AF722486

The Standard of Knowledge for the Aviation, Space, & Defense Industry Quality Practitioner: The AS&D Quality Body of Knowledge (BoK)—Version 1

The Standard of Knowledge for the Aviation, Space, & Defense Industry Quality Practitioner: The AS&D Quality Body of Knowledge (BoK)—Version 1

Michael J. Dreikorn, Ed.D.

Library of Congress Cataloging-in-Publication Data

Dreikorn, Michael J., 1961-

The Standard of Knowledge for the Aviation, Space & Defense Industry Quality Practitioner: The AS&D Quality Body of Knowledge (BoK)—Version 1 / Michael J. Dreikorn

p. cm.

Includes bibliographical references and index.

ISBN 978-1-937974-00-8 (Softcover)

1. Aviation, Space & Defense industry—Professional standards. 2. Education—Aims and objectives. 3. Knowledge Management—Process-based ontological engineering.

I. Title

THE CENTER FOR PROFESSIONAL KNOWLEDGE

a Collective Body of Industry Knowledge

The Center for Professional Knowledge Publishing (The CPK Publishing)

Bokeelia, Florida

Website: www.TheCPK.org

The Aerospace and Defense Learning Institute (ADLI) is the AS&D industry focus unit of The Center for Professional Knowledge. Visit The ADLI at www.TheADLI.com for additional information related to knowledge management standards and learning opportunities for professionals in the AS&D industry.

Front cover graphic credit: NASA JPL; Dark energy and gravity: Yin and Yang of the Universe.

Table of Contents

Chapter 1

Overview to Aviation, Space, and Defense Quality Practitioner Body of Knowledge

Introduction

The industry that produces the products and services that we depend upon for our air transportation, space exploration, and national defense is known as the aviation, space, and defense (AS&D) industry. Some of the products of this industry include aircraft, engines, spacecraft, missiles, ships, tanks, guns, radars, and countless others. The general public may assume that the professionals who perform the business of the industry are well trained and competent in the processes they execute, and by most accounts this is a valid assumption. However, there are a number of factors that may affect the validity of that assumption in the near future, and for good reason.

The first factor is that of an aging workforce. In most industrial nations, the average age of an AS&D professional is fifty-eight (ITA, 2002). And since the AS&D industry has not done a good job of capturing the requisite domain knowledge, the knowledge embedded in the minds of these professionals will leave the industry as they retire. The second factor is an uncertain economy, which has kept many professionals from entering retirement and creating an unknown variable as to when they may exit the industry. The third factor is the global outsourcing of work. With an ever-increasing volume of work leaving its nation of origin, product and technology-based knowledge is also leaving. This means that companies may not have the know-how in the future to develop or service many of the products they deliver today or have in the past. The fourth factor is that process knowledge is not well documented in the AS&D industry. This means that the ability to transfer process knowledge to new professionals cannot be consistently deployed. The fifth factor is that there is no industry-wide body of knowledge (BoK) to train new entrants to the AS&D industry.

Combining all of the above-noted factors within a short amount of time translates into the proverbial *perfect-storm effect* on the intellectual capability of the AS&D industry. The solution is not simple and there is no single silver bullet. To head off this storm, governments, industry, and academia will need to combine their strategic focus and resources to work a systemic approach to sustaining the intellectual capability of the AS&D industry. And in order to do this, foundations must be created to focus such efforts—this includes the definition of an AS&D professional BoK.

A structured BoK provides a profession with the ability to train its new entrants as well as deploy consistent process across an industry. It provides a foundation for training professionals on the execution of standards, methods, and processes, as well as empowers them to engage in accurate decision making. Through this process of knowledge-based decision making, organizations and professionals become more capable in their work.

Decisions are inherent to organizational function. As subsets of society, organizations allocate resources based on decision outcomes. The accuracy of these choices is reflected in the success or failure of the organization. Decision, systems, and knowledge management theorists suggest derived benefits from the application of formalized, systematic, and structured methods to the decision-making process (Fishburn, 1964; Lindgren, 1971; Bertalanffy, 1972; Laszlo, 1972; Hofstadter, 1989).

The collective knowledge of an organization, both tacit and explicit, is referred to as organizational memory (OM). OM, also known as corporate memory, is an element of the larger knowledge management discipline. When structured and documented, OM is the organizational kin to a professional BoK. In general, knowledge management represents the acquisition, creation, usage, representation, and evolution of both types of knowledge—tacit and explicit. Failure to apply systematic and collective management of an organization's memory results in the loss of corporate knowledge and limits decision-making ability. Proposed solutions include a shared conceptualization for representing organizational knowledge. These high-level conceptualizations are called *ontologies*. Fundamentally, ontology represents a consistent vocabulary used to describe a concept. "Every knowledge base, knowledge-based system, or knowledge-level agent is committed to some conceptualization, explicitly or implicitly" (Gruber, 1993). Ontology is an appropriate academic term for a BoK. The challenge for any domain ontology (BoK) is for it to accurately reflect the needs of the industry and profession while minimizing variance in application.

The *AS&D Quality BoK* is the result of developing a new way of engineering domain ontology. The aim of the study was to explore the application of process mapping as a structuring methodology to engineer professional domain ontology. The term for this application is *process-based ontological engineering*. The conceptual framework combined the areas of ontology structuring within the field of knowledge management with that of process mapping within the field of quality improvement. In this initial exploration, the quality processes related to the AS&D industry were identified and analyzed. Over time, other areas of professional domain ontology will be identified and shared utilizing the structuring methodologies described in this book. For such, The Center for Professional Knowledge (CPK) was established for the sharing of process-based knowledge management methodologies among industries, and the Aerospace and Defense Learning Institute (ADLI) for the development and sustainment of domain ontology within the AS&D industry.

To ensure global application and acceptance, a process of verification was carried out to ensure the content of this BoK is consistent with the best practices described in the International Aerospace Quality Group's Supply Chain Management Handbook (SCMH). And the resulting PBOE methodology also forms the foundational framework for the International Aerospace Quality Group's People Capability BoK development standard. The use of this BoK will provide the global AS&D community a common framework to develop and sustain professional knowledge capabilities and reduce variation within process execution expectations and professional decision making. Meaning, there is now a global standard for process execution and related requisite knowledge. The challenge now is for organizations to align process responsibility to individuals (internally and in the supplier base) and ensure people doing the work have the knowledge to get the job done correctly.

As previously described, the urgency for developing domain ontology within the AS&D industry is based on several factors, one being the increasingly aging workforce. As individuals within this workforce retire, their undocumented but often crucial tacit knowledge is lost to the profession. It is frequently the tacit knowledge of an industry that enables the performance of business processes. However, the urgency is not limited solely to the AS&D industry. There are a number of other industries experiencing a loss of intellectual capital, and there is an urgent need for a process-based methodology that will enable consistent deployment of ontology engineering.

A process-based approach assesses the business processes of an organization, industry, or profession in order to identify what is actually going on. Business processes are task activities (e.g., contract review, buying, designing, auditing, and inspection) that are performed to create value in the performance of a service or the creation of a product. Ideally, these processes are executed in a consistent manner. The more consistent a business process, the more variability and waste that can be eliminated. This is especially important in an industry with a global supplier base, such as the AS&D industry.

As macro-business processes are identified within an industry, the tasks associated with performing the business processes can be identified. For example, in the business process of *supplier selection* there are a series of interrelated business micro-processes, such as request-for-quotation development, process assessments, and system audits. Within the microbusiness processes there are a series of tasks that must be performed. For example, the micro-business process of *process assessments* tasks would include the planning of the assessment, performance of the assessment, and documentation of the assessment. Each of these tasks will have specific requirements for knowledge and skills. In the task of *performing assessment*, the individual performing the assessment will require the knowledge of assessment criteria as well as the skills of interviewing and other means of acquiring data. Each task can have multiple knowledge and skill requirements. By applying a structured method, such as Bloom's taxonomy (Anderson and Krathwohl, 2001), of analyzing the tasks to identify the type and depth of knowledge and/or skill, a quantitative measurement can be created to provide greater fidelity to knowledge and skill classification (reference figure 1).

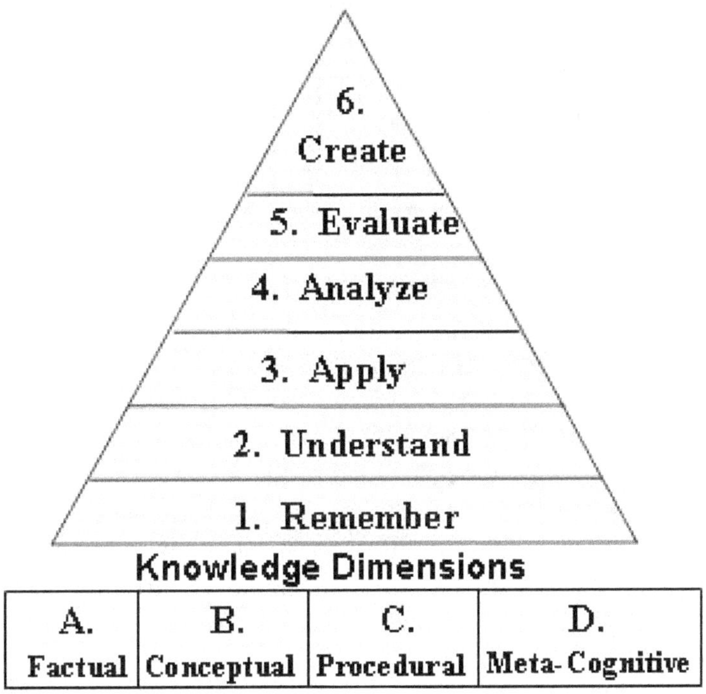

Figure 1, Bloom's taxonomy model, as revised by Anderson and Krathwohl (2001).

For example, when performing a process assessment an individual may require procedural knowledge at a level that will enable the individual to perform the task in accordance with the established criteria. It may only be required for the individual to *apply* the procedural knowledge related to process assessments. Whereas, interviewing skills may be more conceptual in nature, and to effectively apply the procedural requirements the individual must grasp the conceptual skills of interviewing at a higher level, such as *analyze* or *evaluate*.

As an industry's business processes are identified, these are best placed within the logical progression of the process execution. Meaning, business process "C" will follow "B" and "D" will follow "C" (reference figure 2). When these business processes are laid out in such a fashion, it is referred to as process mapping. And when the process mapping reflects the beginning and end of an entire product or service line, it is referred to as the business process lifecycle. As it relates to identifying a comprehensive list of knowledge and skills for an industry, it is important to recognize that tasks performed in one phase of the lifecycle may require different knowledge or skills at a different phase of the lifecycle.

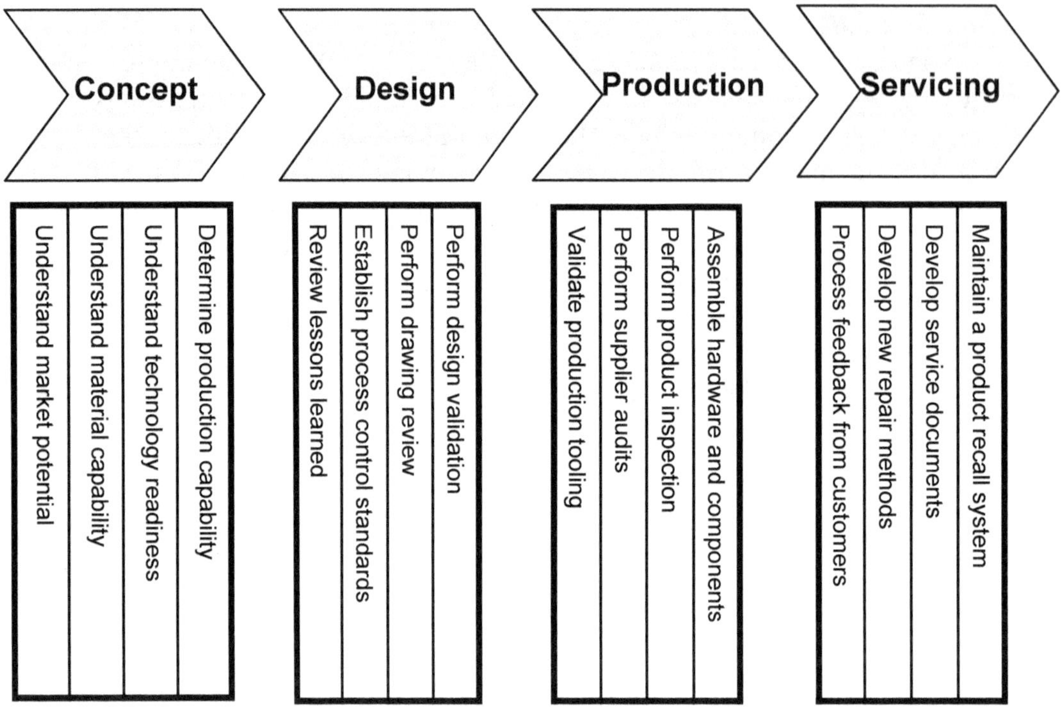

Figure 2, Illustration of major lifecycle phases with sample processes.

In the AS&D industry, the responsibility for task performance is traditionally assigned to the various disciplines within an organization, which can also vary from organization to organization. For example, the responsibility for preparing a request-for-quotation is traditionally assigned to a procurement professional, whereas the responsibility for verifying that the contents of the request-for-quotation are complete and accurate may be assigned to the quality discipline. In this case, there are two disciplines that share responsibility of tasks within a microbusiness process. The procurement discipline would define developing the request-for-quotation as an element of their management processes and quality discipline the verification task as a quality management process (QMP). So, within the overall business process lifecycle of an industry, the various active disciplines will typically have their own set of discipline-related management processes.

Each of these discipline-related management processes is ideally a part of the overall business process lifecycle. Meaning, as the overall definition of the lifecycle business processes change, so do the disciple-related management processes. It is when we disconnect a discipline from the actual business processes of the lifecycle that we see the creation of artificial silos within organizations. These silos frequently lead to fragmentation in business processes, loss of integration, the creation of waste, and can lead to unsafe situations. Functional silos can also result in the establishment of differing and potentially conflicting ontologies for a single process, depending on which professional function is describing the requisite knowledge.

In developing the AS&D Quality BoK, focus has been maintained on the lifecycle business processes of the AS&D industry. Within this focus, the processes that relate to quality were identified as QMP, not to create a discipline-related silo but rather in an attempt to identify the interrelated business processes of an industry.

By identifying the QMPs within the AS&D business process lifecycle, tasks were identified and a process-based ontology for AS&D quality professionals established. With this BoK, the AS&D industry is enabled to structure the development of professional knowledge and skills around the business processes for which quality professionals are responsible. The intended result of a process-based ontology is a more capable and ultimately safer industry. More importantly, the application of a process-based ontology structure underscores the applicability and value of this concept so that it might be utilized in further research when investigating the identification of a professional ontology in emerging and changing industries.

Developing the AS&D Quality BoK

The AS&D Quality BoK was developed through the application a process-based structuring methodology for the rapid development of a domain ontology, and specifically, by the evaluation of process-based structuring of domain tacit knowledge. The widely accepted recognition of knowledge as a social construct (Nonaka 1994; Hassell 2003) combined with the lack of common definitions for both tacit knowledge and organizational memory have resulted in primarily qualitative research on these subjects. In developing the AS&D Quality BoK, the questions of tacit knowledge composition and organizational memory structure were examined through application of a predominantly case study methodology. This approach logically supports the descriptive nature of organizational knowledge creation through observation, personal view, and inductive analysis.

The research design was a multiple-case-study approach with applied process mapping and including semi-structured interviews comprising open-ended questions and researcher observations. The case study and qualitative method are well suited to the research questions in that no commonly accepted methodology for the study of knowledge ontologies exists. Although the ability of qualitative data to explain causal relationships is disputed, the value of qualitative research to elicit understanding of meaning, context, process, and influence (Maxwell, 1996) are elements inherent in the definition of both ontology and knowledge management. The application of subject matter experts (SME) validated the descriptive data for analysis.

Within a purposeful sample of AS&D-related organizations, the business processes that relate to the quality management processes were identified through a process-mapping methodology in a multiple-case study. The organizational sample consists of AS&D organizations of varying sizes. Organizations included a large organization, with approximately ten thousand employees and with an extended supplier base; a medium-size organization, with approximately three thousand employees and a moderate-size supplier population; and a small organization, with approximately one hundred employees and a small supplier base. All of the sample organizations are producers of AS&D products and have both commercial and military product lines. This makes all the sample organizations susceptible to the same regulatory requirements for quality management and product assurance.

Process Mapping the AS&D Lifecycle

Process mapping organizational business process involves the application of process review, procedural review, and interviewing. The lifecycle approach is used to ensure the mapping process has a beginning and an end. The lifecycle approach also ensures that the interrelationship of processes is captured and reduces the possibility of hidden processes going unnoticed. Traditionally, a lifecycle is defined by four major phases: concept, design, production, and servicing. Figure 2 illustrates the lifecycle model with sample processes aligned.

In the Concept phase, the organization assesses its overall capabilities to design, produce, deliver, and service a product or service. The Concept phase also assesses the marketability of the potential products and services. Processes that are typically found in the Concept phase include determination of material capacity, assessment of the state of technology readiness, and determination of production capability.

The Design phase of the lifecycle is where a new product or service is defined in terms of technical requirements. Such definition includes the creation of engineering drawings, process control standards, and material specifications. Processes that typically occur in the Design phase include review of lessons learned from previous designs, drawing review processes, and design validation.

Within the Production phase, the organization converts the design into a tangible product or service. In the Production phase, the organization may be conducting manufacturing processes within its own facilities or may engage a supplier to augment or totally replace its own manufacturing capacity. Processes that are typically found in the Production phase include tooling validation, supplier audits, product inspection, and assembly work.

In the Servicing phase, organizations are focused on maintaining the delivered products. This may take the form of product support and consultation to customer as well as providing scheduled maintenance and repair services. Typical processes within the Servicing phase include repair development, development of service documents, and product recall.

While conducting the initial research for the AS&D Quality BoK, it was quickly determined that the lifecycle phases would need to be redefined to ensure accuracy of process mapping. The first two lifecycle phases remained the same—conceptual and design. The third phase of production was divided into four more defined phases. And since the sample organizations represented manufacturers and in some cases integrated maintenance facilities, the fourth phase, that of Servicing, was integrated into the Production phase. Use and operation of AS&D products, such as that by airlines and warfighters, was not included in the initial development of the AS&D BoK. The resulting six lifecycle phases for this study are illustrated in figure 3.

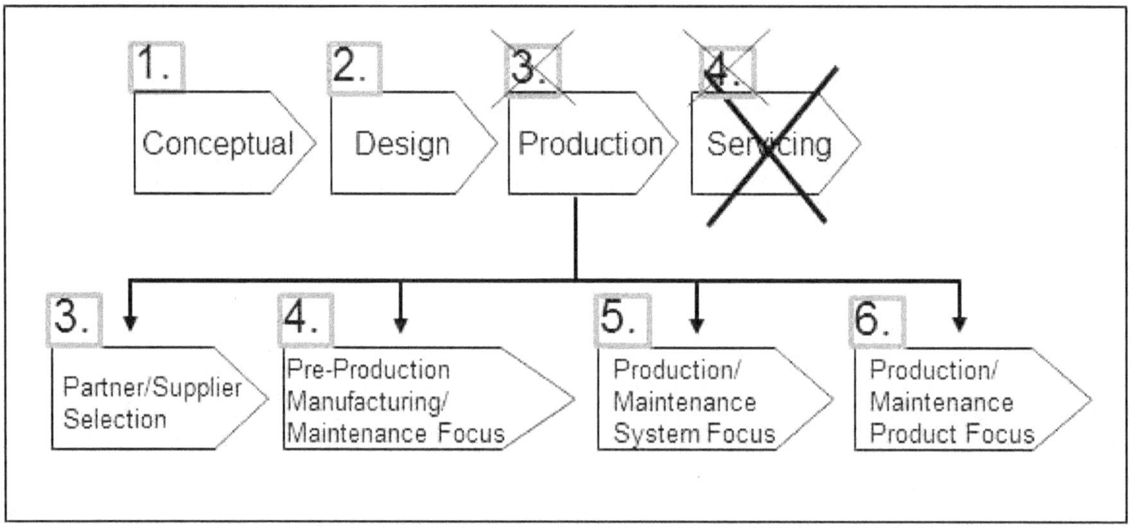

Figure 3, Six phases of the lifecycle for the current AS&D Quality BoK.

The Partner/Supplier Selection phase contains the business process related to ensuring that capable suppliers and partners are selected. By the nature of being enacted prior to engaging a supplier or partner, the processes within this phase are mostly proactive. Examples of processes within this phase include performing producability studies, make-buy decisions, in-house capability/capacity, applying lessons learned, and logistics assessments.

The Pre-Production Manufacturing/Maintenance phase addresses business processes related to proactive activities to ensure process capability prior to serial production or servicing. These processes include pre-award surveys, capability assessments, export controls, and the like.

The Production/Maintenance with System Focus phase of the lifecycle contains the business processes that are not product-related during the serial production and/or maintenance services of an organization. These processes include auditing, training, procedural controls, and others.

The Production/Maintenance with Product Focus phase of the lifecycle contains the processes that directly interact with the product and/or maintenance services the organization is providing. Examples of such processes include inspection, nonconformance handling, root cause analysis, and issuance of governmental forms.

Within each phase there are a series of interconnected business processes. The main value of this process-based BoK is the robust methodology of identifying each business process within a phase. To ignore or misidentify a business process would mean that an area of domain ontology might not be addressed. Thus, applying a process-mapping methodology with a high level of capability is mandatory for reliable results.

Throughout all of the phases of the lifecycle, there are quality related business processes that occur. And there will always be some business processes that are unique to the commodities and services the organization provides or produces. Commodity or organizationally unique processes were identified during the mapping process but were not considered as part of the final domain ontology. To avoid confussio in the intial release of this AS&D Quality BoK, processes that had little generic industry-wide applicability were removed. Future development of the AS&D BoK will provide for more definition into unique business processes and broader aspects of the industry.

Identifying the Tasks within the Business Process

Each business process within the lifecycle may consist of one single task or may require the execution of several tasks. It is at the task level that the knowledge and skill assessment occurs. Bloom's taxonomy (1956), as revised by Anderson and Krathwohl (2001) was used as the standardized knowledge assessment model (reference figure 1). Bloom's taxonomy has been used in various studies (Anderson and Krathwohl, 2001) to provide multi-level competency categorization of subject matter. In developing the AS&D Quality BoK, each business process was analyzed to identify the specific tasks associated with process execution. Then, each task was analyzed to identify and categorize knowledge elements. By applying Bloom's taxonomy, the minimum knowledge requirements were established at the task level.

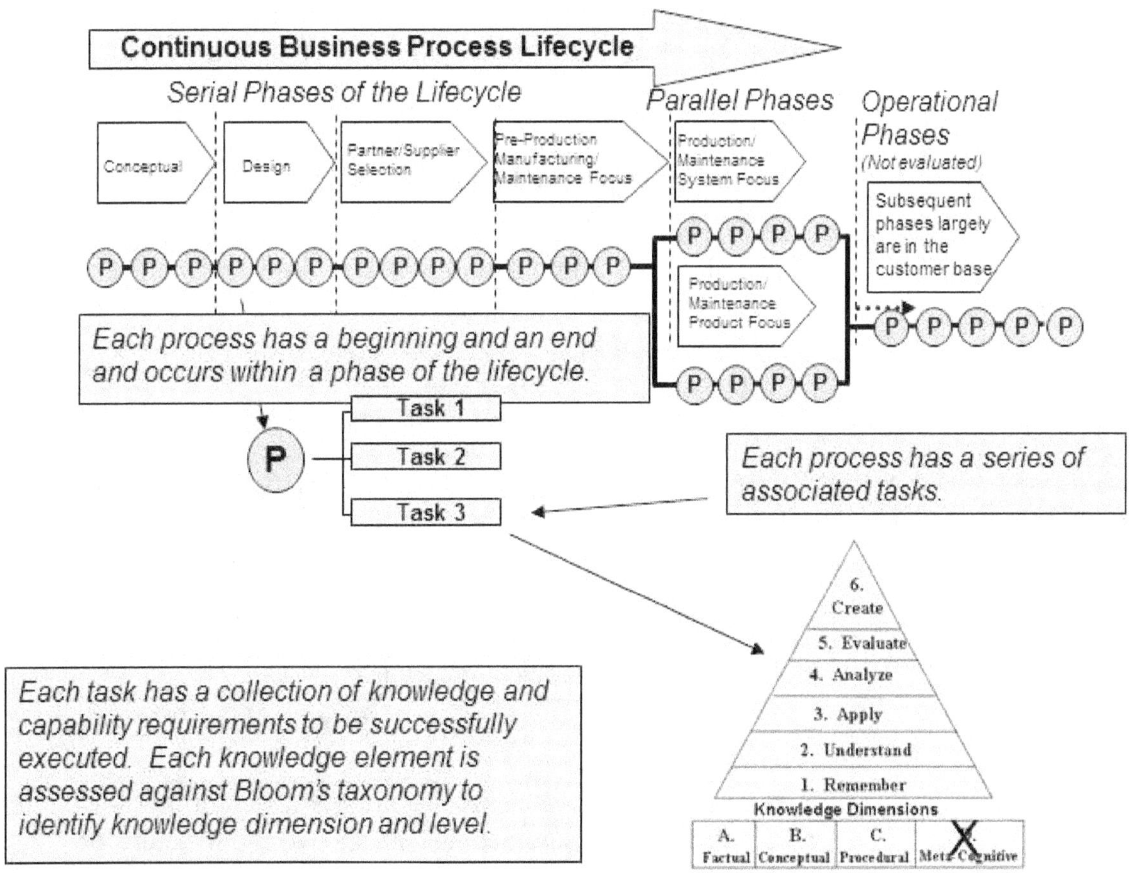

Figure 4, Overview of process-based ontological engineering methodology (Dreikorn, 2009).

Because of the operational nature of the business processes and knowledge expectations of the sample organizations, the knowledge dimension of meta-cognitive was not applied in this research. However, all other dimensions and knowledge levels, as defined by Bloom, were applied to provide measurement expectations of practitioner knowledge.

Throughout the study, subject matter experts (SMEs) were employed to validate the use of terms, scoring of knowledge levels, and to establish consensus of language use. The use of SMEs in the development of the AS&D Quality BoK did not bias the BoK outcome, but rather served to ensure the outcome made sense from an industry perspective. Figure 4 provides an overview of the process-based ontological engineering methodology that was developed in concert with the AS&D Quality BoK.

Briefly on Bloom's

Key to making the AS&D Quality BoK measurable is the application of Bloom's taxonomy. For a detailed understanding on how to use Bloom's taxonomy, read Anderson and Krathwohl (2001). Bloom's taxonomy essentially has two axis, one being knowledge type and the other knowledge level. The four knowledge types are shown in table 1. For the purposes of the AS&D Quality BoK, metacognitive knowledge classification was not utilized.

Procedural Knowledge	How to do something, methods of inquiry, and criteria for using skills, algorithms, techniques, and methods. Knowledge of how to engage criteria, methods, and procedures.
Conceptual Knowledge	The interrelationships among the basic elements within a larger structure that enable them to function together. Ability to understand systems, holistics, and cause and effect.
Factual Knowledge	The basic elements professionals must know to be capable with a discipline or domain and to solve problems.
Metacognitive Knowledge	Knowledge of cognition in general as well as awareness and knowledge of one's own cognition. (*Not used in the AS&D Quality BoK*)

Table 1, The four knowledge types of Bloom's taxonomy.

In addition to the four knowledge types, there are six levels of knowledge, as described by Bloom's taxonomy. Table 2 describes the six levels of knowledge.

1. **Remember**	Retrieve relevant knowledge from long-term memory.
2. **Understand**	Construct meaning from instructional messages, including oral, written, and graphic communication.
3. **Apply**	Carry out or use a procedure or method in a given situation.
4. **Analyze**	Break material into constituent parts and determine how parts relate to one another and to an overall structure or purpose.

5. **Evaluate**	Make judgments based on criteria and standards.
6. **Create**	Put elements together to form a coherent or functional whole; reorganize elements into a new pattern or structure.

Table 2, The six levels of knowledge in Bloom's taxonomy

Limitations

As with any study, there are some limitations. For example, because the focus is the AS&D industry as a whole, the resulting process-based domain ontology is generic in nature. Though this will provide for broad application, it will require augmentation in the technical- and commodity-based (i.e., turbine versus piston engines, airplanes versus helicopters, tanks versus ships, and forgings versus electronics) aspects of knowledge, which are product specific.

Statement of Potential Significance

The process-based ontology engineering methodology enables industries and professions to assess the actual work of the industry/profession and accurately define the requisite knowledge and skills to effectively deploy its business processes. Additionally, the process-based ontology engineering methodology contributes to the knowledge management field by providing a consistent and non-subjective manner in defining ontologies. The resulting standardization of ontology engineering has the potential of significantly reducing variance in the professional fields within common industries and enable accelerated development and sharing of knowledge.

Within the AS&D industry, the resulting AS&D Quality domain ontology ultimately allows for greater standardization of professional capability within the industry. Further qualitative and quantitative research may be applied to understand the influence of elements of specific domain ontology as it relates to performance and other measures. From an operational (business) perspective, the establishment of the AS&D professional quality domain ontology provides the professional with a structured outline of what knowledge and skills are needed to capably execute the processes for which they are responsible.

Utilizing a process-based ontological engineering methodology can also enable organizations to anticipate and plan compliance to new and evolving regulatory requirements. For example, process mapping can be used to identify hazards and to help define the organizational scope needed for the system description for a Safety Management System. And the application of process mapping can also be used by regulatory authorities for both the development of their personnel and for determining equivalencies between authorities/states.

From an industry perspective, if widely deployed in a controlled manner, a process-based ontology will lead to meaningful professional certification of personnel and reduce variance among defined professional activities such as decision making. That is, the variance in knowledge and skills among quality professionals could be reduced within an organization and its extended supplier-base. And confidence in the human-based processes can be increased by applying controls and oversight to knowledge management. Structured and focused learning can be developed and deployed where identifiable gaps are found.

From a theoretical perspective, the science and practice of quality may become more defined than it is at present. It is quite possible that the application of quality tools (such as process mapping in the present study) may prove useful in other sociological inquiries. For example, wherever a process is enacted there are a series of steps and sequential actions that can be mapped and analyzed. The tool of process mapping illustrates the sequential and interrelated nature of processes and allows the researcher to analyze a process within a process while keeping the entire process in perspective. The present case may then provide methodological examples that can be extended to other research studies and other industries and transform the academic to the practical.

By utilizing a process-based ontological engineering methodology, the future state for establishing and sustaining professional ontologies may evolve from being expert-based and riddled with bias to one that requires definition by the industry's business processes.

Chapter 2

AS&D Quality Body of Knowledge—Topical Description

The AS&D Quality BoK was developed independently of any national quality organization and is intended to represent the worldwide AS&D industry. However, in recognition that the American Society for Quality (ASQ) has a long established professional development and certification process for quality professions, the AS&D Quality BoK is being presented in this book utilizing the ASQ categorization of major knowledge topic areas (reference Appendix B).

The following is a topical narrative of the AS&D Quality BoK supplement to the ASQ BoK for Quality Engineers (QE) for those processes that are relevant to the AS&D industry. This supplement is not intended to describe the entire AS&D quality professional BoK, but rather only those additional processes that are not included in the ASQ QE BoK and are considered requisite by the AS&D industry.

Chapter 4 provides the detailed analysis of tasks and taxonomy analysis.

AS&D Relevant Knowledge Areas (Structured to the ASQ QE BoK criteria)

I.G.1. Contracts

Ensure Contract Terminations: To ensure when a contract is terminated with a customer that the terms and conditions of termination are complied with and quality is maintained throughout the process.

Ensure Contract Terminations—Value Stream: To ensure when a contract is terminated with a supplier/partner that the terms and conditions of termination are complied with and quality is maintained throughout the process.

I.J. Knowledge Management

Monitor Changes in Value Stream Personnel: To ensure supplier/partner personnel have the appropriate capability and understanding to meet design and contract requirements. Through surveillance and reporting, monitor supplier/partner personnel changes. Evaluate changes of key personnel to ensure appropriate capabilities are maintained.

Monitor Changes in Organizational Personnel: To ensure organizational personnel have the appropriate capability and understanding to meet design and contract requirements. Through surveillance and reporting, monitor organizational personnel changes. Evaluate changes of key personnel to ensure appropriate capabilities are maintained.

I.M. Material Resource Planning (MRP)

Coordinate Scheduling (MRP): To ensure adequate support is provided (e.g., inspection, testing, etc.), ensure critical points in material flow are recognized and planned. The planning process should be a joint effort to identify timing and work scope. This planning may include the identification of critical skills, special equipment, or logistics considerations.

To ensure action is taken at critical points in material flow, as defined in program planning. Ensure skills, equipment, and other considerations are available when needed. Document the outcomes of planned activities. In the event that conditions prevent the execution of activities as planned, ensure remedial planning is accomplished to ensure tasks are effectively performed.

II.B. Documentation of the Quality System

Participate in Process/Procedure Ballot (Sign-off) Review: To provide input to changes in policies and procedures. Ensure changes to policies and procedures are appropriate with regards to product design, quality requirements (FAI, QMS, standards, FAR), organizational culture, available resources, and equipment.

Ensure Records & Documentation Are Appropriately Maintained: To ensure the documentation required to document traceability, product configuration, and quality acceptance are appropriately maintained. Ensure the retention times and locations of records comply with contract requirements. For electronic records, ensure data is backed-up at appropriate locations and content is protected from derogation, as well as, content is readable utilizing current technology (e.g., absence of floppy disc drives).

Monitor Value Stream Documentation Control Processes: Within the value stream, to ensure documentation control processes are appropriately controlled, to include changes reviewed and approved by authorized persons, changes implemented at the appropriate configuration control level, and implemented into appropriate work documents. Ensure documents are retained in accordance with specific requirements.

Monitor Organizational Documentation Control Processes: Within the organization, to ensure documentation control processes are appropriately controlled, to include changes reviewed and approved by authorized persons, changes implemented at the appropriate configuration control level, and implemented into appropriate work documents. Ensure documents are retained in accordance with specific requirements.

II.D.3. Audit Planning and Implementation

Ensure Supplier Self Audit (QMS): To ensure suppliers/partners perform self-assessments to ensure compliance to their QMS and contract requirements (internal audit). Records of self-assessments must be maintained as quality records. Persons performing the self-assessments must be qualified to perform self-assessments. Records shall contain the scope of the assessment, elements of the assessment that were not performed, and findings and observations discovered during the assessment. Ensure past findings and observations are included in audit scope planning. Ensure records are available for review by external auditors.

Ensure Organizational Self Audit (QMS): To ensure the organization performs self-assessments to ensure compliance to their QMS, contract, and regulatory requirements (internal audit). Records of self-assessments must be maintained as quality records. Persons performing the self-assessments must be qualified to perform self-assessments. Records shall contain the scope of the assessment, elements of the assessment that were not performed, and findings and observations discovered during the assessment. Ensure past findings and observations are included in audit scope planning. Ensure records are available for review by external auditors.

Perform Periodic QMS Evaluations—Value Stream: To ensure suppliers/partners comply with approved QMS and other stated requirements. Plan, schedule, and perform periodic QMS evaluations at supplier/partner locations. Utilize risk-based methods as appropriate. Ensure the results of the audit are documented, to include scope of audit. Where second or third party sources are utilized as a means of determining supplier/partner compliance to their QMS, ensure second and third parties have the appropriate skills and capabilities to perform the scope of the audits.

Perform Full Quality Management System Evaluation: Perform a complete quality management system evaluation of the selected supplier/partner to ensure compliance to all applicable requirements. Ensure the results of the evaluation are appropriately documented, to include any findings or observations.

II.D.4. Audit Reporting and Follow Up

Document QMS Findings: To document noncompliance found during QMS evaluations, ensure such findings and observations are documented to reflect the required condition and the encountered condition and such are documented on appropriate records.

II.H. Control of Customer/Government Material

Ensure Control of Customer Owned Material: To ensure customer-owned material is controlled in accordance with defined requirements. This control shall include identification, proper handling, preservation, and updating as required. May require physical audits of customer-owned materials, understanding of special storage requirements, and traceability and point of usage documentation.

II.I. Control of Customer/Government Equipment/Property

Ensure Control of Customer/Government Furnished Equipment/Materials: To ensure government-owned property and material is controlled in accordance with defined requirements. May require compliance to applicable customer/government standards and/or customer/government contract clauses. This control shall include identification, proper handling, preservation, and updating as required. May require physical audits of customer/government-owned property/materials, understanding of special storage requirements, and traceability and point of usage documentation.

II.J. Capability/Capacity Measurement

Notification of Changes in Capacity and/or Capability: To ensure the organization is aware of any changes (positive or negative) to the capacity and/or capability of a supplier/partner. (Includes all aspects, e.g., engineering, production, maintenance, verification, and facilities.) Ensure there is a notification process in place to make all appropriate parties aware of changes.

Ensure Process Verification (Process Tryout): To ensure processes perform as designed, process verification process is conducted. Ensure processes are adequately defined and operating in a conforming fashion. As processes are submitted for production, ensure a documented process tryout is performed. (Note: The process of process tryout is a Production and/or Servicing process. Only when the process is being applied to verify production/servicing and yields a quality record is it considered a Quality process.)

III.A. Classification of Quality Characteristics

Baseline the Quality and Regulatory Requirements: Combine previous quality requirements review processes performed during the Design Phase and create a baseline quality requirements document. Ensure baseline quality requirements document are accurate and complete. Ensure consolidated requirements are organized, documented, stored and integrated (decompose to system and subsystem requirements) into the requirements hierarchy using a requirements management tool.

III.B. Design Inputs and Review

Define Customer/Program Requirements: To understand the stated and identify the unstated requirements of the customer/program and document such as a preliminary requirements document. To understand the stated and identify the unstated regulatory requirements and document such as a preliminary requirements document.

Determine Design Capability: To understand the present and potential state of engineering capability and to ensure engineering capability can meet the stated and unstated requirements of the customer/program. To ensure mechanisms exist for design consideration of production capabilities and mitigation of predicted nonconformance activity. (Assessment of risk associated with variation to standard-work.) Also, ensure supplier/partner has sufficient resources to accomplish design requirements.

Establish a Preliminary Design and Production/Servicing Execution Plan: To understand where and how the stated and unstated work is to be performed. Establish a program/project execution plan identifying sources for work scope performance.

Determine Quality Requirements—General: Review customer, design, and process requirements and determine quality requirements for the product/program. This review process should include lessons learned material from other similar programs as well as industry source material. These requirements shall be documented in the appropriate organizational format and stored in a database that facilitates requirement hierarchy navigation, configuration control of requirements, and the development of requirement metrics.

Determine Quality Requirements—Regulatory: Review regulatory requirements and determine quality requirements for the product/program. These requirements shall be documented, stored, and integrated into the requirements hierarchy using a requirements management tool. (Examples of such requirements include FAA, EASA, OSHA, DoD, NASA, FMS, and others.)

Determine Quality Requirements—Special Processes: Review design and production requirements and identify special processes associated with program. Determine special process requirements for the product/program. These requirements shall be documented, stored, and integrated into the requirements hierarchy using a requirements management tool.

III.B.1. Workmanship Standards

Determine Material Control Laboratory (MCL) Requirements: Review engineering, process, and production requirements and establish the appropriate material control requirements for laboratory review. This may include destructive and non-destructive evaluation.

Identify Applicable Engineering Source Approval (ESA) Requirements: Review engineering (design), process, and production requirements and determine if there are production processes that require engineering review and approval prior to production release and prior to the implementation of any changes (processes declared unchangeable by engineering without subsequent review and approval).

Provide QA/Engineering Interpretation: Provide interpretation of design and quality requirements to other members of the organization to ensure accurate understanding of requirements. Understand functional differences within the organization. Ensure any formal direction is appropriately documented. Ability to understand technical requirements. Ability to understand organizational challenges and recommendations and engage with engineering, procurement, and other functional disciplines to resolve issues.

Monitor Engineering Source Approval (Process Approval—as required): To ensure processes that are designated as controlled by engineering ("frozen process") are effectively controlled. Evaluate engineering source controlled processes to ensure they are performing as approved. Ensure evaluation results are appropriately documented.

III.B.2. Produceability Review

Perform Manufacturability Review of Design: Review engineering, process, and production requirements and assess such against the known capability of manufacturing/servicing to ensure the design can be produced. Aspects of the review include: ability to produce, cost to produce, and ability to produce with an appropriate level of quality.

III.B.3. Inspectability Review

Perform Inspectability Review of Design: Review engineering, process, and production requirements and assess such against the known capability to inspect/test product at the appropriate point of production to ensure conformity to design. In the event the process is found insufficient apply remediation activities such as request for design changes, process changes, or the development of new methods. Design documentation should contain the appropriate requirements for quality. (e.g., Measurement, Inspection, Key Characteristic Control, and First Article Quality Plans.) Ensure data collection capabilities support the requisite inspection, measurement, and control plans.

III.B.4. Design Validation

Ensure Design Validation (Standard Work): To verify that the design validation process is completed and that there is objective evidence of its completion. The design validation process must ensure the design meets all customer and regulatory requirements. If there are variations in the design that have not been previously produced, risk mitigation plans must be developed (variations to standard work).

III.C.2. Computer Produced Design (CPD) Control

Apply CPD Techniques/Methods (Solid Model): As Computer Produced Designs (CPD) are created, such must be transferable into production and inspection protocol. Ensure such protocol is validated and controlled through all lifecycle aspects. Verify CPD controls include effective control of electronic data to include retention and retrieval.

III.D. Design Verification

Determine Product Validation Laboratory Testing Requirements (Engineering Tests): Review development, deployment, logistic, customer, and usage requirements to determine the appropriate categories and approaches to Product Validation Laboratory Testing.

III.E. Reliability and Maintainability

Determine Production/Servicing Capability: To assess and understand the present and potential state of production/servicing capability of the organization and ensure its capability to produce/service is known and provided as an input to the design process and to ensure the production base is capable of fulfilling the final stated requirements of design.

Determine Production/Servicing Capacity: To understand the present and potential state of production/servicing capacity to ensure the production base has the capacity to produce/service to the requirements of the customer/program.

Perform Maintainability Review of Design: Review the design for ease of maintaining the product while it is in the possession of the customer/end-user.

Perform Design FMEA: Review design to quantify and categorize the potential areas of design risk and ensure risk mitigation plans are developed.

Perform Process FMEA: Review processes to quantify and categorize the potential areas of process risk and ensure risk mitigation plans are developed.

III.G. Configuration Management

Plan for Configuration Management Process: Review the Configuration Management Processes to be used on the program to ensure compliance with established Systems Engineering (SE) and Configuration Management (CM) practices. Ensure the establishment of a change process for program and design changes is being planned for, and coordination and communication are integral parts of the process's design. (Awareness of common CM tools.)

Ensure the Establishment of Configuration Tracking/Status Accounting System: Ensure a system for tracking changes and released configuration status is developed and implemented prior to work authorization.

Ensure Configuration Controls: Ensures a congruence of part configuration with engineering configuration, and that changes to one are adequately coordinated with the other. Implementation of approved changes is planned produceably (e.g., ability to implement the appropriate configuration in production and in servicing) and effectively either in line or through modification kits, which are thoroughly coordinated with the affected internal and/or external organizations.

III.G.1. Configuration Baseline

Ensure Configuration Status and Accounting (FCA): Prior to delivery, ensure configuration reconciliation is performed between Engineering Bill of Materials (EBOM) and Manufacturing Bill of Materials (MBOM), complete with a history of Deviation and Waivers, open and pending changes, and objective evidence of change incorporation as applicable. Ensure the capability to successfully perform a Functional Configuration Audit (FCA), if required.

Ensure Configuration Status and Accounting (PCA): Prior to delivery, ensure configuration reconciliation is performed between Engineering Bill of Materials (EBOM) and Manufacturing Bill of Materials (MBOM), complete with a history of Deviation and Waivers, open and pending changes, and objective evidence of change incorporation as applicable. Ensure the capability to successfully perform a Physical Configuration Audit (PCA) exists if required.

Ensure Engineering Support Is Provided (Organizational): To ensure organizational members have accurate technical/design data to include the assurance of configuration management. To ensure understanding of requirements is accurate by assessing the implementation of technical requirements into documented work processes.

Monitor Engineering Change Request Process: To ensure engineering processes are performed in accordance with system requirements. To ensure engineering changes are implemented effectively and configuration management maintained.

Review Engineering Changes (MCL/SQE/QE) (Quality Review): To ensure quality considerations are included in engineering changes, including produceability, inspectability, and production testing requirements.

Ensure Positive Control of Unvalidated Engineering Changes: To provide a means of deploying un-validated engineering changes into production while maintaining positive recall control of any product or services in which un-validated changes have been applied.

Ensure Ship in Advance of Completion/Approval Process Is Controlled: To provide a means to control product and its production configuration when shipping product without all processes being complete or approved.

III.G.2. Tech Data Management (Incl. Export Control)

Determine Quality Requirements—Export Controls: Review design and production requirements and determine export control considerations such as ITAR and EAR requirements for the product/program. These requirements shall be documented in the appropriate organizational format.

Create Technical Data for Suppliers: In order to effectively communicate to partners/suppliers, create a master file of documents that contain all of the design information required to produce the product/service. Configuration management of applicable revisions must be assured. This may include quality (incl. customer and regulatory), reliability, and other related documents.

Apply Export Control Rules (as required): Review technical data and assess if special handling processes are required to ensure compliance with export requirements.

IV.A.2. First Article Inspections (AS/EN/SJAC 9102)

Review and Approve Development of First Article Inspection Report: To ensure the organization is appropriately planning and is compliant to applicable First Article Inspection (FAI) standards (e.g., AS/EN/SJAC 9102). Ensure FAI plans addresses organization's internal requirements as well as higher-level (customer, prime) requirements. Ensure tooling is included in FAI plan. Ensure supplier and sub-tier supplier activities are included in FAI plan.

Ensure Deliverable Software Conformance—First Article: To ensure deliverable/embedded software meets design requirements and that production systems will assure conformity. Where code can be verified, ensure the review process is structured and the resulting code is maintained for future review.

Ensure NDT/Special Process Conformance—First Article: To ensure NDT and special processes are included in FAI plan and are verified to meet design/specification requirements. Ensure special processes have objective evidence of conformity and that NDTs are performed per requirements. Ensure personnel are qualified to perform special processes and NDT. Ensure personnel at all levels of the supplier/partner base are qualified to perform special processes and NDT.

Ensure Hardware Conformance—First Article: To ensure the geometric and physical characteristics are conformed through the FAI process. Ensure appropriate measurement devices (e.g., calibrated, and accuracy) are utilized when making inspection measurements.

Ensure Deviations Are Reviewed and Approved: To formally document deviations discovered in pre-production processes and to ensure appropriate review and approval is concluded. Ensure deviation authority is defined and controlled. Ensure deviations are re-planned for a follow up FAI.

Approve Product Conformance—First Article Reports: To ensure the entire deliverable product conforms to design and purchasing document requirements and complies with FAI requirements. Verify all lower-level/component-level FAIs are complete and accurate.

IV.A.3. Process/Manufacturing/Servicing/Quality Planning

Verify Production/Servicing Work Instructions: Review production/servicing work instructions and ensure the process of completing work is accurately defined and work definition and verification is appropriately sequenced. Work instruction review will require review of design and process documents and assurance of correlating revisions (configuration management) and the coordination with relevant inspection planning.

Review Product/Servicing Process Planning: Review documented process planning to ensure processes are adequately defined and controlled. Process planning may include statistical process control plans, process control sheets, and other means of process definition. Ensure process planning is controlled by configuration management processes.

Ensure In-Process Inspection/Operator Certification: To ensure inspections are being performed at appropriate points in production/servicing and the inspections are being performed by qualified persons. Where operators are performing inspection functions, ensure a comprehensive operator certification program is in place. If required, ensure organization has approval to perform operator certification. (e.g., Approval may be required by primes and regulatory authorities.) Where special processes are being performed, persons responsible for inspection/verification shall be appropriately qualified and may require special certification (e.g., welding, NDT, chemical processes.)

Ensure Adequacy of In-Process Inspection/Machine Probing: To ensure in-process inspections and machine probing are appropriately planned, sequenced, implemented, and controlled. May require process auditing and documentation review. Ensure inspection personnel are appropriately qualified to perform inspection and are available when inspection is required. Ensure the in-line probing equipment has adequate capability to perform the required inspections. Ensure the use of machine probing is incorporated into organizational procedures and work instructions and utilization is documented in the first article inspection (FAI) reports, as appropriate.

Review Supplier Process Planning: Review documented process planning at suppliers/partners to ensure processes are adequately defined and controlled. Process planning may include statistical process control plans, process control sheets, and other means of process definition. Ensure process planning is controlled by configuration management processes.

Verify Supplier Work Instructions: Review supplier/partner work instructions and ensure the process of completing work is accurately defined. Work instruction review will require review of design and process documents and assurance of correlating revisions (configuration management).

Ensure Process Verification (Process Tryout): To ensure processes perform as designed, a process verification process is conducted. Ensure processes are adequately defined and operating in a conforming fashion. As processes are submitted for production, ensure a documented process tryout is performed, when appropriate. (Note: Unless the output of the verification process yields a quality record, the Production/Servicing process tryout is not a Quality domain process.)

IV.A.3.a. Mandatory Inspections

Establish Delegated Inspection Process: To provide for organizational personnel to perform delegated governmental/customer inspection functions. Delegation approval may include FAA, EASA, NASA, company/organization, prime, and other entities. Ensure delegation process is well defined, means for delegation process oversight provided, and training for inspectors is available.

Perform Supplier Designee Management: To ensure persons who have been delegated inspection authority are capable to execute inspection responsibilities and are in compliance with applicable requirements.

Provide Operator/Inspector Training/Supervision Support: To ensure personnel have accurate understanding of requirements. To ensure personnel have access to organizational decision-making processes. Ability to provide instruction to operators and/or inspectors in the methods of product/service verification.

Ensure Self-Release Program: To provide a means of delegating inspection authority to suppliers/partners and create appropriate quality and shipping documentation. Ensure self-release programs at suppliers/partners are approved, personnel understand requirements, documentation is maintained, and product is conforming.

Monitor Operator Certification Controls: When required, to provide a means of delegating inspection authority to personnel other than quality inspectors. Ensure the integrity of the certification process and the capability of delegated inspections.

IV.B.1. Material Identification, Status, and Traceability

Ensure Traceability Controls: To ensure materials, components, and products are controlled in such a manner that traceability is assured to all specified requirements (Note: Traceability requirements may originate from various sources.). Ensure drawings, process documents, work instructions, and quality records accurately reflect applicable traceability requirements and that traceability is maintained to the level of product definition that requirements specify. Ensure product identification processes are appropriately defined.

Monitor Parts Release Kitting Verification: To ensure parts assembled into kits are effectively identified, documented, and controlled. Parts making up the kit must be identified and controlled with the appropriate revision status.

Monitor Stamp Control: To ensure stamps and signatures utilized to perform acceptance of work processes are effectively controlled and traceable to the individual accepting work. (Includes all media, including electronic.)

IV.B.4. Material Review Board (MRB)

Document Product Nonconformities: To ensure product and process nonconformities are appropriately documented.

Manage Quality Review (MRB Activities)—Product: To ensure appropriate segregation, review, disposition, corrective work, and approvals are made in the event of a product nonconformance. Ensure personnel are authorized (e.g., MRB) personnel and have approved the corrective work activities. (Note: Personnel authorization requirements may be defined by the customer and/or regulatory authorities and such authorizations must be provided in writing.)

Define Material Review Board (MRB) Requirements: Define requirements for material review processes for creation of nonconforming product/services. Ensure the process is documented in a controlled procedure and identifies the responsibilities of the entities involved in the design, procurement, and production of the product/service.

IV.B.4.a. Escape Containment

Perform Escapes (nonconformance to field) Management and Field Notification/Recall Process: To ensure nonconforming materials that have left the quality system are positively controlled in notification/recall processes and/or field management dispositions.

IV.B.4.b. Industry/Government Notification of Escapes and Suspected Escapes

Manage Reportable Nonconformance, Defect, and Malfunction (e.g., FAA CFR 14 Part 21.3) Reporting Process: To ensure discrepancies defined in regulatory requirements (e.g., CFR 14 Part 21.3) are appropriately reported within specified time requirements.

IV.B.4.e. Positive Control of Scrap

Monitor Control of Scrap: To ensure scrap materials are positively controlled.

IV.C. Acceptance Sampling (Non-SPC)

Ensure Sampling Inspection Processes Are Appropriate and Controlled: To ensure the sampling plans established and controlled in compliance with appropriate sampling standards, and plans are appropriately utilized. Ensure personnel utilizing sampling plans understand sampling methods. Verify customer/government approval of sampling methods is/are provided prior to application.

IV.D. Measurement and Test

Perform Source Inspection: To assure product conformity at source of production/servicing. Plan, schedule, and perform product inspection activities and document results in appropriate quality records. Ensure personnel performing source inspection have the appropriate skills and experience to understand and access product/service conformity.

Perform Product Conformity Audits: To ensure the established production/servicing processes continuously produce conforming product. Plan, schedule, perform, and document product conformity audits. Utilize risk-based methods as appropriate.

Monitor Inspection Control Processes: To ensure inspection processes are appropriately planned and performed throughout production/servicing processes by qualified persons. Evaluate effectiveness of inspection/testing processes, ensure authorized personnel perform inspection/test functions, and ensure appropriate documentation.

IV.D.2. Destructive and Nondestructive Tests

Ensure Non-Destructive Test Sequences Are Appropriately Applied: Review work instructions, process planning, and quality records to ensure nondestructive testing is being performed at the appropriate points in production/servicing. The NDT process shall be documented in appropriate records and define the standards of reference and the limits of acceptance. Ensure the documentation of NDT results are accurately defined in applicable work instructions and quality records.

Ensure Hardware Conformance—NDT Inspection: To ensure physical product qualities are verified by appropriate nondestructive evaluation techniques. Ensure supplier/partner NDT processes are performed to requirements, equipment is appropriately maintained (to include calibration and shelf-life), and personnel performing NDT hold appropriate qualifications.

Perform Special Process and/or NDT Evaluations: When special processes and/or nondestructive testing are to be performed within the supplier/partner base, such shall be evaluated to the applicable specifications and standards to ensure capability and compliance. These evaluations will include review of people, process, and equipment.

IV.D.3. Measuring Techniques

Ensure First Piece Setup: To ensure the first article produced for each production run in recurring production processes is appropriately set-up so that it will yield conforming characteristics and meets all stated requirements.

Ensure Hidden Dimension Verification Is Performed (as required): To ensure all required verification characteristics are positively conformed to design and engineering data sets. Ensure required inspections and tests on product characteristics are performed that cannot be verified at a later point. Ensure quality records accurately reflect inspection of hidden characteristics.

Ensure Test and Inspection Using Measurement Software, Equipment, etc. Is Performed (as required): Ensure inspections (NDT and visual) and verification (e.g., testing, software model runs) of product characteristics at appropriate points and with appropriate techniques. Ensure process, people, and equipment are appropriately qualified and controlled. Results of inspection and test activities must be defined and documented.

Ensure Hardware Conformance—Dimensional Inspection—CMM/Layout/Bench: Ensure dimensional inspections are performed utilizing appropriate equipment and gages, utilizing appropriate inspection techniques, utilizing qualified personnel, and are appropriately documented.

Ensure Hardware Conformance—Visual Inspection: To ensure physical product qualities are verified by appropriate visual means. Ensure visual inspection processes are appropriately defined and personnel performing the visual inspection process are qualified to the scope of their responsibilities.

Ensure Internal Receiving Inspection Controls Are Adequate: To ensure receiving inspection processes that receive supplier/partner products and services, including customer/government furnished equipment/materials, are capable of performing required verification processes. Inspection requirements must meet the requirements of the customer, design, and production complexity. Receiving inspection processes may include physical inspection, documentation review, and other means of verifying conformity of design and contract requirements.

IV.E. Metrology

Determine Tool & Gage Design Requirements: Review regulatory, engineering, process, and production/servicing requirements (capability, maintainability, reliability, certification, and human factors) and determine requirements for design, production, and maintenance of tools and gages. This may include calibration and traceability (reference to primary standards) requirements.

Monitor Calibration Control: To ensure the calibration system is in compliance with the approved system. Evaluate equipment maintenance, handling methods, calibration standards, use, and storage.

IV.F. Measurement System Analysis (MSA)

Monitor Gage R&R Control: To ensure reliability of equipment used for product acceptance. To ensure equipment used to accept products are appropriate to the level of accuracy required. When performed, ensure gage R&R studies are documented.

Monitor Correlation of Metallurgical Testing: To ensure the standards and equipment utilized for metallurgical testing are accurate. Ensure documentation of correlation studies is available throughout the supplier/partner base. Special verification and monitoring emphasis on should be applied to spectrometers and associated control standards.

IV.G.1. Shop Assist

Ensure Shop Assist Process: To provide for a temporary supplier relationship to perform work that is not intended to be permanently outsourced. Ensure the scope and use of shop assist is limited to the temporary use of production capacity and is not actually an outsourcing of longer-term production and/or services. Ensure supplier/partner has systems in place to perform work in accordance with organizational procedures and process, including the application of computerized numerical control machines and associated programing. Ensure documentation is controlled.

IV.I. Control of Tooling

Ensure Process Tooling Changes: To ensure tooling utilized in the production/servicing processes are conforming to design requirements, including any changes to such tooling (Note: Including electronic media for tooling and inspection.). Ensure references to tooling requirements are accurately reflected in work instructions, process sheets, and other technical documents.

IV.K.1. Risk Based Acquisition Management

Request Quotes from Suppliers: Ensure a document is created that requests from suppliers/partners formal documents that define their terms, conditions, and pricing to perform work that is stated within the technical data package and defined in a statement of work.

Evaluate Quotes from Suppliers: Evaluate the formal offer packages that potential suppliers/partners have provided in response for quotations. Ensure the responses accurately address all of the technical requirements stated in the scope of work. Evaluate all requested deviations/objections noted in the response from suppliers/partners, and assess the impact on quality and schedule.

Make a Supplier Selection: Based on the criteria established by the organization, and the ability of the potential supplier/partner to deliver to the defined scope of work, select a supplier/partner.

IV.K.3. Flow Down of Requirements (Including Purchase Orders)

Create Purchasing Documents: Create formal documents to procure products or services from suppliers/partners. The documents shall contain all applicable information or references to all applicable information related to the procured scope of work.

Provide QA/Engineering Interpretation: Provide interpretation of design and quality requirements to suppliers and to other members of the organization to ensure accurate understanding of requirements. Understand cultural differences within the supplier/partner base. Ensure any formal direction is appropriately documented. Ability to understand technical requirements. Ability to understand supplier challenges and recommendations and engage with engineering, procurement, and other functional disciplines to resolve issues.

Ensure Engineering Support Is Provided (Value Stream): To ensure suppliers/partners have accurate technical/design data to include the assurance of configuration management. To ensure understanding of requirements is accurate by assessing the implementation of technical requirements into documented work processes.

Identify Flow Down of Unique Customer Requirements: To review all applicable customer requirements documents, assess and determine that any and all applicable customer and/or regulatory requirements are conveyed in the technical data and/or purchasing documents.

Assemble Master Technical Data and Bid Package: To create a master package of technical data, terms and conditions, and other pertinent documents that accurately communicate to the supplier the expectations of the customer.

Apply Unique Requirements to Purchasing Documents: To effectively communicate to the supplier/partner any special or unique requirements. This communication is to be made on applicable purchasing documents and may be included in engineering data.

IV.K.4. Deliverable Data

Issue Purchasing Document: Issue the purchasing documents to the supplier/partner. Ensure the documents are formally controlled within an established system. Purchasing documents shall ensure the correct design configuration is documented.

Verify Master Technical Data and Bid Package: To ensure the contents of the master technical data and bid package are complete and correct.

IV.K.5. Pre-Award/Post-Award Evaluations

Determine Supplier Capability (Assess who has capability): To understand where technical capability resides in the supplier/partner base. For existing suppliers/partners, past performance may demonstrate capability. For new suppliers/partners there should be some objective evidence established to demonstrate technical capability as it relates to the design and production requirements.

Determine Supplier Capacity (Assess who has capacity to take on additional/changed work): To understand where production/servicing capacity resides in the supplier/partner base. This requires recent assessment of supplier/partner workload and an accurate understanding of their total production capacity.

Perform Pre-Award Surveys: For new suppliers/partners, or existing who are potentially assuming work for which they have no previous performance evidence, perform an evaluation at the points where value is added, which may include sub-tier suppliers. This evaluation is to determine if the supplier/partner is potentially qualified to participate in the present procurement. Scope of the evaluation may not be as extensive as full system or process approvals.

Perform Financial Capability Assessment on Supplier/Partner: To the extent a supplier/partner will require extensive resources to perform the stated scope of work, assess the financial capability of the supplier and their sustained ability to conduct business.

Verify Suppliers' Technical Capabilities: To physically evaluate and ensure the supplier/partner can perform to the requirements of the technical data and purchasing documents. This verification shall also ensure the supplier has the ability to interpret the design media provided.

IV.K.6. Supplier Approvals (ASL)

Ensure Organizational Approvals Match Approved Commodity Types: Review the approvals granted to the organization and ensure the approvals match the products or services the organization is capable of providing. The approvals shall note the scope of system approval, product approvals, and special process approvals. (i.e., Unique approvals may include commodities such as forgings, castings, ESD controls, NDT, special processes, etc., as well as, governmental certifications such a production and/or repair approvals.)

Add Supplier to Approved Supplier List: Once a supplier/partner is formally approved, ensure they are listed on the organization's list of approved suppliers. The approved suppliers list should be made available to all points where its use is expected. This may require updating of electronic databases or the reissuance of paper documents. Approved suppliers shall be identified with their specific product and system capabilities (e.g., heat treat, castings, and electronics).

IV.K.7. Sub-Tier Supplier Control

Ensure Supplier Has an Adequate Supplier Control System: Evaluate the supplier/partner's supplier management processes and ensure the supplier/partners have effective control over the work processes sourced at sub-tier suppliers. This evaluation shall ensure that there is effective control at all levels of the supplier-base. Suppliers/Partners must have capability and capacity to ensure all work subcontracted to sub-tier suppliers. Verification may require the performance of physical audits at sub-tier suppliers. Supplier/Partner should have formal monitoring systems in place to assess supplier performance. Supplier control processes shall also possesses controls for managing noncompliance and nonconformities, to include disposition and corrective action. (The general expectation is for the organization to substantiate through evaluation compliance to requirements throughout the entire value stream and the technology capability to transfer product/servicing verification data.)

Ensure Supplier's Utilize Source Approval List for Sub-Tier Selection (as required): To ensure suppliers/partners only utilize approved sources when required by design or other formal document. Use of approved suppliers does not absolve suppliers/partners from the responsibility of sub-tier supplier management.

Monitor Supplier Sub-Tier Control: To ensure suppliers/partners are effectively controlling their supply base by ensuring requirements are flowed through purchasing documents and complied with.

IV.K.9. Direct/Drop Shipments

Ensure Drop-Ship Approvals (Shipments Directly from Supplier to Supplier): To provide for a means to control shipments of product from one supplier to another, which is not directly controlled by the receiving supplier. Ensure appropriate documented approvals are maintained to permit drop-shipments between suppliers. Ensure positive quality control drop-shipments are maintained throughout supplier/partner base.

Ensure Direct-Ship Approvals (Shipments Directly from Supplier to Customer): When the organization is not the end-item producer, to provide for a means to control direct shipments of product from the organization's supplier to the end-item customer. Ensure appropriate documented approvals are maintained to permit direct-shipments to the end-item customer or other defined destinations (e.g., customer and regulatory approvals). Ensure positive quality control is maintained over direct-shipments.

IV.K.10. Supplier Support Resources

Determine Support Resource Needs (IPT, SQA, Labs, etc.): To understand the actual and total resource needs to support outsourcing to suppliers/partners. Based on supplier, design, and production maturity, resource needs may vary. Need to understand differences in regional and international influences on required support resources.

Determine Oversight Resources Needs and Plan Surveillance: To understand the on-going surveillance support resource requirements to sustain the outsourcing to suppliers/partners.

IV.K.11. Monitoring of Suppliers *(RCCAPA in Section V.D.)*

Monitor Supplier Resource Capacity: To ensure the effectiveness of the supplier's/partner's production/servicing capacity. Through review of supplier/partner notification of changes in capacity and/or capability, assess the potential impact of those changes on product and system quality. During on-going assessments and communication with supplier/partners, establish awareness of potential capability/capacity changes.

Notification of Changes in Capacity and/or Capability: To ensure the organization is aware of any changes (positive or negative) to the capacity and/or capability of manufacturing and/or servicing of a supplier/partner, including location of suppliers and significant changes in management/ownership. Ensure there is a notification process in place to make all appropriate parties aware of changes, including required notifications to customers/regulatory authorities.

Monitor Financial Condition of Key Suppliers: To ensure the ability of a supplier/partner to sustain production/servicing. Maintain awareness of the financial condition of suppliers/partners by review public information and discussions within industry. Establish awareness through observation in the supplier base of conditions such as material changes, changes in suppliers, equipment condition, employee turnovers, etc.

Monitor Risk of Ultra-Small Key Suppliers: In the event that the organization engages an ultra-small supplier for services or products that are considered key, ensure risk management plans are in place. (Ultra-small suppliers are organizations that will be adversely affected by the absence of one to two persons.)

IV.K.13. Supplier Quality Plans

Develop Alternate Supplier Approval Process (as required): To provide a means of outsourcing work to a potential supplier that has not been approved under the required processes. Create a quality plan that defines the specific controls to be in place to effectively manage work to a supplier that is not approved under normal processes. If the quality plan is subject to customer approval prior to implementation, ensure such approval is obtained.

IV.M. Special Process Control

Ensure Chemical/ Metallurgical Processes Are Conforming: Ensure chemical and metallurgical processes are conforming to design and process requirements. May require destructive and/or nondestructive evaluation methods. Personnel must be appropriately qualified to perform special processes.

Perform Periodic Special Processes (Weld, Peen, Plating, etc.) Evaluations: To assure control, capability, compliance, and conformity of special processes performed in production/servicing processes. Plan, schedule, perform, and document periodic evaluations of special processes. Utilize risk-based methods as appropriate.

IV.M.1. Second and Third Party Surveillance

Monitor Second and Third Party Surveillance Process: To monitor suppliers/partners who have systems and/or special processes approved by a second or third party (e.g., Nadcap) are maintaining their approval as current. For suppliers who allow their certification to expire, engage an alternative means of surveillance and to assure product/service quality.

IV.N.1. Deliverable Software

Ensure Software Conformance—Deliverable: To ensure deliverable software (software that is embedded in a product and is delivered to a customer) conforms to design requirements. Ensure verification processes are performed (e.g., code comparison, run tests, peer review) to verify deliverable software is conforming. Ensure quality records reflect software configuration, conformity, and acceptance. As required, ensure testing tools are provided with the deliverable code or software program.

Monitor Software Quality Plan: To ensure software quality plans are being applied as required.

IV.O. Packaging, Handling, & Shipping

Ensure Packaging Controls Are in Place: Ensure the organization has processes to protect products and services in packaging to prevent damage during transport. Depending on contractual requirements, may require familiarity with various packaging standards. Ability to understand and interpret packaging specifications and drawing requirements. Ensure shipping documents, to include labeling are legible, applied appropriately, and contain all applicable data required.

Ensure Packaging & Preservation: To prevent deterioration of product during shipping and storage. Evaluate compliance to packaging, handling, and preservation requirements.

IV.O.2. Foreign Object Debris

Monitor Handling Practices/Foreign Object Damage Controls: To ensure products are not damaged during handling and processing. Evaluate material handling process to ensure product is effectively protected during processing. Evaluate work practices and ensure there are positive practices to prevent FOD.

IV.P. Product Pedigree Assurance

Ensure Buy Back Processes Are Controlled: To provide a means to control product to re-enter a production system that has previously been certified as airworthy. Assurances such as state of airworthiness, regulatory authority approvals, customer approvals, and other product and system requirements must be addressed.

IV.R. Certification of Flight Readiness/Airworthiness Certification

Complete Governmental Forms Related to Production/Servicing and/or Design Conformity: To create the appropriate documentation for product design and production approval by governmental agencies. Ensure authorized persons complete the official forms and that supporting documentation is provided.

IV.R.1. Civil Aviation

Issue Governmental Forms for Documentation of Major Repairs and Alteration (e.g., FAA Form 337): Where appropriate, produce governmental documentation to record and approve major repairs and alternations (e.g., FAA Form 337), and that only authorized persons have signed the forms.

Issue FAA Form 8130-series/EASA Form 1 (Authorized Release Certificate/Airworthiness Certificate): When applicable, produce a civil airworthiness certificate and ensure an authorized person has signed the form.

IV.R.2. Military Products

Issue Governmental Product Acceptance Documentation (e.g., DD Form 250 [USA]): When applicable, produce the U.S. Government documentation to accept products/services and transfer ownership (DD 250). An awareness and understanding of governmental policies, procedures, and regulations related to product and system certification (e.g., U.S. Air Force certification requirements are provided in AFPD 62-6 and AFI 62-601.)

V.A. Quality Control Tools

Apply Continuous Improvement Support: To ensure process waste is removed and quality is improved, utilize data and various quality tools to understand process performance and improvement opportunity. Ensure improvement action does not compromise approved production controls.

V.B. Quality Management and Planning Tools

Establish Appropriate Performance Metrics & Data Analysis: To provide visibility of product, process, and organizational performance establish appropriate performance measurements to assess process performance and quality.

V.D. Corrective Action

Monitor/Approve Root Cause Analysis—Product/Process: Ensure appropriate root cause analysis tools are utilized during analysis of documented findings. Ensure authorized (e.g., MRB) personnel have approved the identification of root cause. Ensure product/services previously produced/provided are reviewed for applicability.

Monitor/Approve Corrective & Preventive Action Processes—Product/Process: To ensure effective corrective and preventative action processes are in place and applied to the root cause of noncompliance/nonconformance in a timely fashion. Verify effectiveness after implementation and prior to closure of the corrective action records.

Take Action Based on Data Analysis: Based on analysis of data, take improvement action to achieve objectives.

Monitor QMS Findings: To ensure noncompliance is documented when found during QMS assessments and are managed for root cause and corrective action.

Monitor Root Cause Analysis—QMS (Root Cause Process): To ensure appropriate root cause analysis tools and such are applied during analysis of documented findings.

Approve Root Cause Analysis—QMS (Individual Root Cause Analysis): To ensure the true root cause of QMS noncompliance is identified and appropriate review and approval provided.

Monitor Corrective & Preventive Action Processes—QMS: To ensure effective corrective and preventative action processes are in place and such are applied to the root cause of noncompliance. Ensure process identifies and contains any potential impact on product.

Approve Corrective & Preventive Action—QMS: To ensure the documented corrective and preventative action determinations effectively address the root cause in a timely fashion.

Follow Up of Corrective & Preventative Action: To verify the effectiveness of corrective and preventative actions by assessing implemented solutions after actions had been taken. Establish appropriate interval for effectiveness assessment; audit the implemented solutions to ensure they reflect the agreed-to arrangements; and that the solution prevents recurrence of the root cause.

Monitor Supplier-QMS Corrective and Preventative Action Process: As noncompliance is identified at suppliers/partners, ensure procedures are in place and the process of effective closed-loop corrective and preventative action is deployed. May include managing the implementation timing of corrective action and the management of deviations.

V.G. Lessons Learned

Capture, Document, and Preserve Lessons Learned: To recognize, understand, and document lessons learned during the application of the previous processes in the relevant phase so that such may be used in future programs/projects.

Apply Lessons Learned: Based on the lessons learned in previous programs/projects, apply lessons learned to ensure previous issues are not repeated.

VI.A. Collecting and Summarizing Data

Collect Performance Metrics & Perform Data Analysis: Collect data related to process performance and quality and perform analysis of data. Present data in visual formats for use in the organization.

VI.F. Statistical Process Control (SPC)

Monitor Process Certification /Statistical Process Control: To ensure statistical techniques and inspection sampling plans are appropriately and effectively applied.

Chapter 3

The AS&D Quality Professions

Overview

When combining the global reach, cultural diversity, product diversity, and various business models of the AS&D industry, it is safe to say that there is no one way to assign work within an organization. It should always remain the organization's responsibility to determine how work task responsibility is to be assigned. However, that said it should also be possible to establish a global consensus of core quality professions.

Recognizing that the ASQ has had long standing global recognition of its professional Quality BoK and Quality professions, the AS&D Quality BoK is being described utilizing ASQ's existing framework of professions, while supplementing with AS&D unique groupings of tasks. The following provides a brief description of each of the defined AS&D Quality professions. For certification of AS&D Quality professionals, the ADLI will accept current ASQ certifications, as noted in each description of a profession, as evidence that certain knowledge quality fundamentals have been met. The applicant for ADLI professional certification needs only to present their relevant ASQ certification, meet work experience and reference requirements, and successfully pass a supplemental knowledge examination. AS&D Quality professions who do not hold an existing relevant ASQ certification are required to successfully pass the complete knowledge examination, as well as, meet the work experience and professional references requirements for certification.

The AS&D Quality Professions Described

AS&D Quality Engineer—Level 1 (Practitioner)

The AS&D Quality Engineer profession is defined within this BoK at two levels—Level 1 (Practitioner) and Level 2 (Senior). The two levels of quality engineer are established to recognize that as professional capability is developed and demonstrable, an AS&D Quality Engineer should be able to assume greater work responsibilities. The significant difference between a Level 1 and Level 2 AS&D Quality Engineer is the assignment of tasks with greater complexity. The Level 1 AS&D Quality Engineer is primarily focused on the internal work processes of the organization. Work processes related to the supplier-base are not aligned to the Level 1 AS&D Quality Engineer as *Primary* responsibility.

At both levels, the AS&D Quality analyzes and solves quality problems, develops quality assurance plans and instructions, prepares procedures, mentors inspectors and operators, participates in product and system audits, analyzes quality costs and other quality data, and applies fundamental statistical methods for process control.

In addition to the work tasks and knowledge requirements defined in the AS&D Quality BoK, Level 1 AS&D Quality Engineers may consider the BoK outline of the ASQ Certified Quality Technician as a valuable guideline for fundamental knowledge capabilities.

AS&D Quality Enginee—Level 2 (Senior)

Though inclusive of all the responsibilities and knowledge requirements for the Level 1 Quality Engineer, the Level 2 AS&D Quality Engineer has a much broader scope of responsibility with greater complexity. The Level 2 AS&D Quality Engineer responsibilities include not only the work processes associated with the internal business processes of the organization but also many of those that interface with the organization's supplier-base. The Level 2 is not intended to substitute the profession of the AS&D Supplier Quality Engineer, but rather provide a supplement role to engage internal organizational processes.

In addition to the requirements of the Level 1, the Level 2 AS&D Quality Engineer shall be able to develop and ensure the effective operation of quality assurance systems and safety management systems (reference ICAO Annex 6), apply and conduct analysis of product testing and inspection methods, apply and evaluate metrology and statistical methods to diagnose and correct improper quality verification and validation practices, apply tools related to human factors, conduct analysis of quality cost, apply risk identification practices, effectively communicate quality and risk data, and provide mentorship to the organization in quality and safety concepts, tools, and practices.

In addition to the work tasks and knowledge requirements defined in the AS&D Quality BoK, Level 2 AS&D Quality Engineers may consider the BoK outline of the ASQ Certified Quality Engineer as a valuable guideline for fundamental capabilities.

AS&D Quality Manager

The AS&D Quality Manager must possess the ability to see the big picture, develop strategic and tactical plans, champion the processes related to quality, develop organizational talent related to quality, and ensure the products and services the organization provides meets the requirements of the customer and relevant regulatory authorities. Though the responsibilities of the AS&D Quality Manager are significant, there is a lesser emphasis on statistical capabilities and a greater emphasis on leadership, requirement flows, and effective communication.

In addition to the work tasks and knowledge requirements defined in the AS&D Quality BoK, AS&D Quality Managers may consider the BoK outline of the ASQ Certified Manager of Quality/Organizational Excellence as a valuable guideline for fundamental knowledge capabilities.

AS&D Supplier Quality Engineer

With regard to knowledge requirements, the AS&D Supplier Quality Engineer is similar to the Level 2 AS&D Quality Engineer, however, with a focus on the organization's supplier-base. AS&D Supplier Quality Engineers are responsible for the accurate assessment of supplier capability, identifying potential risks, assurance of effective product and service verification and validation processes, investigation and analysis of product and process issues in the supplier-base, facilitating effective communication between the supplier and the organization, conducting product and system audits, and verification of compliance to requirements, including relevant regulatory requirements.

In addition to the *hard skills* that are identified in the AS&D Quality BoK, the AS&D Supplier Quality Engineer must have the ability to understand the cultural influences related to their specific scope of responsibilities. Developing cultural sensitivity capability is critical within a global industry. However, recognition must be given that there is no single training course that makes one expert in cultural sensitivity. These are skills that are developed over time, exposure and mentorship. It is recommended that when an AS&D Supplier Quality Engineer is assigned a regional scope of responsibility, the organization apply the needed diligence to ensure the professional has the needed cultural capability for the specific assignment.

In addition to the work tasks and knowledge requirements defined in the AS&D Quality BoK, AS&D Supplier Quality Engineers may consider the BoK outline of the ASQ Certified Quality Engineer as a valuable guideline for fundamental knowledge capabilities.

AS&D Software Quality Engineer

The AS&D Quality BoK has not yet mapped out all of the processes related to software quality assurance. However, in the development of the existing AS&D Quality BoK, some critical business processes were identified that are typically aligned to a Software Quality Engineer profession. The knowledge requirements for these processes and associated tasked were evaluated and are provided accordingly. It is the intent of the Aerospace and Defense Learning Institute to conduct research into this area in the future.

A typical AS&D Software Quality Engineer is able to develop and maintain Systems & Software Quality Engineering (SSQE) Common Processes that are aligned with AS9100, Software Engineering Institute's (SEI's) Capability Maturity Model-Integrated (CMMI) program, and various customer and governmental requirements. Software Quality Engineers who are practicing in the AS&D industry may consider the BoK outline of the ASQ Certified Software Quality Engineer as a valuable guideline for fundamental knowledge capabilities.

AS&D Program Quality Engineer

The AS&D Program Quality Engineer has the broadest responsibilities of all the AS&D quality professions, but not necessarily the deepest. The AS&D Program Quality Engineer is typically aligned to a specific program, such as an aircraft model, missile type, or vehicle category; and typically serves as the primary customer Quality interface with both prime and government quality representatives. Some organizations refer to the Program Quality Engineer as the Customer Quality Engineer, with a blending of responsibilities from the Contract Review Quality Engineer. The AS&D Program Quality Engineer has responsibilities in all phases of the lifecycle, with strong emphasis on risk identification and mitigation early in a program’s life. The responsibilities of an AS&D Program Quality Engineer are data rich, and they play a critical role in the organization’s communication processes.

AS&D Program Quality Engineers typically measure quality performance at a program level, whereas other quality professions typically measure quality performance at a unit-level or enterprise-level. As many AS&D organizations budget organizational work at the program-level, AS&D Program Quality Engineers have the critical responsibility of assessing requirements and, collectively with cognizant Quality Managers, ensuring resources are properly aligned. AS&D Program Quality Engineers are also typically assigned the responsibility of ensuring configuration control is maintained.

In addition to the work tasks and knowledge requirements defined in the AS&D Quality BoK, AS&D Program Quality Engineers may consider the BoK outline of the ASQ Certified Quality Engineer and that of the ASQ Certified Manager of Quality/Organizational Excellence as valuable guidelines for fundamental knowledge capabilities.

AS&D Contract Review Quality Engineer

Particularly within the space and defense aspects of the AS&D industry, customer contracts can be very complex and can contain requirements through reference of codes and/or through reference in standards and specifications. The AS&D Contract Review Quality Engineer is expected to be able to review contracts; understand commercial and government acquisition policies, rules, and practices; assess and identify quality and certification requirements; assess organization compliance to requirements; and document and communicate compliance gaps.

The capabilities of AS&D Contract Review Quality Engineer may be employed in review of customer contracts but also in review of contracts between the organization and a supplier or partner. In addition to the work tasks and knowledge requirements defined in the AS&D Quality BoK, AS&D Contract Review Quality Engineers may consider the BoK outline of the ASQ Certified Quality Technician as a valuable guideline for fundamental knowledge capabilities.

AS&D Maintenance, Repair, and Overhaul Quality Engineer

The maintenance, repair, and overhaul (MRO) aspects of the AS&D are also frequently referred to as *aftermarket*. MRO work is that which is performed on product that has already been provided for operational use. For example, an aircraft engine that requires an overhaul is an example of MRO work, as is an aircraft that is being converted from a passenger configuration to a cargo configuration. The skills and knowledge requirements of an AS&D MRO Quality Engineer are focused on the application of work activities that fall within the MRO business. As defined in the AS&D Quality BoK, the AS&D MRO QE is predominately product and/or service focused and may be supported by AS&D Quality Engineers (Level 2) for conducting complex statistical analysis studies.

AS&D MRO Quality Engineers must have a thorough understanding of relevant regulatory requirements, ability to evaluate product and process verification and validation methods, develop and maintain inspection and test planning, manage quality related work, and provide mentorship to the MRO organization in quality and regulatory matters.

In addition to the work tasks and knowledge requirements defined in the AS&D Quality BoK, AS&D MRO Quality Engineers may consider the BoK outline of the ASQ Certified Quality Technician and that of the Certified Quality Inspector as a valuable guideline for fundamental knowledge capabilities.

AS&D Special Process Quality Engineer

The highest potential for latent defects in products and services is through the misapplication and/or loss of control in special processing. Special process are "processes for production and service provision where the resulting output cannot be verified by subsequent monitoring or measurement and, as a consequence, deficiencies become apparent only after the product is in use or the service has been delivered" (Reference AS/EN/SJAC 9100C, paragraph 7.5.2). Special processes include, heat treatment, coatings, bonding, and others.

The responsibilities of the AS&D Special Process Quality Engineer include special processes that are performed within the organization as well as those that are performed within the organization's supplier-base. The AS&D Quality BoK does not address the technical aspects of each special process, but rather only the common work processes related to special processes. The development and certification of AS&D Special Process Quality Engineers must address both these common work processes and clearly define specifically for which special processes the AS&D Special Process Quality Engineer is capable and responsible.

Some responsibilities of the AS&D Special Process Quality Engineer include development of special process control plans, validation of special processes, monitoring the performance of special processes, and conducting investigations of product failures related to special processes. The AS&D Special Process Quality Engineer are expected to maintain their currency in technical standards related to their areas of expertise and maintain awareness of AS&D industry-managed programs such as Nadcap. AS&D Special Process Quality Engineers who specialize in nondestructive evaluation may also consider the American Society for Nondestructive Testing (ASNT) and the Performance Review Institute's (PRI) Nadcap as authoritative knowledge centers.

AS&D Special Process Quality Engineers that include nondestructive evaluation in their portfolio of capabilities should also hold a Level III certification issued from a national NDT society (e.g., ASNT or other national equivalent) or the organization, for the specific methods of expertise. When a company provides Level III certification of NDT personnel, diligence must be applied to ensure appropriate Level III certification and oversight processes are place. Similar is true for AS&D Special Process Quality Engineers who are responsible for welding processes. The American Welding Society (AWS) provides professional education and certification to weld inspectors and AS&D quality professional who are responsible for such processes should be appropriately certified at the method level.

In addition to the common work tasks and knowledge requirements defined in the AS&D Quality BoK, AS&D Special Process Quality Engineers may consider the BoK outline of the ASQ Certified Quality Technician as a valuable guideline for fundamental knowledge capabilities.

AS&D Material Laboratory Quality Engineer

The AS&D Material Laboratory Quality Engineer could be considered the scientist within the quality professions. They work mostly within a laboratory environment, preparing samples for testing and conducting evaluations. The career path for an AS&D Material Laboratory Quality Engineer is typically very long. A foundational engineering degree in metallurgy or material science is typically expected, and advance degrees are sought as increased responsibilities are assigned. This is a highly technical profession and requires significant investment by both the practitioner and organization to ensure sustained professional capability.

Traditionally, the Material Laboratory Quality Engineer has been internally focused on the capabilities and performance of their own laboratory. However, given the continued globalization of the AS&D industry, AS&D Material Laboratory Quality Engineers must be more supplier-based focused. Their global responsibilities require a broader knowledge base of international material standards and of test equipment. One of the greatest risks to traditional AS&D organizations is not having access to a capable AS&D Material Laboratory Quality Engineer to monitor supplier capability and performance. Without the capability of oversight, the organization loses the ability to verify product and/or service and safety is compromised.

The AS&D Quality BoK has not yet mapped the specific technical processes related to the AS&D Material Laboratory Quality Engineer, however, the common work processes are provided. In addition to the common work tasks and knowledge requirements, AS&D Material Laboratory Quality Engineers may consider the BoK outline of the ASQ Certified Quality Technician as a valuable guideline for fundamental knowledge capabilities.

AS&D Production Control Quality Engineer

The AS&D Production Control Quality Engineer is also frequently referred to as the "Shop Quality Engineer." They are the quality engineers that are on the shop floor in the day-to-day activities and are responsible for providing quality expertise, mentorship, and oversight to inspectors and operations personnel. The AS&D Production Control Quality Engineer interacts with production planning in the establishment of inspection protocol, manage the work of quality, engage in investigations related to product and/or system quality, monitor process controls, and communicate to members of the organization.

In addition to the work tasks and knowledge requirements defined in the AS&D Quality BoK, AS&D Production Control Quality Engineers may consider the BoK outline of the ASQ Certified Quality Technician and that of the Certified Quality Inspector as a valuable guideline for fundamental knowledge capabilities.

AS&D Regulatory Quality and Certification Specialist

The AS&D Regulatory Quality and Certification Specialist is the organization's expert in matters related to regulatory requirements for design, production, MRO, and certification of product, personnel, and system. They typically interact with representatives of the civil aviation authorities (e.g., FAA and EASA), maintain an understanding of regulatory requirements, evaluate potential impacts to the organization related to changes in regulations, and mentor the organization in regulatory matters.

The AS&D Regulatory Quality and Certification Specialist is typically responsible for developing and monitoring organizational plans for achieving regulatory approvals for design (TC, STC, TSOA, PMA), production (PC, TSOA, PMA) and repair stations; validating reciprocal design approvals across multiple regulatory authorities (e.g. FAA TSOA in EASA or CAAC); and assuring shipping documentation controls (e.g., FAA Form 8130-3, EASA Form 1, etc.). They are also expected to be expert in the controls for commercial parts; distributor accreditation programs (e.g., FAA Advisory Circular 00-56); and self-disclosure processes (e.g., FAA Advisory Circular 00-58).

To maintain their currency, the AS&D Regulatory Quality and Certification Specialist may be actively participating in industry forums and are constantly monitoring relevant government websites. In addition to the work tasks and knowledge requirements defined in the AS&D Quality BoK, AS&D Regulatory Quality and Certification Specialists may consider the BoK outline of the ASQ Certified Quality Technician as a valuable guideline for fundamental knowledge capabilities. Additionally, civil aviation authorities frequently provide online and classroom based learning at little or no charge.

Chapter 4

AS&D Quality Body of Knowledge—Detailed Description

This section, the expanded analysis of the AS&D Quality BoK is provided. Each of the major topics are aligned with business processes that fall within the topic, and each business process is dissected into it various tasks that are required to execute the process. Each task is then evaluated utilizing Bloom's taxonomy and the tasks are assigned primary (P) and support (S) responsibility to the AS&D quality professions. The task responsibility alignment to the professions is provided as a guideline for standardization of the AS&D industry. The final word on task responsibility shall reside with the organization that is responsible for work execution.

Utilizing the ASQ categorization of major knowledge topic areas, the AS&D Quality BoK is presented supplementing the existing ASQ BoK for QE. To facilitate sorting of the extensive AS&D Quality BoK, a code was assigned at the task level. The code reflects the placement of the task within the lifecycle. Table 3 provides an outline for the prefix code that identifies the tasks placement within the lifecycle. To avoid confusion between what occurs in the organization versus what occurs in the supplier-base, the Pre-Production/Pre-Maintenance Phase was segregated into two categories—Supplier and Organizational focus, increasing the number of lifecycle phases to seven. The supplier category contains processes that are focused on an organization's supplier base. Whereas, the organizational category addresses processes relevant to internal production and/or servicing.

Code	Lifecycle Phase
CP	Conceptual Phase
DP	Design Phase
SS	Partner/Supplier Selection Processes (Pre-approval)
PP	Pre-Production/Pre-Maintenance Processes—Supplier Focus
PO	Pre-Production/Pre-Maintenance Processes—Organizational Focus (Pre-FAI Acceptance)
PM	Production/Maintenance (Post-FAI)—System Focus
PF	Production/Maintenance (Post-FAI)—Product Focus

Table 3, Prefix codes for lifecycle placement.

In addition to the above alpha codes, each process is also identified with a numeric code, referencing its placement within the lifecycle phase. And each task within a business process is identified with a dash number. This coding is primarily for electronic manipulation of the AS&D Quality BoK, however, can also serve as a quick indicator to the BoK user of the placement of the task within the lifecycle. The coding also serves as a structuring methodology when processes are added or revised to the ASD BoK.

Each of the identified AS&D business processes are described by the process' purpose, and the general knowledge and skills required for the successful execution of the business process. Within each business process, the associated tasks are further analyzed and the knowledge types and levels identified. Figure 5 provides a visual example of the task-level knowledge requirements.

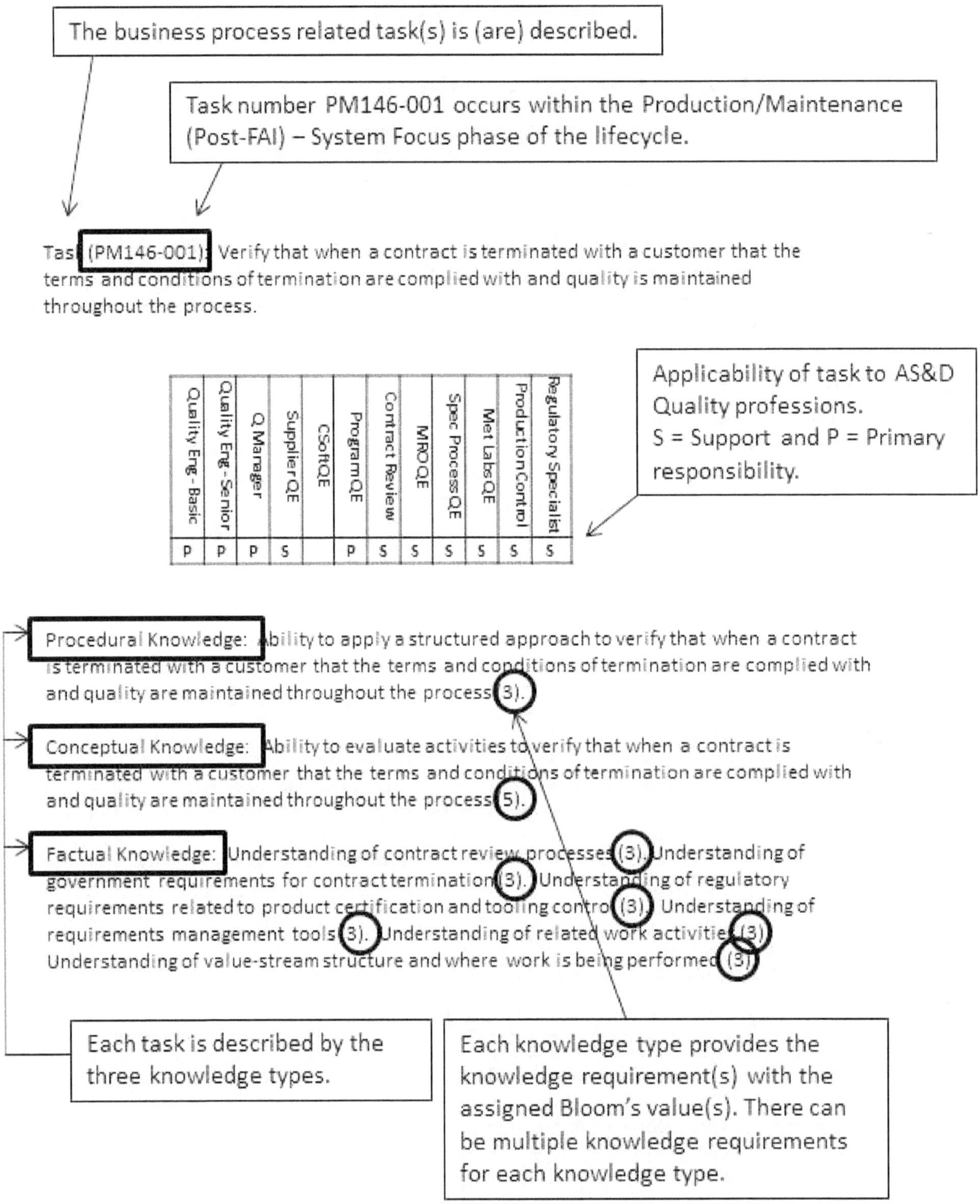

Figure 5, Task-level definition.

Assigning Responsibilities to the AS&D Quality Professions

As a guideline, each task has an associated responsibility matrix assigned. The purpose of the responsibility matrix is to indicate, generally, which AS&D Quality professions are responsible for the execution of a specific task. Within the matrix, there is an "S" or "P" indicated to show if the assignment of task responsibility is a "Support" role or one of "Primary" responsibility. In a support role, the identified profession is not primarily responsible for the execution of the task, and therefore does not require the same level of knowledge and/or skill as a primary responsibility holder. Professions that carry an assignment of primary ("P") responsibility should meet the levels of knowledge and skills identified for the task. Where a task level matrix is blank, it is not typical for that profession to carry a primary or support responsibility for task execution.

The AS&D industry is very diverse and the assignment of task responsibility can vary from organization-to-organization. The provided responsibility matrices are to be used as guidelines for industry-wide professional development and professional certification. As presented in this book, the AS&D Quality BoK should be considered a standard of professionalism at the process and task level. Organizations should be the final determining entities for the assignment of work responsibilities.

However, in the interest of developing global standardization of the AS&D Quality profession, the Aerospace and Defense Learning Institute (ADLI) applies the guidelines provided in this book as its Standard for the development of professional certification. Table 4 provides a listing of the Quality professions that are included in the existing assignment of responsibility matrix. As the AS&D Auditor profession is presently managed by industry, that profession is not included in this matrix.

Profession Abbreviation	Full Professional Title
Quality Eng—Basic	AS&D Quality Engineer—Level 1 (Practitioner)
Quality Eng—Senior	AS&D Quality Engineer—Level 2 (Senior)
Q Manager	AS&D Quality Manager
Supplier QE	AS&D Supplier Quality Engineer
CSoftQE	AS&D Software Quality Engineer
Program QE	AS&D Program Quality Engineer
Contract Review	AS&D Contract Review Quality Engineer
MRO QE	AS&D Maintenance, Repair and Overhaul Quality Engineer
Spec Process QE	AS&D Special Process Quality Engineer
Mat Labs QE	AS&D Material Laboratory Quality Engineer
Production Control	AS&D Production Control Quality Engineer
Regulatory Specialist	AS&D Regulatory Quality and Certification Specialist

Table 4, The major AS&D Quality professions (Auditor not included).

The Detailed AS&D Quality BoK

The following AS&D Quality BoK is presented utilizing the topic categorization developed by the ASQ for its Quality Engineer BoK. The ASQ Quality Engineer BoK is not duplicated in this book.

I.G.1 Contracts

Process: Ensure Contract Terminations—Organizational

Process Purpose: To ensure when a contract is terminated with a customer that the terms and conditions of termination are complied with and quality is maintained throughout the process of production and services to the completion of termination.

Knowledge and Skill Requirements: Understanding of contract review processes. As applicable, understanding of government requirements for contract termination. Ability to verify that when a contract is terminated with a customer that the terms and conditions of termination are complied with and quality is maintained throughout the process. Ability to perform process, product, and QMS audits.

Task (PM146-001): Verify that when a contract is terminated with a customer that the terms and conditions of termination are complied with and quality is maintained throughout the process.

Quality Eng—Basic	Quality Eng—Senior	Q Manager	Supplier QE	CSoftQE	Program QE	Contract Review	MRO QE	Spec Process QE	Met Labs QE	Production Control	Regulatory Specialist
P	P	P	S		P	S	S	S	S	S	S

Procedural Knowledge: Ability to apply a structured approach to verify that when a contract is terminated with a customer that the terms and conditions of termination are complied with and quality is maintained throughout the process (3).

Conceptual Knowledge: Ability to evaluate activities to verify that when a contract is terminated with a customer that the terms and conditions of termination are complied with and quality are maintained throughout the process (5).

Factual Knowledge: Understanding of contract review processes (3). Understanding of government requirements for contract termination (3). Understanding of regulatory requirements related to product certification and tooling control (3). Understanding of requirements management tools (3). Understanding of related work activities (3). Understanding of value stream structure and where work is being performed (3).

I.J. Knowledge Management

Process: Monitor Changes in Value Stream Personnel (People Capability)

Process Purpose: To ensure supplier/partner personnel have the appropriate capability and understanding to meet design and contract requirements. Through surveillance and reporting, monitor supplier/partner personnel changes. Evaluate changes of key personnel to ensure appropriate capabilities are maintained.

Knowledge and Skill Requirements: Ability to identify any specialized skill requirements related to the program, throughout the value stream. Ability to evaluate that supplier/partner and key sub-tier supplier personnel have the appropriate capability and understanding to meet design and contract requirements. Ability to monitor, through monitoring and reporting, supplier/partner and key sub-tier supplier personnel changes. Ability to evaluate changes of key personnel to ensure appropriate capabilities are maintained.

Task 1 (PM210-001): Identify any specialized skill requirements related to the program, throughout the value stream.

Quality Eng—Basic	Quality Eng—Senior	Q Manager	Supplier QE	CSoftQE	Program QE	Contract Review	MRO QE	Spec Process QE	Met Labs QE	Production Control	Regulatory Specialist
S	S	S	P		P			S	S		

Procedural Knowledge: Ability to apply a structured process to identify any specialized skill and/or certification requirements related to the program, throughout the value stream (3).

Conceptual Knowledge: Ability to analyze project/program, design, and production/servicing processes to identify any specialized skill and/or certification requirements related to the program, throughout the value stream (4).

Task 2 (PM210-002): Evaluate that supplier/partner and key sub-tier supplier personnel have the appropriate capability and certifications to meet design and contract requirements.

Quality Eng—Basic	Quality Eng—Senior	Q Manager	Supplier QE	CSoftQE	Program QE	Contract Review	MRO QE	Spec Process QE	Met Labs QE	Production Control	Regulatory Specialist
S	S	S	P		P			S	S		

Procedural Knowledge: Ability to apply a structured approach to assess supplier/partner and key sub-tier supplier personnel have the appropriate capability and certifications to meet design and contract requirements (3).

Conceptual Knowledge: Ability to analyze the capabilities of supplier/partner and key sub-tier supplier personnel and verify that they have the appropriate capability and certifications to meet design and contract requirements (4).

Factual Knowledge: Understanding of regulatory requirements related to personnel qualifications (3). Understanding of contractual and project/program requirements related to personnel qualifications (3). Understanding of requirements management tools (3). Understanding of technical work processes related to the project/program (3). Understanding of industry standards related to applied technologies of the project/program (3). Understanding of personnel skill qualification and certification practices (3). Knowledge of value stream structure and where work is being performed (3).

Task 3 (PM210-003): Monitor, through on-going over-sight and reporting, supplier/partner and key sub-tier supplier personnel changes. Evaluate changes of key personnel to ensure appropriate capabilities are maintained.

Quality Eng—Basic	Quality Eng—Senior	Q Manager	Supplier QE	CSoftQE	Program QE	Contract Review	MRO QE	Spec Process QE	Met Labs QE	Production Control	Regulatory Specialist
S	S	S	P		P			S	S		

Procedural Knowledge: Ability to apply a structured approach to monitor, through on-going over-sight and reporting, changes to supplier/partner and key sub-tier supplier personnel (3).

Conceptual Knowledge: Ability to apply a structured approach to monitor, through on-going over-sight and reporting, changes to supplier/partner and key sub-tier supplier personnel and analyze potential impact to project/program performance (4).

Factual Knowledge: Understanding of regulatory requirements related to personnel qualifications (3). Understanding of requirements management tools (3). Understanding of contractual and project/program requirements related to personnel qualifications (3). Understanding of technical work processes related to the project/program (3). Understanding of industry standards related to applied technologies of the project/program (3). Understanding of personnel skill qualification and certification practices (3). Knowledge of value stream structure and where work is being performed (3).

Task 4 (PM211-001): Identify any specialized skill requirements related to the program, throughout the value stream.

Quality Eng—Basic	Quality Eng—Senior	Q Manager	Supplier QE	CSoftQE	Program QE	Contract Review	MRO QE	Spec Process QE	Met Labs QE	Production Control	Regulatory Specialist
S	S	S	P		P			S	S		

Procedural Knowledge: Ability to apply a structured process to identify any specialized skill and/or certification requirements related to the program, throughout the value stream (3).

Conceptual Knowledge: Ability to analyze project/program, design, and production/servicing processes to identify any specialized skill and/or certification requirements related to the program, throughout the value stream (4).

Factual Knowledge: Understanding of regulatory requirements related to personnel qualifications (3). Understanding of contractual and project/program requirements related to personnel qualifications (3). Understanding of requirements management tools (3). Understanding of technical work processes related to the project/program (3). Understanding of industry standards related to applied technologies of the project/program (3). Understanding of personnel skill qualification and certification practices (3).

Process: Monitor Changes in Organizational Personnel

Process Purpose: To ensure organizational personnel have the appropriate capability and understanding to meet design and contract requirements. Through surveillance and reporting, monitor organizational personnel changes. Evaluate changes of key personnel to ensure appropriate capabilities are maintained.

Knowledge and Skill Requirements: Ability to identify any specialized skill requirements related to the program. Ability to evaluate that organizational personnel have the appropriate capability and understanding to meet design and contract requirements. Ability to monitor, through monitoring and reporting, organizational personnel changes. Ability to evaluate changes of key personnel to ensure appropriate capabilities are maintained.

Task 1 (PM211-002): Evaluate that organizational personnel have the appropriate capability and certifications to meet design, contract, and regulatory requirements.

Quality Eng—Basic	Quality Eng—Senior	Q Manager	Supplier QE	CSoftQE	Program QE	Contract Review	MRO QE	Spec Process QE	Met Labs QE	Production Control	Regulatory Specialist
S	P	P	S		P		P	P	P	P	P

Procedural Knowledge: Ability to apply a structured approach to assess organizational personnel have the appropriate capability and certifications to meet design and contract requirements (3).

Conceptual Knowledge: Ability to analyze the capabilities of organizational personnel and verify that they have the appropriate capability and certifications to meet design and contract requirements (4).

Factual Knowledge: Understanding of regulatory requirements related to personnel qualifications (3). Understanding of contractual and project/program requirements related to personnel qualifications (3). Understanding of requirements management tools (3). Understanding of technical work processes related to the project/program (3). Understanding of industry standards related to applied technologies of the project/program (3). Understanding of personnel skill qualification and certification practices (3). Knowledge of where work is being performed (3).

Task 2 (PM211-003): Monitor, through on-going over-sight and reporting, organizational personnel changes. Evaluate changes of key personnel to ensure appropriate capabilities are maintained.

Quality Eng—Basic	Quality Eng—Senior	Q Manager	Supplier QE	CSoftQE	Program QE	Contract Review	MRO QE	Spec Process QE	Met Labs QE	Production Control	Regulatory Specialist
S	P	P	S		P		P	P	P	P	P

Procedural Knowledge: Ability to apply a structured approach to monitor, through on-going over-sight and reporting, changes to organizational personnel (3).

Conceptual Knowledge: Ability to apply a structured approach to monitor, through on-going over-sight and reporting, changes to organizational personnel and analyze potential impact to project/program performance (4).

Factual Knowledge: Understanding of regulatory requirements related to personnel qualifications (3). Understanding of requirements management tools (3). Understanding of contractual and project/program requirements related to personnel qualifications (3). Understanding of technical work processes related to the project/program (3). Understanding of industry standards related to applied technologies of the project/program (3). Understanding of personnel skill qualification and certification practices (3). Knowledge of where work is being performed (3).

Process: Monitor Supplier Resource Capacity

Process Purpose: To ensure the effectiveness of the supplier's/partner's production/servicing capacity. Through review of supplier/partner notification of changes in capacity and/or capability, assess the potential impact of those changes on product and system quality. During on-going assessments and communication with supplier/partners, establish awareness of potential capability/capacity changes.

Knowledge and Skill Requirements: Understanding of industrial capacity measurement methods. Understanding of value stream mapping processes. Understanding of MRP processes. Ability to evaluate changes in supplier/partner facilities, personnel, and work volume and potential impact on their ability to continuously meet design, quality, and production/maintenance requirements. During on-going assessments and communication with supplier/partners, ability to recognize and quantify potential capability/capacity changes. Ability to effectively communicate changes in value stream capacity measures to management.

Task 1 (PM255-001): Evaluate changes in supplier/partner facilities, personnel, and work volume and potential impact on their ability to continuously meet design, quality, and production/maintenance requirements (includes communication with supplier/partner).

Quality Eng—Basic	Quality Eng—Senior	Q Manager	Supplier QE	CSoftQE	Program QE	Contract Review	MRO QE	Spec Process QE	Met Labs QE	Production Control	Regulatory Specialist
S	S	S	P		P			S	S		

Procedural Knowledge: Ability to apply a semi-structured approach to subject changes in supplier/partner facilities, personnel, and work volume to evaluate the potential impact on their ability to continuously meet design, quality, and production/maintenance requirements (4).

Conceptual Knowledge: Ability to perform evaluation of changes in supplier/partner facilities, personnel, and work volume to determine the potential impact on their ability to continuously meet design, quality, and production/maintenance requirements (5).

Factual Knowledge: Understanding of industrial capacity measurement methods (3). Understanding of value stream mapping processes (3). Understanding of MRP processes (3). Understanding of bottleneck locations in production/servicing flow (3). Understanding of the value stream structure and where work is performed (3).

Task 2 (PM255-002): Ability to effectively communicate changes in value stream capacity measures to management.

Quality Eng—Basic	Quality Eng—Senior	Q Manager	Supplier QE	CSoftQE	Program QE	Contract Review	MRO QE	Spec Process QE	Met Labs QE	Production Control	Regulatory Specialist
S	S	S	P		P			S	S		

Procedural Knowledge: Ability to apply a structured approach to effectively communicate changes in value stream capacity measures to management (3).

Conceptual Knowledge: Ability to effectively communicate changes in value stream capacity measures to management (4).

Factual Knowledge: Understanding of industrial capacity measurement methods (3). Understanding of value stream mapping processes (3). Understanding of MRP processes (3). Understanding of bottleneck locations in production/servicing flow (3). Understanding of the value stream structure and where work is performed (3).

I.M. Material Resource Planning (MRP)

Process: Coordinate Scheduling (MRP)

Process Purpose: To ensure adequate support is provided (e.g., inspection, testing, etc.), ensure critical points in material flow are recognized and planned. The planning process should be a joint effort to identify timing and work scope. This planning may include the identification of critical skills, special equipment, or logistics considerations.

Knowledge and Skill Requirements: Understanding of material resource planning processes. Ability to review MRP and determine the scope of quality related support. Ability to schedule adequate quality/inspection support (e.g., inspection, testing, etc.), and ensure critical points in material flow are recognized and planned. Based on MRP requirements, ability to determine if critical skills, special equipment, or logistics considerations are required. Ability to create quality/inspection support that provides a visual representation of timing and extent of support required.

Task 1 (PM080-001): Review MRP and determine the scope of quality related support.

Quality Eng—Basic	Quality Eng—Senior	Q Manager	Supplier QE	CSoftQE	Program QE	Contract Review	MRO QE	Spec Process QE	Met Labs QE	Production Control	Regulatory Specialist
P	P	P	P		P		P	P	P	P	S

Procedural Knowledge: Ability to apply a structured approach to review MRP requirements and estimate the type and quantity of required quality related support (4).

Conceptual Knowledge: Ability to perform evaluation in review of MRP requirements and schedules to identify quality support requirements (5). Ability to analyze requirements evaluation results and estimate the type, quantity, and timing of required quality related support (4).

Factual Knowledge: Understanding of requirements management system (3). Understanding of MRP schedules (3). Understanding of resource planning practices (3). Understanding of organizational resources related to quality and testing (3).Understanding of where work is performed (3). Understanding of value stream structure (3). Understanding of quality processes related to the project/program (3). Understanding of capabilities of quality support services (3).

Task 2 (PM080-002): Schedule adequate quality/inspection support (e.g., inspection, testing, etc.), and ensure critical points in material flow are recognized and planned.

Quality Eng—Basic	Quality Eng—Senior	Q Manager	Supplier QE	CSoftQE	Program QE	Contract Review	MRO QE	Spec Process QE	Met Labs QE	Production Control	Regulatory Specialist
S	S	P	P		P		P	S	S	P	

Procedural Knowledge: Ability to apply a structured approach to schedule adequate quality/inspection support (e.g., inspection, testing, etc.), and ensure critical points in material flow are recognized and planned (4).

Conceptual Knowledge: Ability to analyze quality and testing resource needs and schedule adequate support (e.g., inspection, testing, etc.) as required (4). Ability to analyze established work activities to ensure critical points in material flow are recognized and planned for quality/inspection support (4).

Factual Knowledge: Understanding of requirements management system (3). Understanding of MRP schedules (3). Understanding of resource planning practices (3). Understanding of organizational resources related to quality and testing (3).Understanding of where work is performed (3). Understanding of value stream structure (3). Understanding of quality processes related to the project/program (3). Understanding of capabilities of quality support services (3).

Task 3 (PM080-003): Determine if critical skills, special equipment, or logistics considerations are required to provide adequate support of MRP requirements (Quality related).

Quality Eng—Basic	Quality Eng—Senior	Q Manager	Supplier QE	CSoftQE	Program QE	Contract Review	MRO QE	Spec Process QE	Met Labs QE	Production Control	Regulatory Specialist
P	P	P	P		P		S	S	S	S	

Procedural Knowledge: In review of resource analysis, the ability to apply a structured approach to identify if critical skills, special equipment, or logistics considerations are required (3).

Conceptual Knowledge: The ability to analyze resource requirements and identify if critical skills, special equipment, or logistics considerations are required (4).

Factual Knowledge: Understanding of requirements management system (3). Understanding of unique project/program requirements related to MRP (3). Understanding of organizational resources related to quality and testing (3).Understanding of where work is performed (3). Understanding of value stream structure (3). Understanding of quality processes related to the project/program (3). Understanding of capabilities of quality support services (3). Understanding of skill requirements related to quality/inspection processes (3).

Task 4 (PM080-004): Develop quality/inspection support planning to provide a visual representation of timing and extent of support required, throughout the value stream.

Quality Eng—Basic	Quality Eng—Senior	Q Manager	Supplier QE	CSoftQE	Program QE	Contract Review	MRO QE	Spec Process QE	Met Labs QE	Production Control	Regulatory Specialist
P	P	P	P		P		S	S	S	S	

Procedural Knowledge: Ability to apply a structured approach to develop quality/inspection support planning to provide a visual representation of timing and extent of support required, throughout the value stream (3).

Conceptual Knowledge: Ability to develop quality/inspection support planning to provide a visual representation of timing and extent of support required, throughout the value stream (3).

Factual Knowledge: Understanding of MRP schedules (3). Understanding of resource planning practices (3). Understanding of organizational resources related to quality and testing (3).Understanding of where work is performed (3). Understanding of value stream structure (3). Understanding of quality processes related to the project/program (3). Understanding of capabilities of quality support services (3). Understanding of schedule planning processes (3). Understanding of effective communication processes (3).

Process: Monitor Scheduling (MRP)

Process Purpose: To ensure action is taken at critical points in material flow, as defined in program planning. Ensure skills, equipment, and other considerations are available when needed. Document the outcomes of planned activities. In the event that conditions prevent the execution of activities as planned, ensure remedial planning is accomplished to ensure tasks are effectively performed.

Knowledge and Skill Requirements: Understanding of material resource planning processes. Ability to review MRP and understand the scope of quality related support. Ability to verify that support action is taken at critical points in material flow, as defined in program planning. Ability to verify that skills, equipment, and other considerations are available when needed. Ability to document the outcomes of planned activities. In the event that conditions prevent the execution of activities as planned, the ability to verify that remedial planning is accomplished to ensure tasks are effectively performed. Ability to validate that the entire supplier base understands how to communicate with the procuring system to ensure accurate authority and direction is provided.

Task 1 (PM085-001): Monitor and verify that quality/inspection support action is taken at critical points in material flow, as defined in program planning.

Quality Eng—Basic	Quality Eng—Senior	Q Manager	Supplier QE	CSoftQE	Program QE	Contract Review	MRO QE	Spec Process QE	Met Labs QE	Production Control	Regulatory Specialist
S	P	P	P		P		P	P	P	P	

Procedural Knowledge: Ability to apply a structured approach to monitor and verify that quality/inspection support action is taken at critical points in material flow, as defined in program planning (3).

Conceptual Knowledge: Ability to monitor and perform analysis to verify that quality/inspection support action is taken at critical points in material flow, as defined in program planning (4).

Factual Knowledge: Understanding of MRP schedules (3). Understanding of resource planning practices (3). Understanding of organizational resources related to quality and testing (3).Understanding of where work is performed (3). Understanding of value stream structure (3). Understanding of quality processes related to the project/program (3). Understanding of capabilities of quality support services (3). Understanding of schedule planning processes (3).

Task 2 (PM085-002): Verify that skills, equipment, and other considerations are available when needed.

Quality Eng—Basic	Quality Eng—Senior	Q Manager	Supplier QE	CSoftQE	Program QE	Contract Review	MRO QE	Spec Process QE	Met Labs QE	Production Control	Regulatory Specialist
S	P	P	P		P		P	P	P	P	

Procedural Knowledge: Ability to apply a structured approach to monitor, assess and verify that skills, equipment, and other considerations are available when needed (3).

Conceptual Knowledge: Ability to monitor, assess and verify that skills, equipment, and other considerations are available when needed (3).

Factual Knowledge: Understanding of MRP schedules (3). Understanding of resource planning practices (3). Understanding of organizational resources related to quality and testing (3).Understanding of where work is performed (3). Understanding of value stream structure (3). Understanding of quality processes related to the project/program (3). Understanding of capabilities of quality support services (3). Understanding of schedule planning processes (3).

Task 3 (PM085-003): In the event that conditions prevent the execution of activities as planned, the ability to verify that remedial planning is accomplished to ensure tasks are effectively performed.

Quality Eng—Basic	Quality Eng—Senior	Q Manager	Supplier QE	CSoftQE	Program QE	Contract Review	MRO QE	Spec Process QE	Met Labs QE	Production Control	Regulatory Specialist
P	P	P	P		P		P	P	P	P	

Procedural Knowledge: In the event that conditions prevent the execution of activities as planned, the ability to apply a structured process to obtain necessary support resources and establish remedial planning to ensure tasks are effectively performed (3).

Conceptual Knowledge: In the event that conditions prevent the execution of activities as planned, the ability to analyze the impact to schedule and coordinate necessary support resources (3). Ability to develop remedial planning to effectively communicate remedial actions (4).

Factual Knowledge: Understanding of MRP schedules (3). Understanding of resource planning practices (3). Understanding of organizational resources related to quality and testing (3).Understanding of where work is performed (3). Understanding of value stream structure (3). Understanding of quality processes related to the project/program (3). Understanding of capabilities of quality support services (3). Understanding of schedule planning processes (3).

II.B. Documentation of the Quality System

Process: Participate in Process/Procedure Ballot (Sign-off) Review

Process Purpose: To provide input to changes in policies and procedures. Ensure changes to policies and procedures are appropriate with regards to product design, quality requirements (FAI, QMS, standards, FAR), organizational culture, available resources, and equipment.

Knowledge and Skill Requirements: Understanding of procedure management systems. Ability to review proposed changes to procedures and provide input to changes in policies and procedures. Ability to verify proposed changes to policies and procedures are appropriate with regards to product design, quality requirements (FAI, QMS, standards, FAR), organizational culture, available resources, and equipment.

Task (PM030-001): Review proposed changes to procedures and provide input to changes in policies and procedures. Verify proposed changes to policies and procedures are appropriate with regards to product design, quality requirements (FAI, QMS, standards, FAR), organizational culture, available resources, and equipment.

Quality Eng—Basic	Quality Eng—Senior	Q Manager	Supplier QE	CSoftQE	Program QE	Contract Review	MRO QE	Spec Process QE	Met Labs QE	Production Control	Regulatory Specialist
P	P	P	P	P	P	P	P	P	P	P	P

Procedural Knowledge: Ability to apply a defined and structured process to review proposed changes to procedures and provide input to changes in policies and procedures (3). Ability to apply a defined and structured process to verify proposed changes to policies and procedures are appropriate with regards to product design, quality requirements (FAI, QMS, standards, FAR), organizational culture, available resources, and equipment (3).

Conceptual Knowledge: Ability to analyze proposed changes to procedures and provide input to changes in policies and procedures (4). Ability to evaluate proposed changes to policies and procedures to ensure changes are appropriate and compliant with regards to product design, quality requirements (FAI, QMS, standards, FAR), organizational culture, available resources, and equipment (5). Ability to evaluate proposed changes to assess the effects on resources and schedules (5).

Factual Knowledge: Knowledge of regulatory requirements related to quality management systems (3). Knowledge of contractual requirements related to quality management systems. Understanding of quality management systems and related standards (e.g., AS/EN/SJAC 9100, 9110) (3). Understanding of procedure balloting processes (3). Understanding of organizational and value stream resources (3). Understanding of cultural issues related to organizational systems (4). Understanding of value stream structure (3). Understanding of where work is performed (3). Understanding of applicable work processes and applied technologies (4).

Process: Ensure Records & Documentation Are Appropriately Maintained

Process Purpose: To ensure the documentation required to document traceability, product configuration, and quality acceptance are appropriately maintained. Ensure the retention times and locations of records complies with contract requirements. For electronic records, ensure data is backed-up at appropriate locations and content is protected from derogation.

Knowledge and Skill Requirements: Understanding of applicable industry standards that require retention of quality records. Ability to ensure the documentation required for documentation of traceability, product configuration, and quality acceptance is appropriately maintained. Ability to verify that retention times and locations of records complies with contract requirements. For electronic records, ability to ensure data is backed-up at appropriate locations and content is protected from derogation. Ability to ensure the documentation required to document traceability, product configuration, and quality acceptance are appropriately maintained throughout the supplier/partner base to include retention and safeguarding of data content. Understanding of the legal requirements related to quality records.

Task 1 (PM055-001): Ensure the documentation required to document traceability, product configuration, and quality acceptance is appropriately maintained, throughout the value stream.

Quality Eng—Basic	Quality Eng—Senior	Q Manager	Supplier QE	CSoftQE	Program QE	Contract Review	MRO QE	Spec Process QE	Met Labs QE	Production Control	Regulatory Specialist
P	P		P		P		P	P	P	P	

Procedural Knowledge: Ability to apply a defined and structured assessment process to verify that the documentation required for documentation of traceability, product configuration, and quality acceptance are appropriately maintained, throughout the value stream (3).

Conceptual Knowledge: Ability to analyze established processes and quality records to verify that the documentation required for documentation of traceability, product configuration, and quality acceptance are appropriately maintained, throughout the value stream (4).

Factual Knowledge: Understanding of regulatory requirements related to documentation control (3). Understanding of contractual requirements related to documentation control (3). Understanding of industry requirements related to documentation control (3). Understanding of where work is performed (3). Understanding of value stream structure (3). Understanding of legal requirements related to documentation control (3).

Task 2 (PM055-002): Verify that retention times and locations of records complies with requirements, throughout the value stream.

Quality Eng—Basic	Quality Eng—Senior	Q Manager	Supplier QE	CSoftQE	Program QE	Contract Review	MRO QE	Spec Process QE	Met Labs QE	Production Control	Regulatory Specialist
P	P		P		P		S	S	S	S	

Procedural Knowledge: Ability to apply a defined and structured process to verify that retention times and locations of records complies with requirements, throughout the value stream (3).

Conceptual Knowledge: Ability to assess at all applicable points that retention times and locations of records complies with requirements, throughout the value stream (3).

Factual Knowledge: Understanding of regulatory requirements related to documentation control (3). Understanding of contractual requirements related to documentation control (3). Understanding of industry requirements related to documentation control (3). Understanding of where work is performed (3). Understanding of value stream structure (3). Understanding of legal requirements related to documentation retention (3).

Task 3 (PM055-003): For electronic records, ensure data is backed-up at appropriate locations and content is protected from degradation, throughout the value stream.

Quality Eng—Basic	Quality Eng—Senior	Q Manager	Supplier QE	CSoftQE	Program QE	Contract Review	MRO QE	Spec Process QE	Met Labs QE	Production Control	Regulatory Specialist
P	P		P		S		S	S	S	S	

Procedural Knowledge: Ability to apply a defined and structured assessment process to verify that electronic records are backed-up at appropriate locations and content is protected from degradation (3).

Conceptual Knowledge: Ability to analyze systems and records to verify that electronic records are backed-up at appropriate locations and content is protected from degradation (4).

Factual Knowledge: Understanding of regulatory requirements related to documentation control (3). Understanding of contractual requirements related to documentation control (3). Understanding of industry requirements related to documentation control (3). Understanding of electronic media control practices (3). Understanding of where work is performed (3). Understanding of value stream structure (3). Understanding of legal requirements related to control of electronic records (3).

Process: Monitor Value Stream Documentation Control Processes

Process Purpose: Within the value stream, to ensure documentation control processes are appropriately controlled, to include changes reviewed and approved by authorized persons, changes implemented at the appropriate configuration control level, and implemented into appropriate work documents. Ensure documents are retained in accordance with specific requirements.

Knowledge and Skill Requirements: Understanding of production and inspection planning processes. Understanding of applicable regulatory and industry standard requirements. Ability to verify that documentation control processes are appropriately controlled, to include changes reviewed and approved by authorized persons, changes implemented at the appropriate configuration control level, and implemented into appropriate work documents. Ability to verify that production and inspection records are retained in accordance with specific requirements. Ability to evaluate the effectiveness of supplier/partner documentation control processes at sub-tier suppliers to ensure compliance to requirements. Ability to verify documentation retention and storage at all tiers of the supplier/partner base.

Task (PM215-002): Evaluate the effectiveness of supplier/partner documentation control processes at sub-tier supplier to ensure compliance to requirements. Verify documentation retention and storage at all tiers of the supplier/partner base.

Quality Eng—Basic	Quality Eng—Senior	Q Manager	Supplier QE	CSoftQE	Program QE	Contract Review	MRO QE	Spec Process QE	Met Labs QE	Production Control	Regulatory Specialist
S	S	S	P		P						

Procedural Knowledge: Ability to apply defined and structured assessment processes throughout the value stream to verify the effectiveness and compliance of supplier/partner documentation control processes at sub-tier suppliers (3). Ability to apply defined and structured assessment processes to verify compliance to documentation retention and storage requirements at all tiers of the supplier/partner base (3).

Conceptual Knowledge: Ability to assess and analyze the effectiveness and compliance of supplier/partner documentation control processes at sub-tier suppliers (4). Ability to assess and analyze the compliance to documentation retention and storage requirements at all tiers of the supplier/partner base (4).

Factual Knowledge: Understanding of requirements management tools (3). Understanding of production/servicing and inspection planning processes (3). Understanding of applicable regulatory and industry standard requirements related to production/servicing and inspection planning (3). Understanding of configuration control processes (3). Understanding of documentation review and approval processes (3). Understanding of value stream structure and where work is being performed (3). Understanding of audit practices (3).

Process: Monitor Organizational Documentation Control Processes

Process Purpose: Within the organization, to ensure documentation control processes are appropriately controlled, to include changes reviewed and approved by authorized persons, changes implemented at the appropriate configuration control level, and implemented into appropriate work documents. Ensure documents are retained in accordance with specific requirements.

Knowledge and Skill Requirements: Understanding of production and inspection planning processes. Understanding of applicable regulatory and industry standard requirements. Ability to verify that documentation control processes are appropriately controlled, to include changes reviewed and approved by authorized persons, changes implemented at the appropriate configuration control level, and implemented into appropriate work documents. Ability to verify that production and inspection records are retained in accordance with specific requirements. Ability to evaluate the effectiveness of supplier/partner documentation control processes at sub-tier suppliers to ensure compliance to requirements. Ability to verify documentation retention and storage within the organization.

Task (PM216-001): Evaluate the effectiveness of organizational documentation control processes at sub-tier supplier to ensure compliance to requirements. Verify documentation retention and storage at all levels of the organization.

Quality Eng—Basic	Quality Eng—Senior	Q Manager	Supplier QE	CSoftQE	Program QE	Contract Review	MRO QE	Spec Process QE	Met Labs QE	Production Control	Regulatory Specialist
P	P	P	S	P	P		S	S	S	S	S

Procedural Knowledge: Ability to apply defined and structured assessment processes to verify the effectiveness and compliance to documentation control processes (3). Ability to apply defined and structured assessment processes to verify compliance to documentation retention and storage requirements at all levels of the organization (3).

Conceptual Knowledge: Ability to assess and analyze the effectiveness and compliance of supplier/partner documentation control processes (4). Ability to assess and analyze the compliance to documentation retention and storage requirements at all levels of the organization (4).

Factual Knowledge: Understanding of requirements management tools (3). Understanding of production/servicing and inspection planning processes (3). Understanding of applicable regulatory and industry standard requirements related to production/servicing and inspection planning (3). Understanding of configuration control processes (3). Understanding of documentation review and approval processes (3). Understanding of where work is being performed (3). Understanding of audit practices (3).

II.D.3. Audit Planning and Implementation

Process: Ensure Supplier Self Audit (QMS)

Process Purpose: To ensure suppliers/partners perform self-assessments to ensure compliance to their QMS and contract requirements (internal audit). Records of self-assessments must be maintained as quality records. Persons performing the self-assessments must be qualified to perform self-assessments. Records shall contain the scope of the assessment, elements of the assessment that were not performed, and findings and observations discovered during the assessment. Ensure past findings and observations are included in audit scope planning. Ensure records are available for review by external auditors.

Knowledge and Skill Requirements: Understanding of quality management systems. Understanding of QMS standards. Understanding of applicable regulatory requirements related to internal audits. Ability to perform QMS evaluations. Ability to verify that suppliers/partners perform self-assessments to ensure compliance to their QMS and contract requirements (internal audit). Ability to verify that records of self-assessments are maintained as quality records. Ability to validate that persons performing the self-assessments are qualified to perform self-assessments. Ability to review and verify that records contain the scope of the assessment, elements of the assessment that were not performed, and findings and observations discovered during the assessment. Ability to verify that appropriate objective evidence is maintained. Ability to verify that past findings and observations are included in audit scope planning. Ability to verify that appropriate root cause/corrective action is undertaken to include appropriate follow up actions. Ability to verify that records are available for review by external auditors.

Task 1 (PM090-001): Verify that suppliers/partners perform self-assessments to ensure compliance to their QMS and contract requirements (internal audit).

Quality Eng—Basic	Quality Eng—Senior	Q Manager	Supplier QE	CSoftQE	Program QE	Contract Review	MRO QE	Spec Process QE	Met Labs QE	Production Control	Regulatory Specialist
		S	P		S						

Procedural Knowledge: Ability to apply a defined and structured assessment process to verify that suppliers/partners perform self-assessments to ensure compliance to their QMS and contract requirements (internal audit) (3).

Conceptual Knowledge: Ability to audit suppliers/partners self-assessment processes to ensure compliance to their QMS and contract requirements (internal audit) (3).

Factual Knowledge: Understanding of requirements management system (3). Understanding of regulatory requirements related to internal audits (3). Understanding of contractual and project/program requirements related to internal audit (3). Understanding of industry standards related to internal audit (3). Understanding of supplier audit practices (3). Understanding of the value stream structure and where work is being executed (3).

Task 2 (PM090-002): Verify that records of self-assessments are maintained as quality records, throughout the value stream.

Quality Eng—Basic	Quality Eng—Senior	Q Manager	Supplier QE	CSoftQE	Program QE	Contract Review	MRO QE	Spec Process QE	Met Labs QE	Production Control	Regulatory Specialist
		S	P		S						

Procedural Knowledge: Ability to apply a defined and structured assessment process to verify that records of self-assessments are maintained as quality records, throughout the value stream (3).

Conceptual Knowledge: Ability to audit supplier/partner internal audit processes to verify that records of self-assessments are maintained as quality records, throughout the value stream (3).

Factual Knowledge: Understanding of regulatory requirements related to internal audits (3). Understanding of contractual and project/program requirements related to internal audit (3). Understanding of industry standards related to internal audit (3). Understanding of supplier audit practices (3). Understanding of the value stream structure and where work is being executed (3).

Task 3 (PM090-003): Validate that persons performing the self-assessments are qualified to perform self-assessments, in the value stream. (Note: Program QE support role is to ensure access to secured locations.)

Quality Eng—Basic	Quality Eng—Senior	Q Manager	Supplier QE	CSoftQE	Program QE	Contract Review	MRO QE	Spec Process QE	Met Labs QE	Production Control	Regulatory Specialist
			P		S						

Procedural Knowledge: Ability to apply a structured approach to review training and work experience to verify that persons performing the self-assessments are qualified to perform self-assessments (internal-audits) (3).

Conceptual Knowledge: Ability to review training and work experience to verify that persons performing the self-assessments are qualified to perform self-assessments (internal-audits) (3). Ability to analyze audit work product to verify that persons performing the self-assessments are qualified to perform self-assessments (internal-audits) (4).

Factual Knowledge: Understanding of regulatory requirements related to internal audits (3). Understanding of contractual and project/program requirements related to internal audit (3). Understanding of industry standards related to internal audit (3). Understanding of supplier audit practices (3). Understanding of the value stream structure and where work is being executed (3). Understanding of skill requirements related to internal auditing (3).

Task 4 (PM090-004): Review and verify that records contain the scope of the assessment, elements of the assessment that were/were not performed, and findings and observations discovered during the assessment, throughout the value stream. Verify that appropriate objective evidence is maintained, throughout the value stream. (Note: Program QE support role is to ensure access to secured locations.)

Quality Eng—Basic	Quality Eng—Senior	Q Manager	Supplier QE	CSoftQE	Program QE	Contract Review	MRO QE	Spec Process QE	Met Labs QE	Production Control	Regulatory Specialist
			P		S						

Procedural Knowledge: Ability to apply a defined and structured assessment process to verify that records contain the scope of the assessment, elements of the assessment that were/were not performed, and findings and observations discovered during the assessment (3). Ability to apply a defined and structured assessment process to verify that appropriate objective evidence is maintained (3).

Conceptual Knowledge: Ability to audit and analyze supplier/partner quality records to verify that records contain the scope of the assessment, elements of the assessment that were/were not performed, and findings and observations discovered during the assessment (3). Ability to audit and analyze supplier/partner quality records to verify that appropriate objective evidence is maintained (4).

Factual Knowledge: Understanding of requirements management systems (3). Understanding of regulatory requirements related to internal audits (3). Understanding of contractual and project/program requirements related to internal audit (3). Understanding of industry standards related to internal audit (3). Understanding of supplier audit practices (3). Understanding of the value stream structure and where work is being executed (3).

Task 5 (PM090-005): Verify that past findings and observations are included in audit scope planning (value stream focused).

Quality Eng—Basic	Quality Eng—Senior	Q Manager	Supplier QE	CSoftQE	Program QE	Contract Review	MRO QE	Spec Process QE	Met Labs QE	Production Control	Regulatory Specialist
			P		S						

Procedural Knowledge: Ability to apply a defined and structured assessment process to verify that past findings and observations are included in audit scope planning (3).

Conceptual Knowledge: Ability to audit supplier/partner internal audit planning to verify that past findings and observations are included in audit scope planning (3).

Factual Knowledge: Understanding of requirements management systems (3). Understanding of regulatory requirements related to internal audits (3). Understanding of contractual and project/program requirements related to internal audit (3). Understanding of industry standards related to internal audit (3). Understanding of supplier audit practices (3). Understanding of the value stream structure and where work is being executed (3). Understanding of audit planning practices (3).

Task 6 (PM090-006): Verify that appropriate root cause/corrective action is undertaken to include appropriate follow up actions, throughout the value stream. Verify that records are available for review by external auditors, throughout the value stream.

Quality Eng—Basic	Quality Eng—Senior	Q Manager	Supplier QE	CSoftQE	Program QE	Contract Review	MRO QE	Spec Process QE	Met Labs QE	Production Control	Regulatory Specialist
S	S		P		S		S	S	S	S	S

Procedural Knowledge: Ability to apply a defined and structured assessment process to verify that appropriate root cause/corrective action is undertaken to include appropriate follow up actions (3). Ability to apply a defined and structured assessment process to verify that records are available for review by external auditors (3).

Conceptual Knowledge: Ability to analyze quality records to verify that appropriate root cause/corrective action is undertaken to include appropriate follow up actions (4). Ability to evaluate deployed corrective actions to verify that appropriate root cause/corrective action is undertaken to include appropriate follow up actions (5). Ability to audit supplier/partner quality records to verify that records are available for review by external auditors (3).

Factual Knowledge: Understanding of requirements management systems (3). Understanding of regulatory requirements related to internal audits (3). Understanding of contractual and project/program requirements related to internal audit (3). Understanding of industry standards related to internal audit (3). Understanding of supplier audit practices (3). Understanding of the value stream structure and where work is being executed (3). Understanding of audit planning practices (3). Understanding of robust root cause corrective action processes (3).

Process: Ensure Organizational Self Audit (QMS)

Process Purpose: To ensure the organization performs self-assessments to ensure compliance to their QMS, contract, and regulatory requirements (internal audit). Records of self-assessments must be maintained as quality records. Persons performing the self-assessments must be qualified to perform self-assessments. Records shall contain the scope of the assessment, elements of the assessment that were not performed, and findings and observations discovered during the assessment. Ensure past findings and observations are included in audit scope planning. Ensure records are available for review by external auditors.

Knowledge and Skill Requirements: Understanding of quality management systems. Understanding of QMS standards. Understanding of applicable regulatory requirements related to internal audits. Ability to perform QMS evaluations. Ability to verify organizational compliance to QMS, contract, and applicable regulatory requirements (internal audit). Ability to verify that records of self-assessments are maintained as quality records. Ability to validate that persons performing the self-assessments are qualified to perform self-assessments. Ability to review and verify that records contain the scope of the assessment, elements of the assessment that were not performed, and findings and observations discovered during the assessment. Ability to verify that appropriate objective evidence is maintained. Ability to verify that past findings and observations are included in audit scope planning. Ability to verify that appropriate root cause/corrective action is undertaken to include appropriate follow up actions. Ability to verify that records are available for review by external auditors.

Task 1 (PM091-001): Verify the organization performs self-assessments to ensure compliance to QMS, contract and regulatory requirements (internal audit).

Quality Eng—Basic	Quality Eng—Senior	Q Manager	Supplier QE	CSoftQE	Program QE	Contract Review	MRO QE	Spec Process QE	Met Labs QE	Production Control	Regulatory Specialist
P	P	P	S		S		S	S	S	S	S

Procedural Knowledge: Ability to apply a defined and structured assessment process to verify the organization performs self-assessments to ensure compliance to QMS, contract requirements, and regulatory requirements (internal audit) (3).

Conceptual Knowledge: Ability to audit organizational self-assessment processes to ensure compliance to their QMS, contract requirements, and regulatory requirements (internal audit) (3).

Factual Knowledge: Understanding of requirements management system (3). Understanding of regulatory requirements related to internal audits (3). Understanding of contractual and project/program requirements related to internal audit (3). Understanding of industry standards related to internal audit (3). Understanding of organizational work processes (3). Understanding of the value stream structure and where work is being executed (3).

Task 2 (PM091-002): Verify that records of self-assessments are maintained as quality records.

Quality Eng—Basic	Quality Eng—Senior	Q Manager	Supplier QE	CSoftQE	Program QE	Contract Review	MRO QE	Spec Process QE	Met Labs QE	Production Control	Regulatory Specialist
P	P	P	S		S		S	S	S	S	S

Procedural Knowledge: Ability to apply a defined and structured assessment process to verify that records of self-assessments are maintained as quality records (3).

Conceptual Knowledge: Ability to audit internal audit processes to verify that records of self-assessments are maintained as quality records (3).

Factual Knowledge: Understanding of regulatory requirements related to internal audits (3).

Understanding of contractual and project/program requirements related to internal audit (3). Understanding of industry standards related to internal audit (3). Understanding of organizational work processes (3). Understanding of the value stream structure and where work is being executed (3).

Task 3 (PM091-003): Validate that persons performing the self-assessments are qualified to perform self-assessments.

Quality Eng—Basic	Quality Eng—Senior	Q Manager	Supplier QE	CSoftQE	Program QE	Contract Review	MRO QE	Spec Process QE	Met Labs QE	Production Control	Regulatory Specialist
P	P	P	S		S		S	S	S	S	S

Procedural Knowledge: Ability to apply a structured approach to review training and work experience to verify that persons performing the self-assessments are qualified to perform self-assessments (internal-audits) (3).

Conceptual Knowledge: Ability to review training and work experience to verify that persons performing the self-assessments are qualified to perform self-assessments (internal-audits) (3). Ability to analyze audit work product to verify that persons performing the self-assessments are qualified to perform self-assessments (internal-audits) (4).

Factual Knowledge: Understanding of regulatory requirements related to internal audits (3). Understanding of contractual and project/program requirements related to internal audit (3). Understanding of industry standards related to internal audit (3). Understanding of internal audit practices (3). Understanding of where work is being executed (3). Understanding of skill requirements related to internal auditing (3). Understanding of organizational work processes (3).

Task 4 (PM091-004): Review and verify that records contain the scope of the assessment, elements of the assessment that were/were not performed, and findings and observations discovered during the assessment. Verify that appropriate objective evidence is maintained.

Quality Eng—Basic	Quality Eng—Senior	Q Manager	Supplier QE	CSoftQE	Program QE	Contract Review	MRO QE	Spec Process QE	Met Labs QE	Production Control	Regulatory Specialist
P	P	P	S		S		S	S	S	S	S

Procedural Knowledge: Ability to apply a defined and structured assessment process to verify that records contain the scope of the assessment, elements of the assessment that were/were not performed, and findings and observations discovered during the assessment (3). Ability to apply a defined and structured assessment process to verify that appropriate objective evidence is maintained (3).

Conceptual Knowledge: Ability to audit and analyze quality records to verify that records contain the scope of the assessment, elements of the assessment that were/were not performed, and findings and observations discovered during the assessment (3). Ability to audit and analyze quality records to verify that appropriate objective evidence is maintained (4).

Factual Knowledge: Understanding of requirements management systems (3). Understanding of regulatory requirements related to internal audits (3). Understanding of contractual and project/program requirements related to internal audit (3). Understanding of industry standards related to internal audit (3). Understanding of internal audit practices (3). Understanding of where work is being executed (3). Understanding of organizational work processes (3).

Task 5 (PM091-005): Verify that past findings and observations are included in audit scope planning.

Quality Eng—Basic	Quality Eng—Senior	Q Manager	Supplier QE	CSoftQE	Program QE	Contract Review	MRO QE	Spec Process QE	Met Labs QE	Production Control	Regulatory Specialist
P	P	P	S		S		S	S	S	S	S

Procedural Knowledge: Ability to apply a defined and structured assessment process to verify that past findings and observations are included in audit scope planning (3).

Conceptual Knowledge: Ability to assess internal audit planning to verify that past findings and observations are included in audit scope planning (3).

Factual Knowledge: Understanding of requirements management systems (3). Understanding of regulatory requirements related to internal audits (3). Understanding of contractual and project/program requirements related to internal audit (3). Understanding of industry standards related to internal audit (3). Understanding of internal audit practices (3). Understanding of where work is being executed (3). Understanding of audit planning practices (3). Understanding of organizational work processes (3).

Task 6 (PM091-006): Verify that appropriate root cause/corrective action is undertaken to include appropriate follow up actions. Verify that records are available for review by external auditors.

Quality Eng—Basic	Quality Eng—Senior	Q Manager	Supplier QE	CSoftQE	Program QE	Contract Review	MRO QE	Spec Process QE	Met Labs QE	Production Control	Regulatory Specialist
P	P	P	S		S		S	S	S	S	S

Procedural Knowledge: Ability to apply a defined and structured assessment process to verify that appropriate root cause/corrective action is undertaken to include appropriate follow up actions (3).

Ability to apply a defined and structured assessment process to verify that records are available for review by external auditors (3).

Conceptual Knowledge: Ability to analyze quality records to verify that appropriate root cause/corrective action is undertaken to include appropriate follow up actions (4). Ability to evaluate deployed corrective actions to verify that appropriate root cause/corrective action is undertaken to include appropriate follow up actions (5). Ability to assess quality records to verify that records are available for review by external auditors (3).

Factual Knowledge: Understanding of requirements management systems (3). Understanding of regulatory requirements related to internal audits (3). Understanding of contractual and project/program requirements related to internal audit (3). Understanding of industry standards related to internal audit (3). Understanding of organization work processes (3). Understanding of where work is being executed (3). Understanding of audit planning practices (3). Understanding of robust root cause corrective action processes (3).

Process: Perform Periodic QMS Evaluations—Value Stream

Process Purpose: To ensure suppliers/partners comply with approved QMS and other stated requirements. Plan, schedule, and perform periodic QMS evaluations at supplier/partner locations. Utilize risk based methods as appropriate. Ensure the results of the audit are documented, to include scope of audit. Where second or third party sources are utilized as a means of determining supplier/partner compliance to their QMS, ensure second and third parties have the appropriate skills and capabilities to perform the scope of the audits.

Knowledge and Skill Requirements: Understanding of quality management systems. Understanding of QMS standards. Understanding of applicable regulatory requirements related to supplier audits. Ability to perform QMS evaluations. Understanding of root cause analysis. Understanding of closed-loop corrective action. Understanding of the business processes and/or technical processes under audit. Ability to plan, schedule, and perform periodic QMS evaluations at supplier/partner locations. Ability to utilize risk based methods as appropriate in establishing schedule and scope of QMS audits. Ability to verify that the results of the QMS audit are documented, to include scope of audit. Where second or third party sources are utilized as a means of determining supplier/partner compliance to their QMS, ability to verify that second and third parties have the appropriate skills and capabilities to perform the scope of the audits.

Task 1 (PM105-001): Plan, schedule, and perform periodic QMS evaluations at supplier/partner locations.

Quality Eng—Basic	Quality Eng—Senior	Q Manager	Supplier QE	CSoftQE	Program QE	Contract Review	MRO QE	Spec Process QE	Met Labs QE	Production Control	Regulatory Specialist
S	S	S	P		S		S	S	S	S	S

Procedural Knowledge: Ability to apply a defined and structured process to plan, schedule, and perform periodic QMS evaluations at supplier/partner locations (3).

Conceptual Knowledge: Ability to analyze supplier/partner performance (product and process) to effectively plan and schedule QMS evaluations (4). Ability to perform QMS audits within the value stream (4).

Factual Knowledge: Understanding of requirements management systems (3). Understanding of regulatory requirements related to audits of suppliers/partners (3). Understanding of contractual and project/program requirements related to audits of suppliers/partners (3). Understanding of industry quality management standards (3). Understanding of QMS audit practices (3). Understanding of supplier quality management systems (3). Understanding of work processes within the supplier/partner base (3). Understanding of the value stream structure and where work is being executed (3).

Task 2 (PM105-002): Utilize risk based methods as appropriate in establishing schedule and scope of QMS audits.

Quality Eng—Basic	Quality Eng—Senior	Q Manager	Supplier QE	CSoftQE	Program QE	Contract Review	MRO QE	Spec Process QE	Met Labs QE	Production Control	Regulatory Specialist
S	S	S	P		S		S	S	S	S	S

Procedural Knowledge: Ability to apply a structured approach to apply risk based methods as appropriate in establishing schedule and scope of QMS audits (3).

Conceptual Knowledge: Ability to analyze supplier/partner performance (product and process) to effectively plan and schedule QMS evaluations (4). Ability to analyze product and process performance at similar suppliers/partners to evaluate potential risk at new suppliers (5). Ability to evaluate risk associated with supplier/partner scope of work and past performance to establish quantitative measures of risk (5).

Factual Knowledge: Understanding of industry quality management standards (3). Understanding of QMS audit practices (3). Understanding of supplier quality management systems (3). Understanding of work processes within the supplier/partner base (3). Understanding of the value stream structure and where work is being executed (3). Understanding of risk measurement processes (4). Understanding of supplier/partner performance (3).

Task 3 (PM105-003): Where second or third party sources are utilized as a means of determining supplier/partner compliance to their QMS, verify that second and third parties have the appropriate skills and capabilities to perform the scope of the audits.

Quality Eng—Basic	Quality Eng—Senior	Q Manager	Supplier QE	CSoftQE	Program QE	Contract Review	MRO QE	Spec Process QE	Met Labs QE	Production Control	Regulatory Specialist
S	S	S	P		S		S	S	S	S	S

Procedural Knowledge: Ability to apply a structured approach to verify that second and third parties performing supplier quality audits have the appropriate skills and capabilities to perform the scope of the audits (3).

Conceptual Knowledge: Ability to evaluate the capabilities of second and third parties to perform supplier quality audits (5).

Factual Knowledge: Understanding of requirements management systems (3). Understanding of regulatory requirements related to audits of suppliers/partners (3). Understanding of contractual and project/program requirements related to audits of suppliers/partners (3). Understanding of industry quality management standards (3). Understanding of QMS audit practices (3). Understanding of supplier quality management systems (3). Understanding of work processes within the supplier/partner base (3). Understanding of the value stream structure and where work is being executed (3).

Process: Perform Full Quality Management System Evaluation

Process Purpose: Perform a complete quality management system evaluation of the selected supplier/partner to ensure compliance to all applicable requirements. Ensure the results of the evaluation are appropriately documented, to include any findings or observations.

Knowledge and Skill Requirements: Based on a review of the quality requirements, ability to determine the scope of supplier/partner QMS evaluation. Ability to perform a complete quality management system evaluation of the selected supplier/partner to ensure compliance to all applicable requirements. Ability to ensure supplier/partner has effective controls in place to ensure their suppliers have compliant and effective QMS. Ability ensure the results of the evaluation are appropriately documented, to include any findings or observations. Ability to add the resulting new knowledge to the organizational knowledge library.

Task 1 (SS115-001): Based on a review of the quality requirements, plan the QMS evaluation of the supplier/partner (to include objectives, scope, schedule, and technical expertise).

Quality Eng—Basic	Quality Eng—Senior	Q Manager	Supplier QE	CSoftQE	Program QE	Contract Review	MRO QE	Spec Process QE	Met Labs QE	Production Control	Regulatory Specialist
S	P	S	P		S			S	S		

Procedural Knowledge: Ability to apply a structured process in review of requirements and plan the audit/evaluation to determine if suppliers/partners have an adequate quality management system (4). Ability to apply a structured process to document a quality management system evaluation document (e.g., guide, checklist, etc.) (3).

Conceptual Knowledge: Ability to analyze project/program requirements and plan the audit/evaluation to determine if suppliers/partners have an adequate quality management system (4). Ability to construct a document that details the evaluation plan of a supplier's/partner's quality management system evaluation document (4).

Factual Knowledge: Understanding of project/program requirements (3). Knowledge of requirements management tools (3). Knowledge of regulatory requirements related to quality management systems (3). Knowledge of standards related to quality management systems (3). Knowledge of audit practices (3). Knowledge of audit planning practices (3). Knowledge of effective communication practices (3).

Task 2 (SS115-002): Perform a complete quality management system evaluation of the selected supplier/partner to ensure compliance to applicable QMS requirements defined in the planned evaluation.

Quality Eng—Basic	Quality Eng—Senior	Q Manager	Supplier QE	CSoftQE	Program QE	Contract Review	MRO QE	Spec Process QE	Met Labs QE	Production Control	Regulatory Specialist
S	P	S	P		S			S	S		

Procedural Knowledge: Ability to apply a structured process to conduct an evaluation of a supplier's/partners quality management system to establish compliance status to applicable QMS requirements defined in the planned evaluation (3).

Conceptual Knowledge: Ability to conduct an evaluation of a supplier's/partners quality management system to establish compliance status to applicable QMS requirements (5). Ability to analyze supplier/partner work product to determine compliance to all applicable requirements (4). Ability to effectively document the results of a supplier/partner quality management system evaluation (3).

Factual Knowledge: Understanding of project/program requirements (3). Understanding of audit practices (4). Understanding of applicable technical processes (3). Understanding of applicable material controls (3). Understanding of applicable quality management standards (3). Understanding of applicable regulatory requirements (3). Understanding of effective communication practices (3). Knowledge of requirements management tools (3).

Task 3 (SS115-003): Ensure supplier/partner has effective controls in place to ensure their suppliers have compliant and effective QMS (sub-tier control).

Quality Eng—Basic	Quality Eng—Senior	Q Manager	Supplier QE	CSoftQE	Program QE	Contract Review	MRO QE	Spec Process QE	Met Labs QE	Production Control	Regulatory Specialist
S	P	S	P		S			S	S		

Procedural Knowledge: Ability to apply a structured process to conduct an evaluation of a supplier's/partners quality management system to determine the adequacy of supplier management controls (3).

Conceptual Knowledge: Ability to conduct an evaluation of a supplier's/partners supplier management system to establish compliance status to all applicable requirements (4). Ability to analyze sub-tier supplier work product to determine compliance to all applicable requirements (4). Ability to effectively document the results of a process evaluation (3).

Factual Knowledge: Understanding of project/program requirements (3). Understanding of audit practices (4). Understanding of applicable technical processes (3). Understanding of applicable material controls (3). Understanding of applicable quality management standards (3). Understanding of applicable regulatory requirements (3). Understanding of effective communication practices (3). Knowledge of requirements management tools (3).

Task 4 (SS115-004): Document the results of the evaluation to include any findings, observations, or assessments of risk.

Quality Eng—Basic	Quality Eng—Senior	Q Manager	Supplier QE	CSoftQE	Program QE	Contract Review	MRO QE	Spec Process QE	Met Labs QE	Production Control	Regulatory Specialist
S	P	S	P		S			S	S		

Procedural Knowledge: Ability to apply a structured process to document and effectively communicate the results of quality management system evaluations to include observations and findings in terms that can be used for effective program management (3). Ability to provide recommendations to program management, as required (3).

Conceptual Knowledge: Ability to determine if data package is complete (4). Ability to document and effectively communicate the results of quality management evaluations in terms that can be used for effective program management (3). Ability to detect trends in performance based on data provided (4).

Factual Knowledge: Knowledge of effective communication practices (3). Understanding of project/program requirements (3). Understanding of applicable technical processes (4). Understanding of risk measurement practices (3). Understanding of audit practices (3). Knowledge of trending analysis (3).

II.D.4. Audit Reporting and Follow Up

Process: Document QMS Findings

Process Purpose: To document noncompliance found during QMS evaluations and ensure such findings and observations are documented to reflect the required condition and the encountered condition and such are documented on appropriate records.

Knowledge and Skill Requirements: Understanding of quality management systems. Understanding of QMS standards. Understanding of applicable regulatory requirements related to internal audits. Ability to perform QMS evaluations. Understanding of root cause analysis. Understanding of closed-loop corrective action. Understanding of the business processes and/or technical processes under audit. Ability to document noncompliance found during QMS evaluations. Ability to ensure findings and observations are documented to reflect the required condition and the encountered condition. Ability to ensure observations and finding are documented on appropriate records.

Task (PM110-001): Document noncompliance found during QMS evaluations and ensure such findings and observations are documented to reflect the required condition and the encountered condition. Ensure observations and findings are documented on appropriate records.

Quality Eng—Basic	Quality Eng—Senior	Q Manager	Supplier QE	CSoftQE	Program QE	Contract Review	MRO QE	Spec Process QE	Met Labs QE	Production Control	Regulatory Specialist
P	P	S	P	P	P	P	P	P	P	P	P

Procedural Knowledge: Ability to apply a defined and structured process to document noncompliance found during QMS evaluations (3).

Conceptual Knowledge: Ability to recognize and accurately document noncompliance found during QMS evaluations (4).

Factual Knowledge: Understanding of requirements management systems (3). Understanding of regulatory requirements related to audits of suppliers/partners (3). Understanding of contractual and project/program requirements related to audits of suppliers/partners (3). Understanding of industry quality management standards (3). Understanding of QMS audit practices (3). Understanding of supplier quality management systems (3). Understanding of work processes within the supplier/partner base (3). Understanding of the value stream structure and where work is being executed (3).

II.H. Control of Customer/Government Material

Process: Ensure Control of Customer Owned Material

Process Purpose: To ensure customer-owned material is controlled in accordance with defined requirements. This control shall include identification, proper handling, preservation, and updating as required. May require physical audits of customer-owned materials, understanding of special storage requirements, and traceability and point of usage documentation.

Knowledge and Skill Requirements: Ability to assess organizational processes to verify customer-owned material is controlled in accordance with defined requirements. Understanding of identification, proper handling, preservation, and updating control requirements. Ability to perform physical audits of organizational facilities to verify that customer-owned materials are being appropriately maintained and there is an understanding of special storage requirements, and traceability and point of usage documentation. Ability to verify that customer-owned material is appropriately maintained throughout the organization.

Task 1 (PO050-001): Assess organizational processes to verify customer-owned material is procedurally controlled in accordance with defined requirements.

Quality Eng—Basic	Quality Eng—Senior	Q Manager	Supplier QE	CSoftQE	Program QE	Contract Review	MRO QE	Spec Process QE	Met Labs QE	Production Control	Regulatory Specialist
S	S				P		S			S	

Procedural Knowledge: Ability to apply a structured process to assess organizational processes to verify customer-owned material is controlled in accordance with defined requirements (3).

Conceptual Knowledge: Ability to review organizational processes and procedures to verify that such procedures adequately define control requirements for customer-owned material (4).

Factual Knowledge: Knowledge of contractual requirements for the control of customer-owned material (3). Knowledge of applicable inventory management processes (3). Knowledge of work instruction control processes (3). Understanding of work is being performed (3).

Task 2 (PO050-002): Schedule and perform physical audits of organizational facilities to verify that customer-owned materials are being appropriately maintained and there is an understanding of special storage requirements, traceability and point of usage documentation.

Quality Eng—Basic	Quality Eng—Senior	Q Manager	Supplier QE	CSoftQE	Program QE	Contract Review	MRO QE	Spec Process QE	Met Labs QE	Production Control	Regulatory Specialist
P	P				S		P			P	

Procedural Knowledge: Ability to apply a defined and structured process to audit organizational facilities to verify that customer-owned materials are being appropriately maintained (3). Ability to apply a structured process to verify that persons responsible for the control of customer-owned materials possess an adequate level of understanding of special storage requirements, traceability and point of usage documentation (3). Ability to apply a structured process to schedule an audit in the organization (3).

Conceptual Knowledge: Ability to audit organizational facilities to verify that customer-owned materials are being appropriately maintained (4). Ability to verify that persons responsible for the control of customer-owned materials possess an adequate level of understanding of special storage requirements, traceability and point of usage documentation (4). Ability to interface with appropriate organizational members to schedule an audit in the organization (3).

Factual Knowledge: Knowledge of contractual requirements for the control of customer-owned material (3). Knowledge of applicable inventory management processes (3). Knowledge of work instruction control processes (3). Knowledge of audit practices (3). Understanding of where work is being performed (3).

Task 3 (PO050-003): Verify that customer-owned material is appropriately maintained throughout the organization.

Quality Eng—Basic	Quality Eng—Senior	Q Manager	Supplier QE	CSoftQE	Program QE	Contract Review	MRO QE	Spec Process QE	Met Labs QE	Production Control	Regulatory Specialist
P	P	S			P		S			S	

Procedural Knowledge: Ability to apply a defined and structured process to conduct audit and verification activities throughout the organization to ensure customer-owned materials are being appropriately maintained and that responsible personnel possess an adequate level of understanding of special storage requirements, traceability and point of usage documentation (3).

Conceptual Knowledge: Ability to conduct audit and verification activities throughout the organization to ensure customer-owned materials are being appropriately maintained and that responsible personnel possess an adequate level of understanding of special storage requirements, traceability and point of usage documentation (4).

Factual Knowledge: Knowledge of contractual requirements for the control of customer-owned material (3). Knowledge of applicable inventory management processes (3). Knowledge of work instruction control processes (3). Knowledge of audit practices (3). Knowledge of where customer-owned materials reside (3). Knowledge of material control processes (3). Knowledge of where work is being performed (3).

Task 4 (PP050-001): Assess organizational and supplier/partner processes to verify customer-owned material is procedurally controlled in accordance with defined requirements.

Quality Eng—Basic	Quality Eng—Senior	Q Manager	Supplier QE	CSoftQE	Program QE	Contract Review	MRO QE	Spec Process QE	Met Labs QE	Production Control	Regulatory Specialist
			P		S						

Procedural Knowledge: Ability to apply a structured process to assess organizational and supplier/partner processes to verify customer-owned material is controlled in accordance with defined requirements (3).

Conceptual Knowledge: Ability to review organizational and supplier/partner processes and procedures to verify that such procedures adequately define control requirements for customer-owned material (4).

Factual Knowledge: Knowledge of contractual requirements for the control of customer-owned material (3). Knowledge of applicable inventory management processes (3). Knowledge of work instruction control processes (3).

Task 5 (PP050-002): Schedule and perform physical audits of organizational and supplier/partner facilities to verify that customer-owned materials are being appropriately maintained and there is an understanding of special storage requirements, traceability and point of usage documentation.

Quality Eng—Basic	Quality Eng—Senior	Q Manager	Supplier QE	CSoftQE	Program QE	Contract Review	MRO QE	Spec Process QE	Met Labs QE	Production Control	Regulatory Specialist
			P		S						

Procedural Knowledge: Ability to apply a defined and structured process to audit organizational and supplier/partner facilities to verify that customer-owned materials are being appropriately maintained (3). Ability to apply a structured process to verify that persons responsible for the control of customer-owned materials possess an adequate level of understanding of special storage requirements, traceability and point of usage documentation (3). Ability to apply a structured process to schedule an audit at supplier/partner locations, to include sub-tier supplier (3).

Conceptual Knowledge: Ability to audit organizational and supplier/partner facilities to verify that customer-owned materials are being appropriately maintained (4). Ability to verify that persons responsible for the control of customer-owned materials possess an adequate level of understanding of special storage requirements, traceability and point of usage documentation (4). Ability to interface with appropriate organizational members to schedule an audit at supplier/partner locations, to include sub-tier supplier (3).

Factual Knowledge: Knowledge of contractual requirements for the control of customer-owned material (3). Knowledge of applicable inventory management processes (3). Knowledge of work instruction control processes (3). Knowledge of audit practices (3).

Task 6 (PP050-003): Verify that customer-owned material is appropriately maintained throughout the value stream.

Quality Eng—Basic	Quality Eng—Senior	Q Manager	Supplier QE	CSoftQE	Program QE	Contract Review	MRO QE	Spec Process QE	Met Labs QE	Production Control	Regulatory Specialist
S	S	S	S		P		S			S	

Procedural Knowledge: Ability to apply a defined and structured process to conduct audit and verification activities throughout the value stream to ensure customer-owned materials are being appropriately maintained and that responsible personnel possess an adequate level of understanding of special storage requirements, traceability and point of usage documentation (3).

Conceptual Knowledge: Ability to conduct audit and verification activities throughout the value stream to ensure customer-owned materials are being appropriately maintained and that responsible personnel possess an adequate level of understanding of special storage requirements, traceability and point of usage documentation (4).

Factual Knowledge: Knowledge of contractual requirements for the control of customer-owned material (3). Knowledge of applicable inventory management processes (3). Knowledge of work instruction control processes (3). Knowledge of audit practices (3). Knowledge of value stream structure (3). Knowledge of where customer-owned materials reside (3). Knowledge of material control processes (3).

II.I. Control of Customer/Government Equipment/Property

Process: Ensure Control of Government Furnished Equipment/Materials

Process Purpose: To ensure government-owned material is controlled in accordance with defined requirements. May require compliance to applicable government standards and/or government contract clauses. This control shall include identification, proper handling, preservation, and updating as required. May require physical audits of government-owned materials, understanding of special storage requirements, and traceability and point of usage documentation.

Knowledge and Skill Requirements: Ability to assess organizational processes to verify government-owned material are controlled in accordance with defined requirements. Understanding of identification, proper handling, preservation, and updating control requirements. Ability to perform physical audits of organizational facilities to verify that government-owned materials are being appropriately maintained and there is an understanding of special storage requirements, traceability and point of usage documentation. Ability to verify that government-owned material are being appropriately maintained throughout the organization. Knowledgeable of applicable government standards and ability to assess applicability to project/program.

Task 1 (PO055-001): Assess organizational processes to verify government-owned material is controlled in accordance with defined requirements.

Quality Eng—Basic	Quality Eng—Senior	Q Manager	Supplier QE	CSoftQE	Program QE	Contract Review	MRO QE	Spec Process QE	Met Labs QE	Production Control	Regulatory Specialist
S	S				P		S			S	

Procedural Knowledge: Ability to apply a structured process to assess organizational processes to verify government-owned material is controlled in accordance with defined requirements (3).

Conceptual Knowledge: Ability to review organizational processes and procedures to verify that such procedures adequately define control requirements for government-owned material (4).

Factual Knowledge: Knowledge of contractual and regulatory requirements for the control of government-owned material (3). Knowledge of applicable inventory management processes (3). Knowledge of work instruction control processes (3). Knowledge of applicable government standards related to government-owned material (3).

Task 2 (PO055-002): Schedule and perform physical audits of organizational facilities to verify that government-owned materials are being appropriately maintained and there is an understanding of special storage requirements, traceability and point of usage documentation.

Quality Eng—Basic	Quality Eng—Senior	Q Manager	Supplier QE	CSoftQE	Program QE	Contract Review	MRO QE	Spec Process QE	Met Labs QE	Production Control	Regulatory Specialist
P	P				S		P			P	

Procedural Knowledge: Ability to apply a defined and structured process to audit organizational facilities to verify that government-owned materials are being appropriately maintained (3). Ability to apply a structured process to verify that persons responsible for the control of government-owned materials possess an adequate level of understanding of special storage requirements, traceability and point of usage documentation (3). Ability to apply a structured process to schedule an audit at organizational locations (3).

Conceptual Knowledge: Ability to audit organizational facilities to verify that customer-owned materials are being appropriately maintained (4). Ability to verify that persons responsible for the control of customer-owned materials possess an adequate level of understanding of special storage requirements, traceability and point of usage documentation (4). Ability to interface with appropriate organizational members to schedule an audit at organizational locations (3).

Factual Knowledge: Knowledge of contractual and regulatory requirements for the control of government-owned material (3). Knowledge of applicable inventory management processes (3). Knowledge of work instruction control processes (3). Knowledge of audit practices (3). Knowledge of applicable government standards related to government-owned material (3). Knowledge of where work is being performed (3).

Task 3 (PO055-003): Verify that government-owned material is appropriately maintained throughout the organization.

Quality Eng—Basic	Quality Eng—Senior	Q Manager	Supplier QE	CSoftQE	Program QE	Contract Review	MRO QE	Spec Process QE	Met Labs QE	Production Control	Regulatory Specialist
P	P	S			P		S			S	

Procedural Knowledge: Ability to apply a defined and structured process to conduct audit and verification activities throughout the organization to ensure government-owned materials are being appropriately maintained and that responsible personnel possess an adequate level of understanding of special storage requirements, traceability and point of usage documentation (3).

Conceptual Knowledge: Ability to conduct audit and verification activities throughout the organization to ensure government-owned materials are being appropriately maintained and that responsible personnel possess an adequate level of understanding of special storage requirements, traceability and point of usage documentation (4).

Factual Knowledge: Knowledge of contractual and regulatory requirements for the control of government-owned material (3). Knowledge of applicable inventory management processes (3). Knowledge of work instruction control processes (3). Knowledge of audit practices (3). Knowledge of where government-owned materials reside (3). Knowledge of applicable government standards related to government-owned material (3). Knowledge of where work is being performed (3).

Task 4 (PP055-001): Assess supplier/partner processes to verify government-owned material is controlled in accordance with defined requirements.

Quality Eng—Basic	Quality Eng—Senior	Q Manager	Supplier QE	CSoftQE	Program QE	Contract Review	MRO QE	Spec Process QE	Met Labs QE	Production Control	Regulatory Specialist
S	S	S	P		S						

Procedural Knowledge: Ability to apply a structured process to assess organizational and supplier/partner processes to verify government-owned material is controlled in accordance with defined requirements (3).

Conceptual Knowledge: Ability to review organizational and supplier/partner processes and procedures to verify that such procedures adequately define control requirements for government-owned material (4).

Factual Knowledge: Knowledge of contractual and regulatory requirements for the control of government-owned material (3). Knowledge of applicable inventory management processes (3). Knowledge of work instruction control processes (3). Knowledge of applicable government standards related to government-owned material (3).

Task 5 (PP055-002): Schedule and perform physical audits of supplier/partner facilities to verify that government-owned materials are being appropriately maintained and there is an understanding of special storage requirements, traceability and point of usage documentation.

Quality Eng—Basic	Quality Eng—Senior	Q Manager	Supplier QE	CSoftQE	Program QE	Contract Review	MRO QE	Spec Process QE	Met Labs QE	Production Control	Regulatory Specialist
			P		S						

Procedural Knowledge: Ability to apply a defined and structured process to audit organizational and supplier/partner facilities to verify that government-owned materials are being appropriately maintained (3). Ability to apply a structured process to verify that persons responsible for the control of government-owned materials possess an adequate level of understanding of special storage requirements, traceability and point of usage documentation (3). Ability to apply a structured process to schedule an audit at supplier/partner locations, to include sub-tier supplier (3).

Conceptual Knowledge: Ability to audit organizational and supplier/partner facilities to verify that customer-owned materials are being appropriately maintained (4). Ability to verify that persons responsible for the control of customer-owned materials possess an adequate level of understanding of special storage requirements, traceability and point of usage documentation (4). Ability to interface with appropriate organizational members to schedule an audit at supplier/partner locations, to include sub-tier supplier (3).

Factual Knowledge: Knowledge of contractual and regulatory requirements for the control of government-owned material (3). Knowledge of applicable inventory management processes (3). Knowledge of work instruction control processes (3). Knowledge of audit practices (3). Knowledge of applicable government standards related to government-owned material (3).

Task 6 (PP055-003): Verify that government-owned material is appropriately maintained throughout the value stream.

Quality Eng—Basic	Quality Eng—Senior	Q Manager	Supplier QE	CSoftQE	Program QE	Contract Review	MRO QE	Spec Process QE	Met Labs QE	Production Control	Regulatory Specialist
S	S	S	P		S						

Procedural Knowledge: Ability to apply a defined and structured process to conduct audit and verification activities throughout the value stream to ensure government-owned materials are being appropriately maintained and that responsible personnel possess an adequate level of understanding of special storage requirements, traceability and point of usage documentation (3).

Conceptual Knowledge: Ability to conduct audit and verification activities throughout the value stream to ensure government-owned materials are being appropriately maintained and that responsible personnel possess an adequate level of understanding of special storage requirements, traceability and point of usage documentation (4).

Factual Knowledge: Knowledge of contractual and regulatory requirements for the control of government-owned material (3). Knowledge of applicable inventory management processes (3). Knowledge of work instruction control processes (3). Knowledge of audit practices (3). Knowledge of value stream structure (3). Knowledge of where government-owned materials reside (3). Knowledge of applicable government standards related to government-owned material (3).

II.J. Capability/Capacity Measurement

Process: Notification of Changes in Capacity and/or Capability

Process Purpose: To ensure the organization is aware of any changes (positive or negative) to the capacity and/or capability of manufacturing and/or servicing of a supplier/partner. Ensure there is a notification process in place to make all appropriate parties aware of changes.

Knowledge and Skill Requirements: Ability to assess organizational processes that ensure the organization is aware of any changes (positive or negative) to the capacity and/or capability of manufacturing and/or servicing of a supplier/partner. Ability to assess existing notification processes to ensure all appropriate parties aware of changes in supplier/partner capability/capacity.

Task (PO010-001): Assess organizational processes to ensure the organization is notified of any changes (positive or negative) to the capacity and/or capability of manufacturing and/or servicing.

Quality Eng—Basic	Quality Eng—Senior	Q Manager	Supplier QE	CSoftQE	Program QE	Contract Review	MRO QE	Spec Process QE	Met Labs QE	Production Control	Regulatory Specialist
S	S	S	S		P		S	S	S	S	S

Procedural Knowledge: Ability to apply a structured process to review and assess processes and procedures to determine adequacy of notification of changes (3).

Conceptual Knowledge: Ability to review various streams of communication flow to determine adequacy of change notification (4). Ability to review configuration management processes and determine compliance of change notification (4).

Factual Knowledge: Knowledge of change notification processes (3). Knowledge of configuration management processes (3). Knowledge of regulatory requirements related to change notification (3).

Process: Ensure Process Verification (Process Tryout)

Process Purpose: To ensure processes perform as designed a process verification process is conducted. Ensure processes are adequately defined and operating in a conforming fashion. As processes are submitted for production, ensure a documented process tryout is performed, as required. (Note: Process Tryout is considered an Operations or Production process and not a Quality process.)

Knowledge and Skill Requirements: Ability to verify organizational documented processes to ensure process verifications (process try-outs) are conducted as required. Ability to assess the adequacy of process verification processes. Ability to verify that process tryout activities are appropriately documented, as required.

Task 1 (PO090-001): Verify organizational documented processes to ensure risk factors identified in process try-out are addressed and process verifications are conducted, as required.

Quality Eng—Basic	Quality Eng—Senior	Q Manager	Supplier QE	CSoftQE	Program QE	Contract Review	MRO QE	Spec Process QE	Met Labs QE	Production Control	Regulatory Specialist
P	P				S		P			P	

Procedural Knowledge: Ability to apply a structured process to identify the risk factors associated with the work process (3). Ability to apply a structured assessment process to verify that the organization has documented processes to perform process try-out activities, as required (3).

Conceptual Knowledge: Ability to assess organizational documented processes to verify compliance to defined requirements for process try-outs (3). Ability to recognize, through work analysis, potential risk factors associated with related work processes (4).

Factual Knowledge: Knowledge of manufacturing/servicing processes related to the project/program (3). Understanding of process try-out practices (3). Understanding of contract requirements related to process try-outs (3). Knowledge of risk factors related to related work processes (4).

Task 2 (PO090-002): Assess the adequacy of process verification processes throughout the organization.

Quality Eng—Basic	Quality Eng—Senior	Q Manager	Supplier QE	CSoftQE	Program QE	Contract Review	MRO QE	Spec Process QE	Met Labs QE	Production Control	Regulatory Specialist
P	P				S		P			P	

Procedural Knowledge: Ability to apply a structured assessment process to determine the adequacy of process verification controls (3).

Conceptual Knowledge: Ability to analyze work flow processes to identify where process verification activities should occur within the value stream (4). Ability to analyze established process verification controls to determine adequacy to measure and verify process capability (4).

Factual Knowledge: Knowledge of manufacturing/servicing processes related to the project/program (3). Understanding of process try-out practices (3). Knowledge of process capability measurement (3). Knowledge of where work is being performed (3).

Task 3 (PO090-003): Verify that process verification activities have been accomplished and are appropriately documented throughout the organization.

Quality Eng—Basic	Quality Eng—Senior	Q Manager	Supplier QE	CSoftQE	Program QE	Contract Review	MRO QE	Spec Process QE	Met Labs QE	Production Control	Regulatory Specialist
P	P				S		P			P	

Procedural Knowledge: Ability to apply a structured process to audit the execution and documentation of process verification activities (3).

Conceptual Knowledge: Ability to analyze the execution and documentation of process verification activities to determine compliance to process procedures (4).

Factual Knowledge: Knowledge of manufacturing/servicing processes related to the project/program (3). Understanding of process try-out practices (3). Knowledge of process capability measurement (3). Knowledge of quality records requirements (3). Knowledge of management practices (3). Knowledge of where work is being performed (3).

III.A. Classification of Quality Characteristics

Process: Baseline the Quality and Regulatory Requirements

Process Purpose: Combine previous quality requirements review processes performed during the Design Phase and create a baseline quality requirements document. Ensure baseline quality requirements document are accurate and complete. Ensure consolidated requirements are organized, documented, stored and integrated (decompose to system and subsystem requirements) into the requirements hierarchy using a requirements management tool.

Knowledge and Skill Requirements: Review the preceding quality requirements definition processes of the Design Phase to ensure adequacy and completeness. Create a consolidated baseline quality requirements document. Ensure consolidated requirements are organized, documented, stored and integrated into the requirements hierarchy using a requirements management tool.

Task 1 (DP080-001): Review the preceding quality requirements definition processes of the Design Phase to ensure adequacy and completeness. Create a consolidated baseline quality requirements document. Ensure consolidated requirements are organized, documented, stored and integrated into the requirements hierarchy using a requirements management tool.

Quality Eng—Basic	Quality Eng—Senior	Q Manager	Supplier QE	CSoftQE	Program QE	Contract Review	MRO QE	Spec Process QE	Met Labs QE	Production Control	Regulatory Specialist
P	P	S	S		P	S		S	S	S	

Procedural Knowledge: Ability to apply a defined and structured process in consolidating the various preceding quality review processes into a baseline quality requirements document (4). Ability to apply a structured process for assessing the consolidated quality requirements document for adequacy and completeness (4).

Conceptual Knowledge: Ability to consolidate the results of preceding quality requirements reviews and develop a baseline quality requirements document (4). Ability to evaluate the consolidated quality requirements document for completeness and adequacy (5). Ability to articulate the identification of any gaps identified during the evaluation of quality requirements (4).

Factual Knowledge: Knowledge of quality requirements definition processes (3). Knowledge of customer requirements (3). Knowledge of regulatory requirements (3). Knowledge of gap assessment methods (3). Knowledge of requirements management tools and processes (3).

Task 2 (DP080-002): Evaluate combined baseline quality requirements and determine if there are any gaps in definition.

Quality Eng—Basic	Quality Eng—Senior	Q Manager	Supplier QE	CSoftQE	Program QE	Contract Review	MRO QE	Spec Process QE	Met Labs QE	Production Control	Regulatory Specialist
P	P	S	S		P	S		S	S	S	

Procedural Knowledge: Ability to apply a structured process for identifying potential gaps in the consolidated quality requirements baseline (3)

Conceptual Knowledge: Ability to review consolidated quality requirements baseline and evaluate against customer and regulatory requirements in identification of potential gaps (4).

Factual Knowledge: Knowledge of customer requirements (3). Knowledge of regulatory requirements (3). Knowledge of quality requirements definition processes (3). Knowledge of requirements management tools and processes (3).

III.B. Design Inputs and Review

Process: Define Customer/Program Requirements

Process Purpose: To understand the stated and identify the unstated requirements of the customer/program and document such as a preliminary requirements document. To understand the stated and identify the unstated regulatory requirements and document such as a preliminary requirements document.

Knowledge and Skills Requirements: Ability to elicit from the customer/program the stated and unstated customer needs. Ability to express and present this information back to the customer and work with the customer to get customer acceptance. Ability to understand the stated and identify the unstated regulatory requirements and document such a preliminary requirements document.

Task 1 (CP005-001): Elicit and verify that all customer requirements are available for review.

Quality Eng—Basic	Quality Eng—Senior	Q Manager	Supplier QE	CSoftQE	Program QE	Contract Review	MRO QE	Spec Process QE	Met Labs QE	Production Control	Regulatory Specialist
					P	P					P

Procedural Knowledge: Apply a structured process for eliciting customer requirements and ensuring agreement between parties (3).

Conceptual Knowledge: Evaluate the completeness of the customer's requirements and ensure the customer understands the scope of the project (5).

Factual Knowledge: Understand that customer requirements must be identified (3). Knowledge of requirements management tools (3).

Task 2 (CP005-002): Review customer requirements (functionality, schedules, functionality, cost, reliability, etc.) to determine customer expectations.

Quality Eng—Basic	Quality Eng—Senior	Q Manager	Supplier QE	CSoftQE	Program QE	Contract Review	MRO QE	Spec Process QE	Met Labs QE	Production Control	Regulatory Specialist
					P	P					P

Procedural Knowledge: Ability to interpret technical specifications (4). Ability to follow a customer requirements definition process (3).

Conceptual Knowledge: Ability to organize and to understand the implications of stated and unstated requirements (5). Example: For government contracts that there are implicit requirements such as oversight and material controls.

Factual Knowledge: Knowledge of resource capabilities (3). Knowledge of applicable industry standards (3). Knowledge of value stream capability (3). Knowledge of applicable regulations (3). Knowledge of requirements management tools (3).

Task 3 (CP005-003): Document customer requirements in an appropriate format (UML, SYSML, DODAF).

Quality Eng—Basic	Quality Eng—Senior	Q Manager	Supplier QE	CSoftQE	Program QE	Contract Review	MRO QE	Spec Process QE	Met Labs QE	Production Control	Regulatory Specialist
					P	P					S

Procedural Knowledge: Ability to follow a formal process for documenting the defined customer requirements (3). Ability to apply the documentation model (4). Understand the documentation approval process (2).

Conceptual Knowledge: Understand the interrelatedness of the processes within the value stream and their relationship to meeting customer requirements (5). Understand all aspects of the product in relationship to customer requirements (5). Note: Requirements will have implications on product features, manufacturing, delivery, and servicing operations.

Factual Knowledge: Know which documentation formats are appropriate (accepted by customer, organization, and value stream) (3). Recognizing that all documentation must be complete to have a valid package (3). Knowledge of requirements management tools (3).

Task 4 (CP005-004): Engage in conversation with customer, organization, and suppliers to elicit requirements and understand capabilities. This process is to ensure that requirements are understood and that undocumented requirements are identified and subsequently documented.

Quality Eng—Basic	Quality Eng—Senior	Q Manager	Supplier QE	CSoftQE	Program QE	Contract Review	MRO QE	Spec Process QE	Met Labs QE	Production Control	Regulatory Specialist
S	S	S	S	S	P	S					S

Procedural Knowledge: Apply appropriate protocol for customer communication (3).

Conceptual Knowledge: Ability to converse with customer in a common language (4). Thorough understanding of the product (5). Ability to understand and define the interrelatedness of processes within the value stream (6).

Factual Knowledge: Knowledge of applicable regulations and standards (3). Knowledge of past programs (3). Knowledge of technical capability (4). Knowledge of resource constraints (4). Knowledge of requirements management tools (3).

Task 5 (CP005-005): Identify the stated and unstated regulatory requirements and document such as a preliminary requirements document.

Quality Eng—Basic	Quality Eng—Senior	Q Manager	Supplier QE	CSoftQE	Program QE	Contract Review	MRO QE	Spec Process QE	Met Labs QE	Production Control	Regulatory Specialist
					P	P					P

Procedural Knowledge: Ability to apply a structured approach to identify the stated and unstated regulatory requirements and document such as a preliminary requirements document (3).

Conceptual Knowledge: Ability to evaluate program variables and analyze regulatory documents to ensure both stated and unstated regulatory requirements are identified and adequately documented within a requirements management tool (5).

Factual Knowledge: Understanding of program requirements (3). Understanding of regulatory documents and structure (3). Understanding of program review practices (3). Understanding of requirements management tools (3).

Process: Determine Design Capability

Process Purpose: To understand the present and potential state of engineering capability and to ensure engineering capability can meet the stated and unstated requirements of the customer/program. To ensure mechanisms exist for design consideration of production capabilities and mitigation of predicted nonconformance activity. (Assessment of risk associated with variation to Standard-work.) Also, ensure supplier/partner has sufficient resources to accomplish design requirements.

Knowledge and Skill Requirements: Knowledge of value stream capabilities and systems related to engineering/design. Familiarity with interface specifications of design and development tools. Knowledge of value stream scheduling and resource limitations. Ability to elicit similar information from suppliers.

Task 1 (CP015-001): Determine customer requirements for specific design models.

Quality Eng—Basic	Quality Eng—Senior	Q Manager	Supplier QE	CSoftQE	Program QE	Contract Review	MRO QE	Spec Process QE	Met Labs QE	Production Control	Regulatory Specialist
S	S				P	P					

Procedural Knowledge: Ability to follow a structured process to elicit customer requirements for specific design models (e.g., 3D modeling, CATIA, Unisys, CAD) (3).

Conceptual Knowledge: Understand the functionality of various engineering design tools and determine the capability of existing tools within the organization and value stream (5). Understand the relationship between models and how they communicate (4).

Factual Knowledge: Know the various types of design tools (3). Know the inter-functionality of design tools (3). Know which design tools the organization and value stream presently deploy (3).

Task 2 (CP015-002): Evaluate organizational design capabilities against specific customer requirements. (Does the organization have the appropriate engineering tools and skills to accomplish required work?)

Quality Eng—Basic	Quality Eng—Senior	Q Manager	Supplier QE	CSoftQE	Program QE	Contract Review	MRO QE	Spec Process QE	Met Labs QE	Production Control	Regulatory Specialist
S	P	S	P		P	S					

Procedural Knowledge: Ability to follow a structured process for determining organizational and value stream design capability (3). Create an side-by-side comparison analysis document illustrating customer design capability requirements and existing design capabilities within the organization and value stream (4).

Conceptual Knowledge: Ability to analyze customer design capability requirements, analyze organizational and value stream design capabilities, and evaluate compatibility of existing capability to customer requirements (5). Ability to understand the applied technologies (4).

Factual Knowledge: Know where the design is to be created (3). Understand terms used in design capability (3). Know the relationship between technical documents and specifications (4).

Task 3 (CP015-003): Evaluate design capabilities within the supplier base against specific customer requirements. (Does the value stream have the appropriate engineering tools and skills to accomplish required work?)

Quality Eng—Basic	Quality Eng—Senior	Q Manager	Supplier QE	CSoftQE	Program QE	Contract Review	MRO QE	Spec Process QE	Met Labs QE	Production Control	Regulatory Specialist
S	P	S	P		P	S					

Procedural Knowledge: Ability to follow a structured process for determining value stream design capability (3). Create a side-by-side comparison analysis document illustrating customer design capability requirements and existing design capabilities within the value stream (4).

Conceptual Knowledge: Ability to evaluate customer design capability requirements, analyze value stream design capabilities, and evaluate compatibility of existing capability to customer requirements (5). Ability to understand the applied technologies (4).

Factual Knowledge: Know where design is to be created (3). Understand terms used in design capability (4). Know the relationship between technical documents and specifications (4). Know the various design tools applied within the value stream (3).

Task 4 (CP015-004): Evaluate customer requirements against the existing definition of engineering standard-work. (Is there a difference between what the customer is requesting and what is already defined in standard-work?)

Quality Eng—Basic	Quality Eng—Senior	Q Manager	Supplier QE	CSoftQE	Program QE	Contract Review	MRO QE	Spec Process QE	Met Labs QE	Production Control	Regulatory Specialist
P	P	S	S	P	P	S		P	P		

Procedural Knowledge: Ability to follow a structured process for comparing customer design requirements against existing design standard-work capability (3). Create a side-by-side comparison analysis document illustrating customer design capability requirements and existing standard-work capabilities within the organization and value stream (4).

Conceptual Knowledge: Ability to analyze customer design capability requirements against existing design standard-work capability, within the organization and value stream (5). Ability to understand the applied technologies (4).

Factual Knowledge: Ability to apply the concepts of standard-work and analyze process capability relates to engineering processes (4).

Task 5 (CP015-005): Define, measure, and evaluate risks associated from deviations from engineering standard-work.

Quality Eng—Basic	Quality Eng—Senior	Q Manager	Supplier QE	CSoftQE	Program QE	Contract Review	MRO QE	Spec Process QE	Met Labs QE	Production Control	Regulatory Specialist
P	P	S	S	P	P	S		P	P		

Procedural Knowledge: Ability to follow the structured methodology for identifying and measuring risks associated with deviations to standard work (3). Ability to document risks associated with deviations to standard work (3).

Conceptual Knowledge: Through evaluation, the ability to recognize and quantify deviations from standard work in review of stated and unstated requirements of the customer (5). Capability to evaluate risks factors related to such deviations (e.g., based on deviations in technical, material, equipment) (5). Ability to communicate risk information to technical and management communities (4).

Factual Knowledge: Understanding of risk measurement methods (4). Understanding of engineering standard work processes (3). Understanding of design and production processes (3). Understanding of communication methods (3).

Process: Establish a Preliminary Design and Production/Servicing Execution Plan

Process Purpose: To understand where and how the stated and unstated work is to be performed. Establish a program/project execution plan identifying sources for work scope performance.

Knowledge and Skill Requirements: Knowledge of contractual requirements. Ability to interpret work scope requirements. Knowledge of quality management practices. Knowledge of existing capabilities and capacity of the value stream. Ability to document execution planning into terms the organization can understand.

Task 1 (CP020-001): Utilizing defined design and production requirements, identify potential sources of where work may be performed. Utilize existing information within the organization and value stream.

Quality Eng—Basic	Quality Eng—Senior	Q Manager	Supplier QE	CSoftQE	Program QE	Contract Review	MRO QE	Spec Process QE	Met Labs QE	Production Control	Regulatory Specialist
S	P	S	P		P			S	S		

Procedural Knowledge: Ability to follow a structured process to identify, through analysis, where work scope may be performed (4).

Conceptual Knowledge: Ability to interpret design and production requirements and evaluate the presently defined sources of work execution to identify where work may be performed. (5) (Identification of potential sources may result in multiple sources being selected.)

Factual Knowledge: Knowledge of existing capabilities within the organization and value stream (3). Knowledge of work execution planning processes (3).

Task 2 (CP020-002): Utilizing defined design and production requirements, identify potential sources of where work may be performed. Elicit preliminary capability and capacity information from the value stream. (Ensuring the existing information is accurate)

Quality Eng—Basic	Quality Eng—Senior	Q Manager	Supplier QE	CSoftQE	Program QE	Contract Review	MRO QE	Spec Process QE	Met Labs QE	Production Control	Regulatory Specialist
S	P	S	P		P			S	S		

Procedural Knowledge: Ability to follow a structured process to identify where work scope may be performed (3). Ability to follow structured process to elicit information from appropriate sources (3).

Conceptual Knowledge: Ability to interpret design and production requirements and evaluate potential sources of work execution as to identify where work may be performed (5). Ability to preliminarily evaluate (not necessarily a physical audit) the abilities of potential sources to effectively execute scope of work (capability and capacity) (5). (Identification of potential sources may result in multiple sources being selected.) Ability to identify risk factors to work execution based on available and obtained information (5).

Factual Knowledge: Knowledge of capability and capacity measurements (4). Knowledge of industrial sources of work (3). Knowledge of work execution planning processes (4). Knowledge of risk management in work execution (4).

Task 3 (CP020-003): Document the preliminary work scope execution plan and provide rational for the decisions made.

Quality Eng—Basic	Quality Eng—Senior	Q Manager	Supplier QE	CSoftQE	Program QE	Contract Review	MRO QE	Spec Process QE	Met Labs QE	Production Control	Regulatory Specialist
S	P	S	P		P			S	S		

Procedural Knowledge: Ability to follow a structured process for creating a work scope execution plan and decision making process (utilizing risk management tools) (3).

Conceptual Knowledge: Ability to translate the identified potential sources of work execution into a documented plan (4). Ability to understand value stream capability and capacity issues and risk factors, based on available and obtained information (4).

Factual Knowledge: Knowledge of work scope execution planning processes (4). Knowledge of industrial capabilities and capacity (4). Knowledge of risk management practices in work scope execution (4).

Process: Determine Quality Requirements—General

Process Purpose: Review customer, design, and process requirements and determine quality requirements for the product/program. This review process should include lessons learned material from other similar programs as well as industry source material. These requirements shall be documented in the appropriate organizational format and stored in a database that facilitates requirement hierarchy navigation, configuration control of requirements, and the development of requirement metrics.

Knowledge and Skill Requirements: Ability to analyze potential quality requirements and determine which requirements are applicable to a given production/servicing program. Ability to analyze production/serving program and determine which quality practices would be appropriate and useful. Ability to author and manage requirements.

Task 1 (Determine appropriate information materials for quality requirements, including customer, best practice quality literature, and lessons learned.)

Quality Eng—Basic	Quality Eng—Senior	Q Manager	Supplier QE	CSoftQE	Program QE	Contract Review	MRO QE	Spec Process QE	Met Labs QE	Production Control	Regulatory Specialist
S	P	S	S	S	P			P	P		P

Procedural Knowledge: Ability to follow a structured process to acquire appropriate source materials. (3)

Conceptual Knowledge: Ability to recognize the interrelationships of quality requirements and seek out applicable sources of information (3).

Factual Knowledge: Knowledge of sources of information (3)

Task 2 (DP010-002): Select and implement appropriate requirements management tool (e.g., DOORS), to include the defined value stream.

Quality Eng—Basic	Quality Eng—Senior	Q Manager	Supplier QE	CSoftQE	Program QE	Contract Review	MRO QE	Spec Process QE	Met Labs QE	Production Control	Regulatory Specialist
S	P	S	S	S	P						S

Procedural Knowledge: Ability to follow a structured process in deployment of a requirements management tool (3).

Conceptual Knowledge: Ability to relate quality requirements to specific tasks within a program (4). Ability to structure quality requirements into a hierarchical structure. (4)

Factual Knowledge: Knowledge of various requirements management tools (3).

Task 3 (DP010-003): Extract quality management best practices from industry source material, identify customer quality requirements, and determine organizational quality requirements. Place best practices and requirements in requirements management tool. Review completeness of requirements to ensure best practices and requirements are appropriately aligned with identified processes.

Quality Eng—Basic	Quality Eng—Senior	Q Manager	Supplier QE	CSoftQE	Program QE	Contract Review	MRO QE	Spec Process QE	Met Labs QE	Production Control	Regulatory Specialist
S	P	S	S	S	P	S	S	S	S	S	S

Procedural Knowledge: Ability to follow a structured process for scanning industry best-practices and making determination for program applicability (3).

Conceptual Knowledge: Ability to assess industry best-practices and conceptualize application for intended program. (4) Ability to analyze technologies and processes to determine where best-practices may apply. (4)

Factual Knowledge: Knowledge of where best-practice may be found. (2) Knowledge of proposed technical processes (3).

Process: Determine Quality Requirements—Regulatory

Process Purpose: Review regulatory requirements and determine quality requirements for the product/program. These requirements shall be documented, stored and integrated into the requirements hierarchy using a requirements management tool. (Examples of such requirements include FAA, EASA, OSHA, DoD, NASA, FMS, and others.)

Knowledge and Skill Requirements: Ability to recognize applicable regulatory requirements during review of quality requirements. Ability to research regulatory resource libraries to capture specific regulatory requirements as they relate to the program. Ability to organize regulatory requirements into a matrix (or other appropriate organizational document) that can be integrated into the requirements management tool.

Task 1 (DP015-001): Identify applicable regulatory requirements during review of program quality requirements.

Quality Eng—Basic	Quality Eng—Senior	Q Manager	Supplier QE	CSoftQE	Program QE	Contract Review	MRO QE	Spec Process QE	Met Labs QE	Production Control	Regulatory Specialist
P	P	P	P	P	P	P		S	S		P

Procedural Knowledge: Ability to follow a structured process to define applicable regulatory requirements (3). Ability to review regulatory requirements to determine applicability (3).

Conceptual Knowledge: Ability to analyze design requirements and identify applicable regulatory requirements (4). Ability to conceptualize the design process throughout the entire lifecycle (3).

Factual Knowledge: Knowledge of applicable regulatory requirements (3). Knowledge of design process (3). Knowledge of design certification process (3).

Task 2 (DP015-002): Research regulatory resource libraries to capture specific regulatory requirements as they relate to the program.

Quality Eng—Basic	Quality Eng—Senior	Q Manager	Supplier QE	CSoftQE	Program QE	Contract Review	MRO QE	Spec Process QE	Met Labs QE	Production Control	Regulatory Specialist
S	P	S	S	S	P	P		S	S		P

Procedural Knowledge: Ability to follow a structured process to determine applicable regulatory resource libraries (3).

Conceptual Knowledge: Ability to recognize applicability of regulatory resource libraries (3).

Factual Knowledge: Knowledge of where regulatory research libraries reside (3).

Task 3 (DP015-003): Organize regulatory requirements into a matrix (or other appropriate organizational document) that can be integrated into requirements management tool.

Quality Eng—Basic	Quality Eng—Senior	Q Manager	Supplier QE	CSoftQE	Program QE	Contract Review	MRO QE	Spec Process QE	Met Labs QE	Production Control	Regulatory Specialist
P	P	S	S	S	P	P		S	S		P

Procedural Knowledge: Ability to follow a structured process to organize applicable regulatory requirements into a matrix (3).

Conceptual Knowledge: Ability to recognize regulatory applicability in the design phase and relate such to specific tasks within a matrix (4).

Factual Knowledge: Knowledge of requirements matrix formats (2).

Process: Determine Quality Requirements—Special Processes

Process Purpose: Review design and production requirements and identify special processes associated with program. Determine special process requirements for the product/program. These requirements shall be documented, stored and integrated into the requirements hierarchy using a requirements management tool.

Knowledge and Skill Requirements: Ability to review applicable design and program planning documents and identify special processes that require validation, verification, certification, and control. Ability to research industry resource libraries to analyze potential options available in the performance and control of special processes. Ability to organize, document, store and integrate into the requirements hierarchy using a requirements management tool.

Task 1 (DP020-001): Review applicable design and program planning documents and identify special processes that require validation, verification, certification, and control.

Quality Eng—Basic	Quality Eng—Senior	Q Manager	Supplier QE	CSoftQE	Program QE	Contract Review	MRO QE	Spec Process QE	Met Labs QE	Production Control	Regulatory Specialist
P	P	S	P		P	S		P	P		

Procedural Knowledge: Ability to follow a structured process to identify special processes that require validation, verification, certification, and control (3).

Conceptual Knowledge: Ability to analyze applicable design and program planning documents to identify special processes that require validation, verification, certification, and control (4). Ability to conceptualize the design process throughout the entire lifecycle (4).

Factual Knowledge: Knowledge of special process control requirements (3). Knowledge of design process (3). Knowledge of design certification process (3).

Task 2 (DP020-002): Research industry resource libraries to analyze potential options available in the execution and control of special processes.

Quality Eng—Basic	Quality Eng—Senior	Q Manager	Supplier QE	CSoftQE	Program QE	Contract Review	MRO QE	Spec Process QE	Met Labs QE	Production Control	Regulatory Specialist
S	S	S	S		S			P	P		

Procedural Knowledge: Ability to access special process resource libraries (3).

Conceptual Knowledge: Ability to analyze industry resource libraries to analyze potential options available in the execution and control of special processes (4). Ability to determine appropriate special process controls for the design and program (4).

Factual Knowledge: Knowledge of special process controls (3). Knowledge of where special process resource libraries reside (3).

Task 3 (DP020-003): Document, store and integrate special process quality requirements into the requirements hierarchy using a requirements management tool.

Quality Eng—Basic	Quality Eng—Senior	Q Manager	Supplier QE	CSoftQE	Program QE	Contract Review	MRO QE	Spec Process QE	Met Labs QE	Production Control	Regulatory Specialist
S	P	S	S		P			P	P		

Procedural Knowledge: Ability to apply a structured process for integrating into the requirements management tool, the requirements for special process controls (3).

Conceptual Knowledge: Ability to relate special process controls to specific program tasks (3).

Factual Knowledge: Knowledge of special process controls (2). Knowledge of requirements management tools (2).

III.B.1. Workmanship Standards

Process: Determine Material Control Laboratory (MCL) Requirements

Process Purpose: Review engineering, process, and production requirements and establish the appropriate material control requirements for laboratory review. This may include destructive and non-destructive evaluation.

Knowledge and Skill Requirements: Ability to review applicable design and program planning documents and identify material control requirements that require validation, verification, certification, and control. Ability to research industry resource libraries to analyze potential options available in the execution of production processes and control of materials. Ability to assess the defined value stream and identify material control considerations. Ability to organize, document, store and integrate into the requirements hierarchy using a requirements management tool.

Task 1 (DP040-001): Review applicable design and program planning documents and identify material control requirements that require validation, verification, certification, and control.

Quality Eng—Basic	Quality Eng—Senior	Q Manager	Supplier QE	CSoftQE	Program QE	Contract Review	MRO QE	Spec Process QE	Met Labs QE	Production Control	Regulatory Specialist
P	P	P	P		P		S	S	S	S	S

Procedural Knowledge: Ability to follow a structured process to identify material control requirements that require validation, verification, certification, and control (3). Ability to access applicable industry standards related to material controls (3).

Conceptual Knowledge: Ability to analyze applicable design and program planning documents and identify material control requirements that require validation, verification, certification, and control (4).

Factual Knowledge: Knowledge of material control specifications (3). Knowledge of material control methods (3). Knowledge of material control certification process (3).

Task 2 (DP040-002): Research industry resource libraries to analyze potential options available in the execution of production processes and control of materials. Document results of analysis.

Quality Eng—Basic	Quality Eng—Senior	Q Manager	Supplier QE	CSoftQE	Program QE	Contract Review	MRO QE	Spec Process QE	Met Labs QE	Production Control	Regulatory Specialist
S	P	S	S		P			P	P		

Procedural Knowledge: Ability to access industry resource libraries (2) Ability to follow a structured process to perform the analysis and documentation of potential options for production processes and control of materials (3).

Conceptual Knowledge: Ability to analyze industry resource libraries in review of program requirements to determine alternative options for production processes and control of materials (5).

Factual Knowledge: Knowledge of industry sources related to production processes and control of materials (3).

Task 3 (DP040-003): Evaluate and identify material control considerations. Organize, document, store and integrate findings into the applicable organizational records.

Quality Eng—Basic	Quality Eng—Senior	Q Manager	Supplier QE	CSoftQE	Program QE	Contract Review	MRO QE	Spec Process QE	Met Labs QE	Production Control	Regulatory Specialist
P	P	P	P		P		S	P	P	S	

Procedural Knowledge: Ability to apply a structured process to evaluate material control requirements against the existing capabilities of the organization and value stream (3). Identify concerns, and potential sources of risk related to material control (4). Document results of evaluation into the applicable records of the organization (3).

Conceptual Knowledge: Ability to evaluate the interrelated processes of the organization and value stream in consideration of material control requirements and capabilities (5). Identify potential areas of risk and concern and articulate such into terms useable by management (4).

Factual Knowledge: Knowledge of material control requirements related to program (3). Knowledge of sources of material control information (2). Knowledge of where work is to be executed (3). Knowledge of risk management processes (3). Knowledge of historical performance of applicable suppliers, technologies, and processes (4).

Process: Identify Applicable Engineering Source Approval (ESA) Requirements

Process Purpose: Review engineering (design), process, and production requirements and determine if there are production processes that require engineering review and approval prior to production release and prior to the implementation of any changes (processes declared unchangeable by engineering without subsequent review and approval).

Knowledge and Skill Requirements: Ability to review engineering (design), process, and production requirements and determine if there are production processes that require engineering review (ESA) and approval prior to production release and prior to the implementation of any changes (processes declared unchangeable by engineering without subsequent review and approval).

Task (DP055-001): Review engineering (design), process, and production requirements and determine if there are production processes that require engineering review and approval prior to production release and prior to the implementation of any changes (processes declared unchangeable by engineering without subsequent review and approval). Organize, document, store and integrate ESA requirements into the requirements management tool.

Quality Eng—Basic	Quality Eng—Senior	Q Manager	Supplier QE	CSoftQE	Program QE	Contract Review	MRO QE	Spec Process QE	Met Labs QE	Production Control	Regulatory Specialist
P	P	P	S		P		S	P	P	S	

Procedural Knowledge: Ability to apply a structured process in review of engineering (design), process, and production requirements to determine, identify, and document if there are production processes that require engineering review and approval prior to production release and prior to the implementation of any changes (processes declared unchangeable by engineering without subsequent review and approval) (3). Ability to incorporate the identification of engineering source approval requirements into the requirements management tool (3).

Conceptual Knowledge: Ability to analyze program engineering (design), process, and production requirements to determine, identify, and document if there are production processes that require engineering review and approval prior to production release and prior to the implementation of any changes (processes declared unchangeable by engineering without subsequent review and approval) (4). Ability to organize the identification of ESA requirements into a hierarchical structure for integration into the requirements management tool (3).

Factual Knowledge: Knowledge of ESA processes (3). Knowledge of program-specific technologies and materials (4). Knowledge of where work is to be performed (3). Knowledge of requirements management tools (3).

Process: Provide QA/Engineering Interpretation

Process Purpose: Provide interpretation of design and quality requirements to other members of the organization to ensure accurate understanding of requirements. Understand functional differences within the organization. Ensure any formal direction is appropriately documented. Ability to understand technical requirements. Ability to understand organizational challenges and recommendations and engage with engineering, procurement, and other functional disciplines to resolve issues.

Knowledge and Skill Requirements: Ability to provide interpretation of design and quality requirements to other members of the organization to ensure accurate understanding of requirements. Understand functional differences within the organization. Ability to document formal direction, as appropriate. Ability to understand technical requirements. Ability to understand organizational challenges and recommendations and engage with engineering, procurement, and other functional disciplines to resolve issues. Ability to schedule and verify through follow up actions that interpretation provided was accurately understood.

Task 1 (PO115-001): Provide interpretation of design, quality, and regulatory requirements to other members of the organization to ensure accurate understanding of requirements.

Quality Eng—Basic	Quality Eng—Senior	Q Manager	Supplier QE	CSoftQE	Program QE	Contract Review	MRO QE	Spec Process QE	Met Labs QE	Production Control	Regulatory Specialist
P	P	S	S		S		S	P	P	S	P

Procedural Knowledge: Ability to apply defined processes in providing organizational members with interpretative data (3). Ability to apply defined processes for accessing organizational personnel in providing interpretative information (3).

Conceptual Knowledge: Ability to understand and evaluate unique organizational inquiries (5) Ability to determine appropriate support requirements for interpretation and route to appropriate organization members for evaluation (4).

Factual Knowledge: Understanding of contractual, regulatory, design, and quality requirements as they relate to program/project requirements (3). Knowledge of technical support resources within the organization (3). Knowledge of functional/professional differences as they relate to technical communication (3).

Task 2 (PO115-002): Document formal direction, as appropriate. Publish communication.

Quality Eng—Basic	Quality Eng—Senior	Q Manager	Supplier QE	CSoftQE	Program QE	Contract Review	MRO QE	Spec Process QE	Met Labs QE	Production Control	Regulatory Specialist
P	P	S	S		S		S	P	P	S	P

Procedural Knowledge: Ability to apply a structured process to document formal direction to organizational members (3). Ability to transmit in a closed-loop fashion, the formal direction to the organization, as required (3).

Conceptual Knowledge: Ability to understand the importance of formally documenting direction to organizational members and the purpose of closed-loop transmittal (2).

Factual Knowledge: Knowledge of documentation processes (3). Knowledge of closed-loop data transmittal processes (3).

Task 3 (PO115-003): Schedule and verify through follow up actions that interpretation provided was accurately understood and applied.

Quality Eng—Basic	Quality Eng—Senior	Q Manager	Supplier QE	CSoftQE	Program QE	Contract Review	MRO QE	Spec Process QE	Met Labs QE	Production Control	Regulatory Specialist
P	P	S	S		S		S	P	P	S	P

Procedural Knowledge: Ability to apply a structured process to schedule follow-up assessments to verify that formal interpretative direction was correctly applied (3). Ability to apply a structured process to assess application of formal interpretative direction (3).

Conceptual Knowledge: Ability to understand workflow and affectivity of formal direction so that a follow up assessment can be scheduled to verify appropriate application (4). Ability to analyze work processes at supplier/partner to determine compliance with formal direction (4).

Factual Knowledge: Understanding of workflow processes (3). Understanding of implementation affectivity processes (3). Understanding of applicable technical processes (3). Understanding of system and process audit practices (3). Understanding of where work processes are being performed (3).

Process: Monitor Engineering Source Approval (Process Approval—as required)

Process Purpose: To ensure processes that are designated as controlled by engineering ("frozen process") are effectively controlled. Evaluate engineering source controlled processes to ensure they are performing as approved. Ensure evaluation results are appropriately documented.

Knowledge and Skill Requirements: Understanding of technical program requirements, including approved design. Ability to verify processes that are designated as controlled by engineering ("frozen process") are effectively controlled. Ability to audit and evaluate engineering source controlled processes to ensure they are performing and documented as approved/required. Ability to verify evaluation results are appropriately documented.

Task 1 (PF060-001): From a quality management system perspective, verify processes that are designated as controlled by engineering ("frozen process") are effectively controlled.

Quality Eng—Basic	Quality Eng—Senior	Q Manager	Supplier QE	CSoftQE	Program QE	Contract Review	MRO QE	Spec Process QE	Met Labs QE	Production Control	Regulatory Specialist
P	P	P	P		P		P	P	P	P	P

Procedural Knowledge: Ability to apply a structured assessment process to verify that processes that are designated as controlled by engineering ("frozen process") are effectively controlled (3).

Conceptual Knowledge: Ability to analyze design processes to ensure they adequately control frozen processes, when required (4). Ability to evaluate design and production/servicing work activities to determine if processes that are designated as controlled by engineering ("frozen process") are effectively controlled (5).

Factual Knowledge: Understanding of regulatory requirements related to design control (3). Understanding of contractual and project/program requirements related to design control (3).Understanding of engineering source controlled processes (3). Understanding of frozen processes practices (3). Understanding of audit practices (3). Understanding of the value stream structure and where work is performed (3). Understanding of requirements management tools (3).

Task 2 (PM240-001): From a product perspective, verify that processes that are designated as controlled by engineering ("frozen process") are effectively controlled as applied to the product or service.

Quality Eng—Basic	Quality Eng—Senior	Q Manager	Supplier QE	CSoftQE	Program QE	Contract Review	MRO QE	Spec Process QE	Met Labs QE	Production Control	Regulatory Specialist
P	P	P	P		S		P	P	P	P	P

Procedural Knowledge: Ability to apply a structured assessment process to verify that processes that are designated as controlled by engineering ("frozen process") are effectively controlled (3).

Conceptual Knowledge: Ability to analyze design processes to ensure they adequately control frozen processes, when required (4). Ability to evaluate design and production/servicing work activities to determine if processes that are designated as controlled by engineering ("frozen process") are effectively controlled (5).

Factual Knowledge: Understanding of regulatory requirements related to design control (3). Understanding of contractual and project/program requirements related to design control (3). Understanding of requirements management tools (3). Understanding of engineering source controlled processes (3). Understanding of frozen processes practices (3). Understanding of audit practices (3). Understanding of the value stream structure and where work is performed (3).

III.B.2. Produceability Review

Process: Perform Manufacturability Review of Design

Process Purpose: Review engineering, process, and production requirements and assess such against the known capability of manufacturing/servicing to ensure the design can be produced. Aspects of the review include: ability to produce, cost to produce, and ability to produce with an appropriate level of quality.

Knowledge and Skill Requirements: Ability to review engineering, process, and production requirements and assess such against the known capability of manufacturing/servicing to ensure the design can be produced. Ability to evaluate the value stream's ability to produce, quality cost to produce, and ability to produce with an appropriate level of quality. Ability to evaluate potential alternative production methods that might require changes to the design.

Task 1 (DP050-001): Review engineering, process, and production requirements and assess such against the known capability of manufacturing/servicing to ensure the design can be produced. Organize, document, store and integrate findings into the appropriate organizational records.

Quality Eng—Basic	Quality Eng—Senior	Q Manager	Supplier QE	CSoftQE	Program QE	Contract Review	MRO QE	Spec Process QE	Met Labs QE	Production Control	Regulatory Specialist
P	P	S	S		P		S	P	S	S	

Procedural Knowledge: Ability to apply a structured process to review engineering, process, and production requirements and assess such against the known capability of manufacturing/servicing to ensure the design can be produced (3). Ability to document findings into the appropriate organizational records (3).

Conceptual Knowledge: Ability to analyze the engineering, process, and production capability requirements of the program and express such in measurable terms (4).

Factual Knowledge: Knowledge of process capability measurement processes (3). Knowledge of program-specific production/servicing processes (3). Knowledge of design definition processes (3).

Task 2 (DP050-002): Evaluate the ability to produce, quality cost to produce, and ability to produce with an appropriate level of quality.

Quality Eng—Basic	Quality Eng—Senior	Q Manager	Supplier QE	CSoftQE	Program QE	Contract Review	MRO QE	Spec Process QE	Met Labs QE	Production Control	Regulatory Specialist
P	P	P	S		P						

Procedural Knowledge: Ability to apply a structured process to measure the ability to produce, quality cost to produce, and ability to produce with an appropriate level of quality (3).

Conceptual Knowledge: Ability to analyze the production/servicing capability (4). Ability to evaluate quality cost to produce to meet program requirements (4). Ability to express the production/servicing capability in terms that are measurable (4). Ability to compare the measured capability against the capability requirements of the program (4).

Factual Knowledge: Knowledge of quality costing methods (3). Knowledge of value stream management principles (3). Knowledge of process capability measurement processes (3).

Task 3 (DP050-003): Identify and evaluate potential alternative production/servicing methods that might require changes to the design.

Quality Eng—Basic	Quality Eng—Senior	Q Manager	Supplier QE	CSoftQE	Program QE	Contract Review	MRO QE	Spec Process QE	Met Labs QE	Production Control	Regulatory Specialist
S	P		S		P		S	S	S	S	

Procedural Knowledge: Ability to apply a structured process to identify potential alternative production/servicing methods (3). Ability to apply a structured process to evaluate the risks and benefits associated with potential production/servicing alternatives (3). Ability to follow a structured process to effectively communicate the results of evaluation in terms useful for management (3).

Conceptual Knowledge: Ability to scan the industry to seek out potential alternative processes and methods for production/servicing to improve quality, cost, and/or efficiency (5). Ability to evaluate potential alternatives and establish potential risks and benefits in application (5). Ability to articulate risks and benefits in terms useful to management (4).

Factual Knowledge: Knowledge of industrial processes utilized in production/servicing (4). Knowledge in quality costing practices (3). Knowledge in risk assessment processes (4). Knowledge of effective communication practices (3).

III.B.3. Inspectability Review

Process: Perform Inspectability Review of Design

Process Purpose: Review engineering, process, and production requirements and assess such against the known capability to inspect/test product at the appropriate point of production to ensure conformity to design. In the event the process is found insufficient apply remediation activities such as request for design changes, process changes, or the development of new methods. Design documentation should contain the appropriate requirements for quality. (e.g., Measurement, Inspection, Key Characteristic Control, and First Article Quality Plans.) Ensure data collection capabilities support the requisite inspection, measurement and control plans.

Knowledge and Skill Requirements: Ability to review engineering, process, and production requirements and assess such against the known capability to inspect/test product at the appropriate point of production to ensure conformity to design. Ability to ensure that measurement, inspection, key characteristic control, and first article inspection are included in applicable design documents and/or other program control documents. Ability to evaluate process inadequacies and establish recommendations for remediation activities such as request for design changes, process changes, or the development of new methods. Ability to organize, document, store and integrate inspection requirements into the requirements hierarchy using a requirements management tool.

Task 1 (DP065-001): Review engineering, process, and production requirements and assess such against the known capability to inspect/test product at the appropriate point of production to ensure conformity to design. Identify and document any process inadequacies. Organize, document, store and integrate inspection requirements into the requirements hierarchy using a requirements management tool.

Quality Eng—Basic	Quality Eng—Senior	Q Manager	Supplier QE	CSoftQE	Program QE	Contract Review	MRO QE	Spec Process QE	Met Labs QE	Production Control	Regulatory Specialist
P	P	S	S		P		S	P	P	S	

Procedural Knowledge: Ability to apply a defined and structured process in the review of engineering, process, and production requirements and assess such against the known capability to inspect/test product at the appropriate point of production to ensure conformity to design (3). Ability to document the results of the definition process in the appropriate forms and format of the organization (3).

Conceptual Knowledge: Ability to perform analysis of engineering, process, and production requirements and assess such against the known capability to inspect/test product at the appropriate point of production to ensure conformity to design (4). Ability to articulate the results of inspectability analysis (3).

Factual Knowledge: Knowledge of design processes (3). Knowledge of production processes (3). Knowledge of inspection and test methods (3). Knowledge of value stream inspection/test capability (3). Knowledge of data collection methods (3).

Task 2 (DP065-002): Ensure that measurement, inspection, key characteristic control, and first article inspection are included in applicable design documents and/or other program control documents.

Quality Eng—Basic	Quality Eng—Senior	Q Manager	Supplier QE	CSoftQE	Program QE	Contract Review	MRO QE	Spec Process QE	Met Labs QE	Production Control	Regulatory Specialist
P	P	S	S		P		S	P	P	S	

Procedural Knowledge: Ability to apply a defined and structured process to ensure that measurement, inspection, key characteristic control, and first article inspection are included in applicable design documents and/or other program control documents (3). Ability to apply the requirements of applicable inspection standards (e.g., AS9102) to applicable design documents and/or other program control documents (3).

Conceptual Knowledge: Ability to contextualize the requirements for measurement, inspection, key characteristic control, and first article inspection to a specific program and ensure the requirements are included in applicable design documents and/or other program control documents (4).

Factual Knowledge: Knowledge of design processes (3). Knowledge of production processes (3). Knowledge of inspection and test methods (3). Knowledge of value stream inspection/test capability (3). Knowledge of industry and governmental standards applicable to measurement, inspection, key characteristic control, and first article inspection (3).

Task 3 (DP065-003): Evaluate process inadequacies and establish recommendations for remediation activities such as request for design changes, process changes, or the development of new methods.

Quality Eng—Basic	Quality Eng—Senior	Q Manager	Supplier QE	CSoftQE	Program QE	Contract Review	MRO QE	Spec Process QE	Met Labs QE	Production Control	Regulatory Specialist
P	P	S	S		P		S	P	P	S	

Procedural Knowledge: Ability to apply a defined and structured process to perform the evaluation of inspection/test process inadequacies (3). Ability to apply a structured process for assessing potential risk associated with inspection/test process inadequacies (3). Ability to apply a structured process for documenting potential inspection/test process inadequacies (3). Ability to document, in an appropriate manner, recommendations for remediation of potential inspection/test inadequacies (4).

Conceptual Knowledge: Ability to assess inspection and test processes in evaluation of process capability (4). Ability to evaluate and measure potential risk associated with inspection/test process inadequacies (4). Ability to articulate potential inspection/test process inadequacies in an appropriate manner (4). Ability to develop recommendations for remediation of potential inspection/test inadequacies (5).

Factual Knowledge: Knowledge of design processes (3). Knowledge of production processes (3). Knowledge of inspection and test methods (3). Knowledge of value stream inspection/test capability (3). Knowledge of industry and governmental standards applicable to measurement, inspection, key characteristic control, and first article inspection (3). Knowledge of risk evaluation methods (3).

III.B.4. Design Validation

Process: Ensure Design Validation (Standard Work)

Process Purpose: To verify that the design validation process is completed and that there is objective evidence of its completion. The design validation process must ensure the design meets all customer and regulatory requirements. If there are variations in the design that have not been previously produced, risk mitigation plans must be developed (variations to standard work).

Knowledge and Skill Requirements: Ability to ensure design validation process meets all customer and regulatory requirements. Ability to assess the design validation process and verify that it is completed and that there is objective evidence of its completion. Ability to recognize variations in design standard work. Ability to ensure that variations in design standard work have risk mitigation plans.

Task 1 (DP100-001): Ensure design validation process meets all customer and regulatory requirements.

Quality Eng—Basic	Quality Eng—Senior	Q Manager	Supplier QE	CSoftQE	Program QE	Contract Review	MRO QE	Spec Process QE	Met Labs QE	Production Control	Regulatory Specialist
P	P	P	P		P		S	P	P	S	P

Procedural Knowledge: Ability to apply a structured assessment to determine if all customer and regulatory requirements have been met in the design validation process (3).

Conceptual Knowledge: Ability to evaluate design validation process to determine if all customer and regulatory requirements have been met (4).

Factual Knowledge: Knowledge of applicable customer and regulatory requirements related to design validation (3). Knowledge of design validation methods and processes (3). Knowledge of requirements management tools (3).

Task 2 (DP100-002): Assess the implementation of the design validation process and verify that it is completed and that there is objective evidence of its completion.

Quality Eng—Basic	Quality Eng—Senior	Q Manager	Supplier QE	CSoftQE	Program QE	Contract Review	MRO QE	Spec Process QE	Met Labs QE	Production Control	Regulatory Specialist
P	P	P	P		P		S	P	P	S	P

Procedural Knowledge: Ability to apply a structured assessment to determine if all customer and regulatory requirements have been met in the execution of design validation activities (3).

Conceptual Knowledge: Ability to evaluate design validation activities to determine if all customer and regulatory requirements have been met (4).

Factual Knowledge: Knowledge of applicable customer and regulatory requirements related to design validation (3). Knowledge of design validation methods and processes (3). Knowledge of requirements management tools (3).

Task 3 (DP100-003): Evaluate proposed design data (objective evidence) to ensure compliance to existing engineering standard work.

Quality Eng—Basic	Quality Eng—Senior	Q Manager	Supplier QE	CSoftQE	Program QE	Contract Review	MRO QE	Spec Process QE	Met Labs QE	Production Control	Regulatory Specialist
P	P	S	S	P	P		S	S	S	S	

Procedural Knowledge: Ability to apply a defined process to access the compliance of proposed engineering data to existing engineering standard work (3).

Conceptual Knowledge: Ability to review and evaluate proposed engineering data to existing engineering standard work to determine compliance with existing engineering standards (4).

Factual Knowledge: Knowledge of engineering standard work definition practices (3). Understanding of the proposed engineering methods, materials, and processes (4).

Task 4 (DP100-004): Ensure that risk mitigation plans are established to manage variations in the design, which have not been previously produced (variations to standard work), using control processes as appropriate.

Quality Eng—Basic	Quality Eng—Senior	Q Manager	Supplier QE	CSoftQE	Program QE	Contract Review	MRO QE	Spec Process QE	Met Labs QE	Production Control	Regulatory Specialist
P	P	P	P	P	P		S	S	S	S	

Procedural Knowledge: Ability to apply a structured process to assess and determine that risk mitigation plans have been established to manage variations in the design, which have not been previously produced (variations to standard work), using control processes as appropriate (3).

Conceptual Knowledge: Ability to evaluate risk mitigation plans to effectively manage risk assigned to variations in the design, which have not been previously produced (variations to standard work) (4). Ability to evaluate control plans to ensure adequacy of risk mitigation (4).

Factual Knowledge: Understanding of design standard work concepts and control processes (3). Understanding of risk mitigation methods, concepts, and processes (3).

Task 5 (DP100-005): Based on the positive results of validating new design requirements, update engineering standard work.

Quality Eng—Basic	Quality Eng—Senior	Q Manager	Supplier QE	CSoftQE	Program QE	Contract Review	MRO QE	Spec Process QE	Met Labs QE	Production Control	Regulatory Specialist
S	P	S	S	P	P		S	S	S	S	

Procedural Knowledge: Ability to apply a structured and defined process to ensure engineering standard work is revised to reflect the known capabilities of the new design methods, materials, and processes (3).

Conceptual Knowledge: Ability to integrate the new knowledge captured during risk mitigation and design validation into the existing engineering standard work definition to reflect the known capabilities of the new design methods, materials, and processes (5).

Factual Knowledge: Understanding of design standard work concepts and control processes (3). Understanding of risk mitigation methods, concepts, and processes (3).

III.C.2. Computer Produced Design (CPD) Control

Process: Apply CPD Techniques/Methods (Solid Model)

Process Purpose: As Computer Produced Designs (CPD) are created, such must be transferable into production and inspection protocol. Ensure such protocol is validated and controlled through all lifecycle aspects.

Knowledge and Skill Requirements: Ability to understand the application of CPD systems as they relate to the program. Ability to identify risks associated with CPD deployment in the value stream. Ability to verify that CPD data transfer process are capable and controlled. Understanding of CPD interfaces to production and inspection processes. Ability to evaluate the capability of CPD tools. Ability to identify issues related to usage of different CPD tools within the value stream. Ability to assess value stream integration capability with production and inspection processes. Ability to organize, document, store and integrate findings into the requirements hierarchy using a requirements management tool.

Task 1 (DP035-001): Identify the application of CPD and electronic-controlled production/inspection systems as they relate to the program.

Quality Eng—Basic	Quality Eng—Senior	Q Manager	Supplier QE	CSoftQE	Program QE	Contract Review	MRO QE	Spec Process QE	Met Labs QE	Production Control	Regulatory Specialist
P	P	P	P		P		S	S		S	

Procedural Knowledge: Ability to follow a structured process to identify the application CPD and electronic-controlled production/inspection systems as they relate to the program (3).

Conceptual Knowledge: Ability to analyze design and program documents to identify CPD and electronic-controlled production/inspection system requirements. (4)

Factual Knowledge: Knowledge of CPD and electronic-controlled production/inspection systems (3).

Task 2 (DP035-002): Identify risks associated with CPD deployment in the value stream with regards to data transfer and compatibility among the value stream.

Quality Eng—Basic	Quality Eng—Senior	Q Manager	Supplier QE	CSoftQE	Program QE	Contract Review	MRO QE	Spec Process QE	Met Labs QE	Production Control	Regulatory Specialist
P	P	P	P		P		S			S	

Procedural Knowledge: Ability to follow a structured process for determination of potential risks associated with deployment of CPD in the value stream (3).

Conceptual Knowledge: Ability to analyze design and program requirements to identify potential sources of deployment risk in the value stream (4). Ability to apply foresight, based on performance in previous programs, to recognize potential sources of risk in deployment of CPD and electronic-controlled production/inspection systems in the value stream (5).

Factual Knowledge: Knowledge of risk management processes (3). Knowledge of CPD and electronic-controlled production/inspection systems (3).

Task 3 (DP035-003): Assess the capability of the value stream to integrate electronic production and inspection processes.

Quality Eng—Basic	Quality Eng—Senior	Q Manager	Supplier QE	CSoftQE	Program QE	Contract Review	MRO QE	Spec Process QE	Met Labs QE	Production Control	Regulatory Specialist
S	P	S	P		P						

Procedural Knowledge: Ability to follow a structured process to assess and document value stream integration capability with production and inspection processes (3).

Conceptual Knowledge: Ability to evaluate the capabilities of the value stream to effectively integrate electronic production and inspection processes based on delivered design protocol (5).

Factual Knowledge: Knowledge of CPD and electronic-controlled production/inspection system requirements (3). Knowledge of value stream electronic control capabilities (3).

Task 4 (DP035-004): Verify that CPD data transfer processes are capable and controlled.

Quality Eng—Basic	Quality Eng—Senior	Q Manager	Supplier QE	CSoftQE	Program QE	Contract Review	MRO QE	Spec Process QE	Met Labs QE	Production Control	Regulatory Specialist
S	P	S	P		P						

Procedural Knowledge: Ability to follow a structured process to verify that CPD data transfer processes are capable and controlled (3).

Conceptual Knowledge: Ability to assess organizational and value stream electronic systems to determine capability and extent of control during data transfer (5).

Factual Knowledge: Knowledge of CPD and electronic-controlled production/inspection system requirements (3). Knowledge of electronic data management systems (3).

III.D. Design Verification

Process: Determine Product Validation Laboratory Testing Requirements (Engineering Tests)

Process Purpose: Review development, deployment, logistic, customer and usage requirements to determine the appropriate categories and approaches to Product Validation Laboratory Testing.

Knowledge and Skill Requirements: Knowledge of requirements and requirement tools. Ability to analyze requirements and determine tests that can be used to ensure product meets requirements. Ability to communicate with supplier. Ability to link requirements to tests within a requirements management tool to ensure requirements coverage.

Task 1 (DP045-001): Identify requirements that should drive Product Validation Laboratory Testing.

Quality Eng—Basic	Quality Eng—Senior	Q Manager	Supplier QE	CSoftQE	Program QE	Contract Review	MRO QE	Spec Process QE	Met Labs QE	Production Control	Regulatory Specialist
P	P	P	S		P			P	P		

Procedural Knowledge: Ability to follow a structured process to identify and document requirements for product validation laboratory testing (3).

Conceptual Knowledge: Ability to analyze program requirements and design data to determine product characteristics that require product validation laboratory testing (5).

Factual Knowledge: Knowledge of design review process (3). Knowledge of product validation laboratory testing requirements and processes (3). Knowledge of applicable customer, regulatory and industry requirements related to product validation laboratory testing (3).

Task 2 (DP045-002): Analyze requirements previously identified and define tests that are appropriate for validating requirements.

Quality Eng—Basic	Quality Eng—Senior	Q Manager	Supplier QE	CSoftQE	Program QE	Contract Review	MRO QE	Spec Process QE	Met Labs QE	Production Control	Regulatory Specialist
P	P	P	S		P			P	P		

Procedural Knowledge: Ability to apply a structured process to define required tests to meet validation requirements (3).

Conceptual Knowledge: Ability to evaluate validation requirements in assessment of identified characteristics, and define appropriate validation testing protocol (5).

Factual Knowledge: Knowledge of validation test planning (3). Knowledge of technologies applied within the program/project (5).

Task 3 (DP045-003): Document tests in a Test Plan. Store the Test Plan in a tool that facilitates linking to the requirements database. Link the tests to the requirements.

Quality Eng—Basic	Quality Eng—Senior	Q Manager	Supplier QE	CSoftQE	Program QE	Contract Review	MRO QE	Spec Process QE	Met Labs QE	Production Control	Regulatory Specialist
P	P	P	S		S			P	P		

Procedural Knowledge: Ability to apply a structured process in developing a validation test planning document (3). Ability to incorporate test planning requirements into the requirements management tool (3).

Conceptual Knowledge: Ability to describe test protocol in terms that can be executed by the organization (4). Ability to plan validation test at points where assurance can be best performed (4). Ability to align test planning into hierarchical structure within the requirements management tool (4).

Factual Knowledge: Knowledge of test planning processes (3). Knowledge of program specific technologies and materials (4). Knowledge of requirements management tools (3).

Task 4 (DP045-004): Ensure correct requirements are utilized to establish their product validation tests.

Quality Eng—Basic	Quality Eng—Senior	Q Manager	Supplier QE	CSoftQE	Program QE	Contract Review	MRO QE	Spec Process QE	Met Labs QE	Production Control	Regulatory Specialist
P	P	P	S		P	S		P	P		S

Procedural Knowledge: Ability to follow a structured process to substantiate that appropriate requirements for product validation testing have been identified (3).

Conceptual Knowledge: Ability to review project/program requirements and evaluate compliance of documented/planned product validation requirements (4). Ability to evaluate the adequacy of product validation planning to project/program requirements (5).

Factual Knowledge: Knowledge of project/program product validation testing requirements (3). Knowledge of product validation planning methods (3). Knowledge of quality records requirements

(3). Knowledge of applicable regulatory requirements for product validation testing (3). Understanding of requirements management tools (3).

Task 5 (DP045-005): Review product validation test/requirements linkage to ensure adequate requirements coverage. Review proposed product validation tests to ensure that they adequately cover requirements.

Quality Eng—Basic	Quality Eng—Senior	Q Manager	Supplier QE	CSoftQE	Program QE	Contract Review	MRO QE	Spec Process QE	Met Labs QE	Production Control	Regulatory Specialist
P	P	P	S		P	S		P	P		S

Procedural Knowledge: Ability to follow a structured process to evaluate the adequacy of product validation test protocol to ensure applicable project/program requirements are addressed (3).

Conceptual Knowledge: Ability to evaluate the adequacy of product validation test protocol to ensure applicable project/program requirements are addressed and linkage to requirements are stated (4).

Factual Knowledge: Knowledge of project/program product validation testing requirements (3). Knowledge of product validation planning methods (3). Knowledge of quality records requirements (3). Knowledge of applicable regulatory requirements for product validation testing (3). Understanding of requirements management tools (3). Knowledge of product past performance (3). Knowledge of applicable test methods (3).

III.E. Reliability and Maintainability

Process: Determine Production/Servicing Capability

Process Purpose: To assess and understand the present and potential state of production/servicing capability of the organization and ensure its capability to produce/service is known and provided as an input to the design process, and to ensure the production base is capable of fulfilling the final stated requirements of design.

Knowledge and Skill Requirements: Ability to assess and understand the present and potential state of production/servicing capability of the organization and ensure its capability to produce/service is known and provided as an input to the design process, and to ensure the production base is capable of fulfilling the final stated requirements of design. Knowledge of value stream capability to include raw materials.

Task 1 (CP025-001): Determine Production capability. (Determine the production capability within a value stream).

Quality Eng—Basic	Quality Eng—Senior	Q Manager	Supplier QE	CSoftQE	Program QE	Contract Review	MRO QE	Spec Process QE	Met Labs QE	Production Control	Regulatory Specialist
P	P	S	P	P	P			P	P		

Procedural Knowledge: Ability to apply a structured process for determining production/servicing capability. (3) (Follow defined procedures to measure production/servicing capability.)

Conceptual Knowledge: Understand the relationship of machines, people, materials, and logistics, as such are applied to production/servicing capability to evaluate existing production/servicing capability (5).

Factual Knowledge: Ability to apply the concepts of production capability (3). An understanding of the production capability of the various equipment that might be applied during the execution of work scope (3). An understanding of the various logistics requirements and options applicable to the program/project (3). An understanding of material control requirements for raw materials that are applicable to the program/project (3).

Task 2 (CP025-002): Evaluate the production/service capability against the work requirements.

Quality Eng—Basic	Quality Eng—Senior	Q Manager	Supplier QE	CSoftQE	Program QE	Contract Review	MRO QE	Spec Process QE	Met Labs QE	Production Control	Regulatory Specialist
P	P	S	P	P	P			P	P		

Procedural Knowledge: While following a structured process, ability to create side-by-side production/servicing capability comparison documents (5). While following prescribe methods of communication, the ability to elicit production/servicing capability from the value stream (4).

Conceptual Knowledge: Ability to interpret customer requirements and convert such into defined and measurable production/servicing capability requirements (5). Based on available and obtained information, the ability evaluate the production/servicing capability of the value stream to ensure the capability is appropriately quantified (5).

Factual Knowledge: Understanding of production/servicing capability measures and analysis (4). Know where the work is to be potentially performed (3).

Task 3 (CP025-003): Evaluate production requirements against the current state of production/servicing standard-work.

Quality Eng—Basic	Quality Eng—Senior	Q Manager	Supplier QE	CSoftQE	Program QE	Contract Review	MRO QE	Spec Process QE	Met Labs QE	Production Control	Regulatory Specialist
P	P	S	P	P	P			P	P		

Procedural Knowledge: Ability to create side-by-side production/servicing capability comparison documents (5). Ability to elicit production/servicing standard-work from the value stream (4). Ability to follow a structured process to establish comparison document (3).

Conceptual Knowledge: Ability to interpret customer requirements and convert such into production/servicing capability requirements (4). Understanding of present state of standard-work within the organization and throughout the value stream (4). Understanding of the relationship between equipment, people, materials, and logistics as they relate to production/service capability (4). The ability to perform an evaluation of the production requirements in relation to the current state of production/servicing standard work, and create a side-by-side analysis of capability alignment (5).

Factual Knowledge: Know the production/service capability of specific equipment and processes as they relate to the program/project (3). Know who the are the providers of equipment, machines, and tools (3). Understand and have the ability to apply the concepts of standard-work (4).

Task 4 (CP025-004): Document gaps in present standard-work against new production/service requirements. Measure the gap as measures of production/service execution risk. Evaluate the present level of production/servicing standard work as it relates to production/servicing capability.

Quality Eng—Basic	Quality Eng—Senior	Q Manager	Supplier QE	CSoftQE	Program QE	Contract Review	MRO QE	Spec Process QE	Met Labs QE	Production Control	Regulatory Specialist
P	P	S	P	P	P			P	P		

Procedural Knowledge: Ability to follow a structured methodology for documenting production/servicing capability (3). Ability to analyze and illustrate risks associated with gaps between production/servicing capability and standard-work or required capability in accordance with appropriate procedures (4).

Conceptual Knowledge: Ability to analyze customer requirements and determine required production/servicing capability (5). Ability to evaluate existing production/servicing operations and determine capability (5). Ability to evaluate differences between known capability, as defined in standard work, and create appropriate documents that define the identified program/project risks associated with production/servicing processes (6).

Factual Knowledge: Based on available and obtained information, know how to analyze the production/servicing capability of the value stream (4). Know the production/servicing capability of existing standard-work (3). Understand the vocabulary and assignments of measure associated with risk measurement (4). Know the various risk measurement processes available to evaluate production/service capability as it aligns to standard work (4).

Task 5 (CP025-005): Determine which production/servicing processes are to be applied to ensure effective execution of work.

Quality Eng—Basic	Quality Eng—Senior	Q Manager	Supplier QE	CSoftQE	Program QE	Contract Review	MRO QE	Spec Process QE	Met Labs QE	Production Control	Regulatory Specialist
S	P	P	P	P	P			P	P		

Procedural Knowledge: Ability to follow a structured process to identify which production/servicing processes have sufficient capability to effectively execute customer work requirements (3). Ability to document production/servicing capability in terms that production/servicing sources can be selected (3).

Conceptual Knowledge: Understand the state-of-the art in regards to production/servicing capability (in existing organization, value stream, and throughout industry) and effectively communicate capability in terms that allow source selections to be made (5). Ability to analyze the results of capability assessments to determine which sources provide the best solution for execution of work (4).

Factual Knowledge: Know the existing structure of the organization's value stream and the associated production/servicing capability (3). Know the state-of-the-art with regards to production/servicing processes (3). Know the state-of-the-art with regards to raw material processing (3).

Task 6 (CP025-006): Evaluate and identify unique and/or special processes are to be applied to ensure effective execution of work.

Quality Eng—Basic	Quality Eng—Senior	Q Manager	Supplier QE	CSoftQE	Program QE	Contract Review	MRO QE	Spec Process QE	Met Labs QE	Production Control	Regulatory Specialist
S	S	S	S		P			S	S		

Procedural Knowledge: Ability to follow a structured process to identify which unique and/or special processes are required and which have sufficient capability to effectively execute customer work requirements (3). Ability to document production/servicing capability in terms that production/servicing sources can be selected (3).

Conceptual Knowledge: Ability to elicit unique and/or special processing requirements from the customer and/or customer specifications (4). Understand the state-of-the art in regards to unique and/or special processes and determine capability (in existing organization, value stream, and throughout industry) and effectively communicate capability in terms that allow source selections to be made (5). Ability to perform capability assessments (4).

Factual Knowledge: Know the existing structure of the organization's value stream and the associated production/servicing capability with regards to unique and/or special processes (3). Know the state-of-the-art with regards to production/servicing processes (3). Know the state-of-the-art with regards to raw material processing (3). Know the sources of technical information (3).

Process: Determine Production/Servicing Capacity

Process Purpose: To understand the present and potential state of production/servicing capacity to ensure the production base has the capacity to produce/service to the requirements of the customer/program.

Knowledge and Skill Requirements: Detailed understanding of desired product/service requirements. Ability to determine the organization's capacity to deliver the required products/services given production resources and constraints. Ability to elicit detailed information from the organization. Planning/scheduling capability.

Task 1 (CP030-001): Understand the production/servicing capacity and material requirements and the resources needed to deliver the product/service. Review results of the product/servicing capability evaluation process. Make preliminary determination of production/servicing/logistics approach with alternatives. Understand and document program constraints.

Quality Eng—Basic	Quality Eng—Senior	Q Manager	Supplier QE	CSoftQE	Program QE	Contract Review	MRO QE	Spec Process QE	Met Labs QE	Production Control	Regulatory Specialist
S	P	P	P		P		S	S	S	S	S

Procedural Knowledge: Ability to follow a structured process in review of customer requirement documents as they relate to capacity and material availability requirements (3). Ability to apply a structured process for determining existing production/servicing capacity and eliciting information from organizations (3).

Conceptual Knowledge: Understand and evaluate the relationship of machines, people, materials, and logistics, as such are applied to production/servicing and as they relate capacity requirements of the program/project (5). Understanding of production/servicing/logistics approaches and ability to select the most capable approach(es) (5). Through evaluation of existing and obtained information, identify potential resource constraints (5).

Factual Knowledge: Understand and have the ability to evaluate production capacity (3). Possess an understanding of the reliability and availability of the various equipment and tools that may be required for the program/project (4). Understand the various requirements related to logistics and the potential options available (4). Know the requirements for raw material controls (4).

Task 2 (CP030-002): Based on physical evaluation of production/servicing processes, determine the production resources required to deliver the desired product/service. (Including evaluation of other actual or planned work that suppliers may be planning to take on.)

Quality Eng—Basic	Quality Eng—Senior	Q Manager	Supplier QE	CSoftQE	Program QE	Contract Review	MRO QE	Spec Process QE	Met Labs QE	Production Control	Regulatory Specialist
S	P	P	P		P		S	S	S	S	S

Procedural Knowledge: Ability to apply a structured process in the review of existing and/or proposed sources of production/servicing to determine production/service capacity (5). Ability to elicit and understand the implications of management's approach for delivering product/service (5). Ability to understand resource limitations and quality requirements as they relate to execution risk (4).

Conceptual Knowledge: Ability to interpret customer requirements and convert such into defined and measurable production/servicing capacity requirements (5). Ability to evaluate existing and/or proposed production/servicing capacity of the value stream to ensure the capacity is appropriately quantified (measurements applied) (5). Ability to elicit future workload information from management (4).

Factual Knowledge: Understanding of, and ability to apply production/servicing capacity measures (4). Know where the work is to be performed or potentially performed (3). Based on analysis of industrial information, an understanding of industrial capacity projections and potential shortages of labor, equipment, and raw material (4).

Task 3 (CP030-003): Elicit from management information about production/servicing plans and resources. Through evaluation, determine the effectiveness of alternative approaches. Identify and characterize risks associated with capacity planning.

Quality Eng—Basic	Quality Eng—Senior	Q Manager	Supplier QE	CSoftQE	Program QE	Contract Review	MRO QE	Spec Process QE	Met Labs QE	Production Control	Regulatory Specialist
S	P	P	P	P	P		S	S	S	S	S

Procedural Knowledge: Ability to follow a structured methodology to evaluate and define possible alternative approaches to production/servicing processing (5).

Conceptual Knowledge: Ability to understand various data inputs to evaluate the viability of alternative production/servicing approaches (5). Ability to evaluate potential risks associated with alternative methods (5). Ability to assess and characterize risks. Ability to effectively communicate the results of the analysis (4). Ability to evaluate the resource requirements of the alternative approaches and create new alternatives to reduce risk (6).

Factual Knowledge: Posses and understanding of the various alternative production/servicing approaches and how such may be applied to the program/project (4). Know the inherent risks associated with alternative approaches, as they relate to program/project resources (3). Know the resource requirements associated with alternative approaches (3).

Task 4 (CP030-004): Document capacity plans and expectations. Review with organization management. Determine projected schedule and resource constraints. (Analysis performed in previous tasks)

Quality Eng—Basic	Quality Eng—Senior	Q Manager	Supplier QE	CSoftQE	Program QE	Contract Review	MRO QE	Spec Process QE	Met Labs QE	Production Control	Regulatory Specialist
S	P	S	P	P	P		S	S	S	S	S

Procedural Knowledge: Ability to follow a structured methodology to develop a capacity plan (3).

Conceptual Knowledge: Ability to evaluate data provided from previous resource analysis tasks and create resource capacity plan (5). Ability to communicate the capacity plan in terms the organization can understand (3).

Factual Knowledge: Know the concepts and requirements of capacity planning (4). Know the appropriate communication and documentation methods related to capacity planning (4).

Task 5 (CP030-005): Perform physical evaluation of the actual production/servicing capacity to perform the expected scope of work.

Quality Eng—Basic	Quality Eng—Senior	Q Manager	Supplier QE	CSoftQE	Program QE	Contract Review	MRO QE	Spec Process QE	Met Labs QE	Production Control	Regulatory Specialist
S	P	S	P	P	P		S	S	S	S	S

Procedural Knowledge: Ability to follow a structured methodology (checklists, guides, procedures) to evaluate and establish actual production/servicing capacity (4).

Conceptual Knowledge: Through physical evaluation, the ability to asses and measure the actual capacity (5). (Physical evaluation may include assessment of facilities, labor, knowledge base, equipment, control of supplier base, and maturity of the quality management system.)

Factual Knowledge: Knowledge of program/project related tooling and equipment capacity and capability (4). Knowledge of capability evaluation processes (5). Knowledge of inspection/production/servicing techniques (5). Knowledge of quality management systems (4). Knowledge of supplier management controls (4).

Process: Perform Maintainability Review of Design

Process Purpose: Review the design for ease of maintaining the product while it is in the possession of the customer/end-user.

Knowledge and Skill Requirements: Ability to review design documents and customer requirements to establish maintainability requirements. Ability to evaluate the proposed design (e.g., use cases) for ease of maintaining (to include human factors) the product while it is in the possession of the customer/end-user. Ability to organize, document, store and integrate maintainability requirements into the requirements hierarchy using a requirements management tool.

Task 1 (DP070-001): Review design documents and customer requirements to establish maintainability requirements. Determine the cost/benefits analysis of maintenance. Organize, document, store and integrate maintainability requirements into the requirements hierarchy using a requirements management tool.

Quality Eng—Basic	Quality Eng—Senior	Q Manager	Supplier QE	CSoftQE	Program QE	Contract Review	MRO QE	Spec Process QE	Met Labs QE	Production Control	Regulatory Specialist
S	P	S	S		P	S					

Procedural Knowledge: Ability to apply a defined and structured process to evaluate design and customer requirements to identify and define maintainability requirements (3). Ability to conduct a cost/benefits analysis on the proposed maintenance activities (3).

Conceptual Knowledge: Ability to analyze customer requirements and design documents to determine and define maintainability requirements (4) Ability to evaluate the cost and benefits of performing maintenance and establishing the appropriate maintenance practices (4).

Factual Knowledge: Knowledge of customer requirements documents (3). Knowledge of design definition requirements (3). Knowledge of maintenance practices (4). Knowledge of cost/benefits analysis methods (3).

Task 2 (DP070-002): Evaluate the proposed design for ease of maintaining (to include human factors) the product while it is in the possession of the customer/end-user. Ability to organize, document, store and integrate maintainability requirements into the requirements hierarchy using a requirements management tool.

Quality Eng—Basic	Quality Eng—Senior	Q Manager	Supplier QE	CSoftQE	Program QE	Contract Review	MRO QE	Spec Process QE	Met Labs QE	Production Control	Regulatory Specialist
S	P	S	S		P	S					

Procedural Knowledge: Ability to organize the methods of maintenance for evaluation (3). Ability to apply a defined and structured process in the evaluation of design documents to ensure the ease of maintenance (4). Ability to document the results of a maintainability analysis in an appropriate form and fashion (3).

Conceptual Knowledge: Ability to evaluate design documents to ensure the ease of maintenance (4). Ability to apply the science of human factors in the determination of maintainability of a specific design (5).

Factual Knowledge: Knowledge of customer requirements documents (3). Knowledge of design definition requirements (3). Knowledge of maintenance practices (4). Knowledge of human factors practices in maintenance (3).

III.E.4. Reliability/Safety/Hazard Assessment Tools (Includes all FMEA)

Process: Perform Design FMEA

Process Purpose: Review design to quantify and categorize the potential areas of design risk and ensure there are risk mitigation plans developed.

Knowledge and Skill Requirements: Ability to apply the concepts and methods of DFMEA to the product design in order to quantify and categorize the potential areas of design risk and ensure there are risk mitigation plans developed, documented and deployed.

Task (DP085-001): Apply the concepts and methods of DFMEA to the product design in order to quantify and categorize the potential areas of design risk and ensure there are risk mitigation plans developed, documented and deployed.

Quality Eng—Basic	Quality Eng—Senior	Q Manager	Supplier QE	CSoftQE	Program QE	Contract Review	MRO QE	Spec Process QE	Met Labs QE	Production Control	Regulatory Specialist
S	P	S	S		P			S	S		

Procedural Knowledge: Ability to apply a structured process in applying a DFMEA in the program/project design (3). Ability to apply a structured process to develop and document mitigation plans required from the results of the DFMEA (3).

Conceptual Knowledge: Ability to evaluate program/project design in application of DFMEA in order to quantify and categorize the potential areas of design risk (5). Ability to create risk mitigation plans in response to DFMEA findings (5).

Factual Knowledge: Knowledge of design practices (3). Knowledge of production processes (3). Knowledge of DFMEA concepts and processes (3). Knowledge of risk mitigation planning methods (3).

Process: Perform Process FMEA

Process Purpose: Review processes to quantify and categorize the potential areas of process risk and ensure there are risk mitigation plans developed.

Knowledge and Skill Requirements: Ability to apply the concepts and methods of Process FEMA to the production/servicing processes in order to quantify and categorize the potential areas of work execution risk and ensure there are risk mitigation plans developed. Ability to ensure PFEMA analysis are organized, documented, stored and integrated into the requirements hierarchy using a requirements management tool.

Task 1 (PO095-001): Apply the concepts and methods of Process FEMA to the production/servicing processes in order to quantify and categorize the potential areas of work execution risk and ensure there are risk mitigation plans developed.

Quality Eng—Basic	Quality Eng—Senior	Q Manager	Supplier QE	CSoftQE	Program QE	Contract Review	MRO QE	Spec Process QE	Met Labs QE	Production Control	Regulatory Specialist
P	P				P		S			S	

Procedural Knowledge: Apply a structured Process FMEA for identifying, quantifying, and categorizing potential areas of work execution risk and ensure, where appropriate, there are documented risk mitigation plans developed (3).

Conceptual Knowledge: Ability to apply Process FMEA methods and concepts to evaluate production/servicing processes in order to quantify and categorize potential areas of work execution risk and ensure risk mitigation plans are developed (4).

Factual Knowledge: Knowledge of Process FMEA methods and concepts (3). Understanding of risk management methods (3).

Task 2 (PO095-002): Ensure PFEMA analysis are organized, documented, stored and integrated into the requirements hierarchy using a requirements management tool.

Quality Eng—Basic	Quality Eng—Senior	Q Manager	Supplier QE	CSoftQE	Program QE	Contract Review	MRO QE	Spec Process QE	Met Labs QE	Production Control	Regulatory Specialist
P	P				P		S			S	

Procedural Knowledge: Ability to apply a structured process to assess established PFMEA processes to determine the adequacy of such processes to organize, document, store and integrate identified risk and defined mitigation plans into the requirements hierarchy using a requirements management tool (3).

Conceptual Knowledge: Ability to evaluate established PFMEA processes to determine the adequacy of such processes to organize, document, store and integrate identified risk and defined mitigation plans into the requirements hierarchy using a requirements management tool (4).

Factual Knowledge: Knowledge of PFMEA methods and concepts (3). Knowledge of risk management methods and concepts (4). Knowledge of requirements management tools (3).

Process: Perform Process FMEA

Process Purpose: Review processes to quantify and categorize the potential areas of process risk and ensure there are risk mitigation plans developed.

Knowledge and Skill Requirements: Ability to apply the concepts and methods of Process FEMA to the production/servicing processes in order to quantify and categorize the potential areas of work execution risk and ensure there are risk mitigation plans developed. Ability to ensure PFEMA analysis are organized, documented, stored and integrated into the requirements hierarchy using a requirements management tool.

Task 1 (PP095-001): Apply the concepts and methods of Process FEMA to the production/servicing processes (in the value stream) in order to quantify and categorize the potential areas of work execution risk and ensure there are risk mitigation plans developed.

Quality Eng—Basic	Quality Eng—Senior	Q Manager	Supplier QE	CSoftQE	Program QE	Contract Review	MRO QE	Spec Process QE	Met Labs QE	Production Control	Regulatory Specialist
			P					S	S		

Procedural Knowledge: Apply a structured Process FMEA for identifying, quantifying, and categorizing potential areas of work execution risk and ensure, where appropriate, there are documented risk mitigation plans developed (3).

Conceptual Knowledge: Ability to apply Process FMEA methods and concepts to evaluate production/servicing processes in order to quantify and categorize potential areas of work execution risk and ensure risk mitigation plans are developed (4).

Factual Knowledge: Knowledge of Process FMEA methods and concepts (3). Understanding of risk management methods (3).

Task 2 (PP095-002): Ensure PFEMA analysis are organized, documented, stored and integrated into the requirements hierarchy using a requirements management tool (in the value stream).

Quality Eng—Basic	Quality Eng—Senior	Q Manager	Supplier QE	CSoftQE	Program QE	Contract Review	MRO QE	Spec Process QE	Met Labs QE	Production Control	Regulatory Specialist
			P					S	S		

Procedural Knowledge: Ability to apply a structured process to assess established PFMEA processes to determine the adequacy of such processes to organize, document, store and integrate identified risk and defined mitigation plans into the requirements hierarchy using a requirements management tool (3).

Conceptual Knowledge: Ability to evaluate established PFMEA processes to determine the adequacy of such processes to organize, document, store and integrate identified risk and defined mitigation plans into the requirements hierarchy using a requirements management tool (4).

Factual Knowledge: Knowledge of PFMEA methods and concepts (3). Knowledge of risk management methods and concepts (4). Knowledge of requirements management tools (3).

III.G. Configuration Management

Process: Plan for Configuration Management Process

Process Purpose: Review the Configuration Management Processes to be used on the program to ensure compliance with established Systems Engineering (SE) and Configuration Management (CM) practices. Ensure the establishment of a change process for program and design changes is being planned for, and coordination and communication are integral parts of the process's design. (Awareness of common CM tools.)

Knowledge and Skill Requirements: Ability to review the Configuration Management Processes to be used on the program and ensure compliance with established Systems Engineering (SE) and Configuration Management (CM) practices. Ability to assess the establishment and adequacy of a change process for program. Ability to evaluate the organization's and value stream's capability and capacity to manage design changes, and ensure the coordination, communication, and appropriate approval processes are integral parts of the design CM processes. (Awareness of common CM tools) Ability to access contractual documents to ensure configuration management control requirements are adequately noted.

Task 1 (DP090-001): Review applicable contractual documents and verify that requirements for configuration management are adequately noted. Review the Configuration Management Processes to be used on the program and ensure compliance with established Systems Engineering (SE) and Configuration Management (CM) practices.

Quality Eng—Basic	Quality Eng—Senior	Q Manager	Supplier QE	CSoftQE	Program QE	Contract Review	MRO QE	Spec Process QE	Met Labs QE	Production Control	Regulatory Specialist
S	P	P	S		P	P		S	S		P

Procedural Knowledge: Ability to access contractual documents to ensure configuration management requirements are adequately noted (3). Ability to apply a structured process to evaluate compliance of program/project-related configuration management process against established SE and CM practices (3). Ability to apply a structured process to evaluate compliance of program/project-related configuration management process against customer and/or regulatory requirements (4).

Conceptual Knowledge: Ability to interpret customer and regulatory requirements for configuration management (4), Ability to evaluate compliance of organizational configuration management processes to ensure compliance to customer and regulatory requirements (4). Ability to assess deployed configuration management process to ensure compliance to customer, regulatory and organizational requirements (4).

Factual Knowledge: Knowledge of regulatory requirements for configuration management (3). Knowledge of contract review processes (3). Knowledge of systems engineering processes as they relate to configuration management (3). Knowledge of regulatory requirements related to configuration management (3).

Task 2 (DP090-002): Evaluate the organization's capability and capacity to manage design changes, and ensure the coordination, communication, and appropriate approval processes are integral parts of the design CM processes. (Awareness of common CM tools)

Quality Eng—Basic	Quality Eng—Senior	Q Manager	Supplier QE	CSoftQE	Program QE	Contract Review	MRO QE	Spec Process QE	Met Labs QE	Production Control	Regulatory Specialist
S	P	P	S		P	P		S	S		P

Procedural Knowledge: Ability to follow a structured process in the assessment of organizational capability and capacity to effectively manage the design and design change process (4). (May include the utilization of assessment checklists and guides.)

Conceptual Knowledge: The ability to evaluate the systems and processes an organization has in place to effectively manage the timely implementation of design changes into the value stream (5). Ability to evaluate the systems and processes an organization has in place to effectively manage the coordination of appropriate approvals prior to implementation of design changes into the value stream (4). Ability to assess the design management tools within the value stream to ensure design changes are effectively coordinated, communicated, and implemented throughout the value stream (5).

Factual Knowledge: Knowledge of regulatory requirements related to configuration management (3). Knowledge of industry standards related to configuration management (3). Knowledge of configuration management tools (4). Knowledge of audit practices (3). Knowledge of design change implementation practices (4).

Process: Ensure the Establishment of Configuration Tracking/Status Accounting System

Process Purpose: Ensure a system for tracking changes and released configuration status is developed and implemented prior to work authorization.

Knowledge and Skill Requirements: Ability to ensure the organization's and value stream's system for tracking changes and released configuration status is developed, capable and implemented prior to engineering release. Ability to assess non-design responsible suppliers to ensure processes for tracing Manufacturing Bill of Materials (MBOM) usage, PO Changes, and Change Implementation exists and are controlled for Status Accounting prior to work authorization. Ability to access contractual documents to ensure configuration management control requirements are adequately noted.

Task 1 (DP095-001): Ensure the organization's (organization with design authorization) system for tracking changes and released configuration status is developed, capable and implemented prior to work authorization.

Quality Eng—Basic	Quality Eng—Senior	Q Manager	Supplier QE	CSoftQE	Program QE	Contract Review	MRO QE	Spec Process QE	Met Labs QE	Production Control	Regulatory Specialist
P	P	P	P		P	S		S	S		S

Procedural Knowledge: Ability to follow a structured process for assessing suppliers' configuration management system in determination that the system is capable of tracking engineering changes and released configuration status and implemented prior to work authorization (3). Ability to follow a structured process for assessing suppliers' configuration management system in determination that the system for tracking engineering changes and released configuration status in compliance with customer requirements, industry standards and regulatory requirements (3).

Conceptual Knowledge: Ability to evaluate suppliers' configuration management system to determine if the system is capable of tracking engineering changes and released configuration status and implemented prior to work authorization (4). Ability to evaluate suppliers' configuration management system to determine if the system for tracking engineering changes and released configuration status is in compliance with customer requirements, industry standards and regulatory requirements (4).

Factual Knowledge: Knowledge of project/program requirements for configuration management (3). Knowledge of configuration management systems (3). Knowledge of customer/industry/regulatory requirements for configuration management (3).

Task 2 (DP095-002): Assess non-design responsible suppliers to ensure processes for tracing Manufacturing Bill of Materials (MBOM) usage, PO Changes, and Change Implementation exist and are controlled for Status Accounting prior to work authorization.

Quality Eng—Basic	Quality Eng—Senior	Q Manager	Supplier QE	CSoftQE	Program QE	Contract Review	MRO QE	Spec Process QE	Met Labs QE	Production Control	Regulatory Specialist
S	S	S	P		P			S	S		

Procedural Knowledge: Ability to follow a structured process to assess non-design responsible suppliers to ensure processes for tracing Manufacturing Bill of Materials (MBOM) usage, PO Changes, and Change Implementation exist and are controlled for Status Accounting (revision control/configuration management) prior to work authorization (3).

Conceptual Knowledge: Ability to evaluate the systems and processes that suppliers' have established to ensure processes for tracing Manufacturing Bill of Materials (MBOM) usage, PO Changes, and Change Implementation exist and are controlled for Status Accounting (revision control/configuration management) prior to work authorization (4).

Factual Knowledge: Knowledge of configuration management requirements (3). Knowledge of production/servicing work instruction planning methods (3).

Task 3 (DP095-003): Ensure contractual documents from the organization to the supplier, and/or supplier to its suppliers, include adequately noted configuration management flow down requirements.

Quality Eng—Basic	Quality Eng—Senior	Q Manager	Supplier QE	CSoftQE	Program QE	Contract Review	MRO QE	Spec Process QE	Met Labs QE	Production Control	Regulatory Specialist
S	S	S	P		P	S					S

Procedural Knowledge: Ability to follow a structured and defined process to determine the compliance of configuration management flow down requirements in contractual documents (compliance assessment occurs throughout the value stream) (3).

Conceptual Knowledge: Ability to evaluate the configuration management flow down processes and procedures throughout the value stream to determine adequacy of controls and assurance of proper flow down (4).

Factual Knowledge: Knowledge of configuration management processes, procedures, and systems (3). Knowledge of compliance auditing practices (3). Understanding of requirements management tools (3). Knowledge of contractual document structures (3).

Process: Ensure Configuration Controls

Process Purpose: Ensures a congruence of part configuration with engineering configuration, and that changes to one are adequately coordinated with the other. Implementation of approved changes are planned produceably (e.g., ability to implement the appropriate configuration in production and in servicing) and effectively either in line or through modification kits that are thoroughly coordinated with the affected internal and/or external organizations.

Knowledge and Skill Requirements: Ability to verify congruence of part configuration with engineering configuration, and that changes to one are adequately coordinated with the other. Ability to verify the implementation of approved changes are planned produceably and effectively either in line or through modification kits that are thoroughly coordinated with the affected internal and/or external organizations.

Task 1 (PO080-001): Verify congruence of part configuration with engineering configuration, and that changes to one are adequately coordinated with the other.

Quality Eng—Basic	Quality Eng—Senior	Q Manager	Supplier QE	CSoftQE	Program QE	Contract Review	MRO QE	Spec Process QE	Met Labs QE	Production Control	Regulatory Specialist
P	P				S		P			P	

Procedural Knowledge: Ability to apply a structured process to assess the congruence of part configuration with engineering configuration, and that changes to one are adequately coordinated with the other (3). Ability to apply a structured process to perform a product audit to determine configuration conformance (3).

Conceptual Knowledge: Ability to identify the appropriate configuration revision for a given part, process, and/or service (4). Ability to review a configuration management system and determine the likely areas for noncompliance (5). Ability to assess congruence of part configuration with engineering configuration, and that changes to one are adequately coordinated with the other (4). Ability to determine configuration compliance throughout a value stream (4).

Factual Knowledge: Understanding of drawing systems (3). Understanding of work instruction systems (3). Understanding of configuration management systems (4). Understanding of project/program requirements related to configuration management (3). Understanding of audit practices (3).

Task 2 (PO080-002): Verify the implementation of approved changes are planned produceably and effectively either in line or through modification kits that are thoroughly coordinated with the affected internal and/or external organizations.

Quality Eng—Basic	Quality Eng—Senior	Q Manager	Supplier QE	CSoftQE	Program QE	Contract Review	MRO QE	Spec Process QE	Met Labs QE	Production Control	Regulatory Specialist
S	P				P						

Procedural Knowledge: Ability to apply a structured process to determine that approved changes are planned produceably and effectively, either in line or through modification kits that are thoroughly coordinated with the affected internal and/or external organizations (3).

Conceptual Knowledge: Ability to review approved changes to verify that they are planned produceably and effectively, either in line or through modification kits that are thoroughly coordinated with the affected internal and/or external organizations (5). Ability to recognize and understand the implications of the uses of various configuration management systems within a value stream (4).

Factual Knowledge: Understanding of design drawing systems (3). Understanding of work instruction systems (3). Understanding of configuration management systems (4). Understanding of audit practices (3). Understanding of value stream structure (3). Understanding of project/program requirements related to configuration management (3).

Task 3 (PP080-001): Verify congruence of part configuration with engineering configuration, and that changes to one are adequately coordinated with the other.

Quality Eng—Basic	Quality Eng—Senior	Q Manager	Supplier QE	CSoftQE	Program QE	Contract Review	MRO QE	Spec Process QE	Met Labs QE	Production Control	Regulatory Specialist
S	P				P			S	S		

Procedural Knowledge: Ability to apply a structured process to assess the congruence of part configuration with engineering configuration, and that changes to one are adequately coordinated with the other (3). Ability to apply a structured process to perform a product audit to determine configuration conformance (3).

Conceptual Knowledge: Ability to identify the appropriate configuration revision for a given part, process, and/or service (4). Ability to review a configuration management system and determine the likely areas for noncompliance (5). Ability to assess congruence of part configuration with engineering configuration, and that changes to one are adequately coordinated with the other (4). Ability to determine configuration compliance throughout a value stream (4).

Factual Knowledge: Understanding of drawing systems (3). Understanding of work instruction systems (3). Understanding of configuration management systems (4). Understanding of project/program requirements related to configuration management (3). Understanding of audit practices (3).

Task 4 (PP080-002): Verify the implementation of approved changes are planned produceably and effectively either in line or through modification kits that are thoroughly coordinated with the affected internal and/or external organizations.

Quality Eng—Basic	Quality Eng—Senior	Q Manager	Supplier QE	CSoftQE	Program QE	Contract Review	MRO QE	Spec Process QE	Met Labs QE	Production Control	Regulatory Specialist
	S		S		P						

Procedural Knowledge: Ability to apply a structured process to determine that approved changes are planned produceably and effectively, either in line or through modification kits that are thoroughly coordinated with the affected internal and/or external organizations (3).

Conceptual Knowledge: Ability to review approved changes to verify that they are planned produceably and effectively, either in line or through modification kits that are thoroughly coordinated with the affected internal and/or external organizations (5). Ability to recognize and understand the implications of the uses of various configuration management systems within a value stream (4).

Factual Knowledge: Understanding of design drawing systems (3). Understanding of work instruction systems (3). Understanding of configuration management systems (4). Understanding of audit practices (3). Understanding of value stream structure (3). Understanding of project/program requirements related to configuration management (3).

III.G.1. Configuration Baseline

Process: Ensure Configuration Status and Accounting—Functional Configuration Audit (FCA)

Process Purpose: Prior to delivery, ensure configuration reconciliation is performed between Engineering Bill of Materials (EBOM) and Manufacturing Bill of Materials (MBOM), complete with a history of Deviation and Waivers, open and pending changes, and objective evidence of change incorporation as applicable. Ensure the capability to successfully perform an Functional Configuration Audit (FCA), if required.

Knowledge and Skill Requirements: Understanding of contract requirements related to configuration management. Understanding of configuration management processes. Ability to verify that prior to delivery, configuration reconciliation is performed between Engineering Bill of Materials (EBOM) and Manufacturing Bill of Materials (MBOM), complete with a history of Deviation and Waivers, open and pending changes, and objective evidence of change incorporation as applicable. Ability to verify that capability to successfully perform an Functional Configuration Audit (FCA), if required.

Task 1 (PF095-001): Verify that prior to delivery, configuration reconciliation is performed between Engineering Bill of Materials (EBOM) and Manufacturing Bill of Materials (MBOM), complete with a history of Deviation and Waivers, open and pending changes, and objective evidence of change incorporation as applicable.

Quality Eng—Basic	Quality Eng—Senior	Q Manager	Supplier QE	CSoftQE	Program QE	Contract Review	MRO QE	Spec Process QE	Met Labs QE	Production Control	Regulatory Specialist
P	P	S	P	P	P		P	S	S	P	

Procedural Knowledge: Ability to apply a defined and structured process to verify that prior to delivery, configuration reconciliation is performed between Engineering Bill of Materials (EBOM) and Manufacturing Bill of Materials (MBOM), complete with a history of Deviation and Waivers, open and pending changes, and objective evidence of change incorporation as applicable (3).

Conceptual Knowledge: Ability to analyze quality records, and applicable engineering documents to verify that prior to delivery, configuration reconciliation is performed between Engineering Bill of Materials (EBOM) and Manufacturing Bill of Materials (MBOM), complete with a history of Deviation and Waivers, open and pending changes, and objective evidence of change incorporation as applicable (4).

Factual Knowledge: Understanding of contract requirements related to configuration management (3). Understanding of configuration management processes (3). Understanding of requirements management tools (3). Understanding of regulatory requirements related to approved designs (3). Understanding of contractual and project/program requirements related to approved design (3). Understanding of concessions and deviations processes (3).

Task 2 (PF095-002): Verify that capability exists within the value stream to successfully perform an Functional Configuration Audit (FCA), if required.

Quality Eng—Basic	Quality Eng—Senior	Q Manager	Supplier QE	CSoftQE	Program QE	Contract Review	MRO QE	Spec Process QE	Met Labs QE	Production Control	Regulatory Specialist
P	P	S	P	P	P		P	S	S	P	

Procedural Knowledge: Ability to apply a structured process to verify that capability exists within the value stream to successfully perform an Functional Configuration Audit (FCA), as required.

Conceptual Knowledge: Ability to analyze the capabilities of persons assigned the responsibility to perform Functional Configuration Audits (FCA), to ensure they have the skills required for the task (4).

Factual Knowledge: Understanding of contract requirements related to configuration management (3). Understanding of configuration management processes (3). Understanding of requirements management tools (3). Understanding of regulatory requirements related to approved designs (3). Understanding of contractual and project/program requirements related to approved design (3). Understanding of concessions and deviations processes (3). Understanding of skill requirements as they relate to FCA (3). Understanding of personnel training records (3).

Process: Ensure Configuration Status and Accounting—Physical Configuration Audit (PCA)

Process Purpose: Prior to delivery, ensure configuration reconciliation is performed between Engineering Bill of Materials (EBOM) and Manufacturing Bill of Materials (MBOM), complete with a history of Deviation and Waivers, open and pending changes, and objective evidence of change incorporation as applicable. Ensure the capability to successfully perform an Physical Configuration Audit (PCA) exists if required.

Knowledge and Skill Requirements: Understanding of contract requirements related to configuration management. Understanding of configuration management processes. Ability to verify that prior to delivery, configuration reconciliation is performed between Engineering Bill of Materials (EBOM) and Manufacturing Bill of Materials (MBOM), complete with a history of Deviation and Waivers, open and pending changes, and objective evidence of change incorporation as applicable. Ability to verify that the capability to successfully perform an Physical Configuration Audit (PCA) exists, if required.

Task 1 (PF100-001): Verify that prior to delivery, configuration reconciliation is performed between Engineering Bill of Materials (EBOM) and Manufacturing Bill of Materials (MBOM), complete with a history of Deviation and Waivers, open and pending changes, and objective evidence of change incorporation with appropriate approvals, as applicable.

Quality Eng—Basic	Quality Eng—Senior	Q Manager	Supplier QE	CSoftQE	Program QE	Contract Review	MRO QE	Spec Process QE	Met Labs QE	Production Control	Regulatory Specialist
P	P	S	P	P	P		P	S	S	P	

Procedural Knowledge: Ability to apply a defined and structured process to verify that prior to delivery, configuration reconciliation is performed between Engineering Bill of Materials (EBOM) and Manufacturing Bill of Materials (MBOM), complete with a history of Deviation and Waivers, open and pending changes, and objective evidence of change incorporation with appropriate approvals, as applicable (3).

Conceptual Knowledge: Ability to analyze quality records, and applicable engineering documents to verify that prior to delivery, configuration reconciliation is performed between Engineering Bill of Materials (EBOM) and Manufacturing Bill of Materials (MBOM), complete with a history of Deviation and Waivers, open and pending changes, and objective evidence of change incorporation with appropriate approvals, as applicable (4).

Factual Knowledge: Understanding of contract requirements related to configuration management (3). Understanding of configuration management processes (3). Understanding of requirements management tools (3). Understanding of regulatory requirements related to approved designs (3). Understanding of contractual and project/program requirements related to approved design (3). Understanding of concessions and deviations processes (3).

Task 2 (PF100-002): Verify that the capability to successfully perform an Physical Configuration Audit (PCA) exists, if required.

Quality Eng—Basic	Quality Eng—Senior	Q Manager	Supplier QE	CSoftQE	Program QE	Contract Review	MRO QE	Spec Process QE	Met Labs QE	Production Control	Regulatory Specialist
P	P	S	P	P	P		P	S	S	P	

Procedural Knowledge: Ability to apply a structured process to verify that the capability exists within the value stream to successfully perform an Physical Configuration Audit (PCA), as required (3).

Conceptual Knowledge: Ability to analyze the capabilities of persons assigned the responsibility to perform Physical Configuration Audits (PCA), to ensure they have the skills required for the task (4).

Factual Knowledge: Understanding of contract requirements related to configuration management (3). Understanding of configuration management processes (3). Understanding of requirements management tools (3). Understanding of regulatory requirements related to approved designs (3). Understanding of contractual and project/program requirements related to approved design (3).

Understanding of concessions and deviations processes (3). Understanding of skill requirements as they relate to PCA (3). Understanding of personnel training records (3).

Process: Ensure Engineering Support is Provided (Organizational)

Process Purpose: To ensure organizational members have accurate technical/design data to include the assurance of configuration management. To ensure understanding of requirements is accurate by assessing the implementation of technical requirements into documented work processes.

Knowledge and Skill Requirements: Understanding of configuration management practices. Understanding of design change processes. Understanding of applicable regulatory requirements related to design changes. Understanding of the relationship between design documents and production planning. Ability to verify that organizational product work processes have accurate technical/design data to include the assurance of configuration management. Ability to verify that organizational members possess an understanding of requirements and such is accurate by assessing the implementation of technical requirements into documented work processes.

Task 1 (PM161-001): Verify that organizational members have access to accurate technical/design data to include the assurance of configuration management.

Quality Eng—Basic	Quality Eng—Senior	Q Manager	Supplier QE	CSoftQE	Program QE	Contract Review	MRO QE	Spec Process QE	Met Labs QE	Production Control	Regulatory Specialist
P	P	P	S		P		P	P	P	P	

Procedural Knowledge: Ability to apply a structure approach to verify that organizational members have access to accurate technical/design data to include the assurance of configuration management (3).

Conceptual Knowledge: Ability to analyze established processes to verify that organizational members have access to accurate technical/design data to include the assurance of configuration management (4). Ability to audit and perform analysis of work product to verify that organization work processes provide for accurate technical/design data to include the assurance of configuration management (4).

Factual Knowledge: Understanding of configuration management practices (3). Understanding of organizational design data processes (3). Understanding of where work is being performed (3).

Task 2 (PM161-002): Verify that organizational members possess an understanding of requirements and such is accurate by assessing the implementation of technical requirements into documented work processes.

Quality Eng—Basic	Quality Eng—Senior	Q Manager	Supplier QE	CSoftQE	Program QE	Contract Review	MRO QE	Spec Process QE	Met Labs QE	Production Control	Regulatory Specialist
P	P	P	S		P		P	P	P	P	

Procedural Knowledge: Ability to apply a structured approach to verify that organizational members possess an understanding of requirements and such is accurate by assessing the implementation of technical requirements into documented work processes (3).

Conceptual Knowledge: Ability to analyze the implementation of business processes to verify that organizational members possess an understanding of requirements and such is accurate by assessing the implementation of technical requirements into documented work processes (4).

Factual Knowledge: Understanding of configuration management practices (3). Understanding of organizational design data processes (3). Understanding of where work is being performed (3). Understanding of audit practices (3). Understanding of requirements management tools (3).

Task 3 (PM161-003): Verify that engineering support is provided to organizational members, as required.

Quality Eng—Basic	Quality Eng—Senior	Q Manager	Supplier QE	CSoftQE	Program QE	Contract Review	MRO QE	Spec Process QE	Met Labs QE	Production Control	Regulatory Specialist
P	P	P	S		P		P	P	P	P	

Procedural Knowledge: Ability to apply a structured approach to verify that engineering support is provided, as required (3).

Conceptual Knowledge: Ability to analyze organizational work processes and work product to determine the appropriate level of engineering support is provided, when required (4).

Factual Knowledge: Understanding of configuration management practices (3). Understanding of organizational design data processes (3). Understanding of where work is being performed (3). Understanding of audit practices (3). Understanding of organizational communication practices (3).

Process: Monitor Engineering Change Request Process

Process Purpose: To ensure engineering processes are performed in accordance with system requirements. To ensure engineering changes are implemented effectively and configuration management maintained.

Knowledge and Skill Requirements: Understanding of configuration management practices. Understanding of design change processes. Understanding of applicable regulatory requirements related to design changes. Understanding of the relationship between design documents and production planning. Ability to verify that engineering processes are performed in accordance with system requirements. Ability to verify that engineering changes are implemented effectively and configuration management maintained. Ability to verify that engineering change control processes and configuration management are effectively deployed throughout the value stream.

Task (PM165-001): Verify that engineering processes are performed in accordance with system requirements, in the organization and throughout the value stream.

Quality Eng—Basic	Quality Eng—Senior	Q Manager	Supplier QE	CSoftQE	Program QE	Contract Review	MRO QE	Spec Process QE	Met Labs QE	Production Control	Regulatory Specialist
P	P	P	P	P	P		P	P	P	P	

Procedural Knowledge: Ability to apply a defined and structured process to verify that engineering processes are performed in accordance with system requirements (3).

Conceptual Knowledge: Ability to analyze engineering processes to verify that such are performed in accordance with system requirements (4).

Factual Knowledge: Understanding of regulatory requirements related to approved designs (3). Understanding of requirements management tools (3). Understanding of contractual and project/program requirements related to changes in design (3). Understanding of organization design control processes (3). Understanding of value stream structure and where work is performed (3).

Process: Review Engineering Changes (MCL/SQE/QE) (Quality Review)

Process Purpose: To ensure quality considerations are included in engineering changes, including produceability, inspectability, and production testing requirements.

Knowledge and Skill Requirements: Understanding of configuration management practices. Understanding of design change processes. Understanding of applicable regulatory requirements related to design changes. Understanding of the relationship between design documents and production planning. Understanding of quality laboratory processes. Understanding of contract review process. Understanding of purchase order processes. Ability to verify that quality considerations are included in engineering changes, including produceability, inspectability, and production testing requirements.

Task (PM170-001): Verify that quality considerations are included in engineering changes, including produceability, inspectability, and production testing requirements.

Quality Eng—Basic	Quality Eng—Senior	Q Manager	Supplier QE	CSoftQE	Program QE	Contract Review	MRO QE	Spec Process QE	Met Labs QE	Production Control	Regulatory Specialist
P	P	P	P	P	P		P	P	P	P	

Procedural Knowledge: Ability to apply a structured approach to verify that quality considerations are included in engineering changes, including produceability, inspectability, and production testing requirements (3).

Conceptual Knowledge: Ability to evaluate engineering changes, to ensure quality considerations are appropriately addressed including produceability, inspectability, and production testing requirements (5).

Factual Knowledge: Understanding of regulatory requirements related to approved designs (3). Understanding of contractual and project/program requirements related to changes in design (3). Understanding of requirements management tools (3). Understanding of organization design control processes (3). Understanding of value stream structure and where work is performed (3).

Process: Ensure Positive Control of Unvalidated Engineering Changes

Process Purpose: To provide a means of deploying un-validated engineering changes into production while maintaining positive recall control of any product or services in which un-validated changes have been applied.

Knowledge and Skill Requirements: Understanding of configuration management practices. Understanding of design change processes. Understanding of applicable regulatory requirements related to design changes. Understanding of the relationship between design documents and production planning. Understanding of quality laboratory processes. Understanding of contract review process. Understanding of purchase order processes. Ability to verify that processes are in place to control the deployment of un-validated engineering changes into production while maintaining positive recall control of any product or services in which un-validated changes have been applied. Ability to evaluate that the controls established are adequate to control the deployment of un-validated engineering changes into production while maintaining positive recall control of any product or services in which un-validated changes have been applied.

Task (PM180-001): Verify that processes are in place to control the deployment of un-validated engineering changes into production while maintaining positive recall control of any product or services in which un-validated changes have been applied.

Quality Eng—Basic	Quality Eng—Senior	Q Manager	Supplier QE	CSoftQE	Program QE	Contract Review	MRO QE	Spec Process QE	Met Labs QE	Production Control	Regulatory Specialist
P	P	P	P	P	P		P	S	S	P	S

Procedural Knowledge: Ability to apply a defined and structured process to verify that processes are in place to control the deployment of un-validated engineering changes into production while maintaining positive recall control of any product or services in which un-validated changes have been applied (3).

Conceptual Knowledge: Ability to analyze the adequacy and compliance of processes to control the deployment of un-validated engineering changes into production while maintaining positive recall control of any product or services in which un-validated changes have been applied (4). Ability to analyze the effectiveness of processes to control the deployment of un-validated engineering changes through the review of work product (4).

Factual Knowledge: Understanding of requirements management tools (3). Understanding of regulatory requirements related to approved designs (3). Understanding of contractual and project/program requirements related to changes in design (3). Understanding of organization design control processes (3). Understanding of production planning processes (3). Understanding of positive recall processes (3). Understanding of value stream structure and where work is performed (3).

Process: Ensure Ship in Advance of Completion Process Is Controlled

Process Purpose: To provide a means to control product and its production configuration when shipping product without all processes being complete or approved.

Knowledge and Skill Requirements: Understanding of special processing, inspection, testing, and installation of parts and components. Understanding of configuration management processes. Understanding of production and inspection records. Ability to verify that processes are established that ensure the effective management of product that is being shipped from one location to another without all of the required processes being completed.

Task (PM185-001): Verify that processes are established that ensure the effective management of product that is being shipped from one location to another without all of the required processes being completed.

Quality Eng—Basic	Quality Eng—Senior	Q Manager	Supplier QE	CSoftQE	Program QE	Contract Review	MRO QE	Spec Process QE	Met Labs QE	Production Control	Regulatory Specialist
P	P	P	P	P	P		P	S	S	P	S

Procedural Knowledge: Ability to apply a structure approach to verify that processes are established that ensure the effective management of product that is being shipped from one location to another without all of the required processes being completed (3).

Conceptual Knowledge: Ability to apply a structure approach to verify that processes are established that ensure the effective management of product that is being shipped from one location to another without all of the required processes being completed (3).

Factual Knowledge: Ability to evaluate processes that have been established to ensure the effective management of product that is being shipped from one location to another without all of the required processes being completed, to verify they are in compliance with requirements and provide positive recall abilities (5). Ability to analyze individual work pages of product/services released prior to final acceptance and/or completion to ensure they are in compliance with applicable control procedures (4).

III.G.2. Tech Data Management (Incl. Export Control)

Process: Determine Quality Requirements—Export Controls

Process Purpose: Review design and production requirements and determine export control considerations such as ITAR and EAR requirements for the product/program. These requirements shall be documented in the appropriate organizational format.

Knowledge and Skill Requirements: Ability to review applicable design and program planning documents and identify potential export control requirements Ability to research sources of information related to export control requirements. Ability to assess the defined value stream and identify potential export control considerations. Ability to organize, document, store and integrate into the requirements hierarchy using a requirements management tool.

Task 1 (DP025-001): Review applicable design and program planning documents and identify potential export control requirements. Organize, document, store and integrate findings into the requirements hierarchy using a requirements management tool.

Quality Eng—Basic	Quality Eng—Senior	Q Manager	Supplier QE	CSoftQE	Program QE	Contract Review	MRO QE	Spec Process QE	Met Labs QE	Production Control	Regulatory Specialist
P	P	P	P	P	P	P		S	S		P

Procedural Knowledge: Access applicable regulatory requirements for export controls (4). Ability to follow a structured process to review applicable design and program planning documents and identify potential export control requirements (4). Ability to follow a structured process to organize, document, store and integrate findings into the requirements hierarchy using a requirements management tool (3).

Conceptual Knowledge: Ability to recognize export control implications throughout the value stream (4).

Factual Knowledge: Knowledge of export control rules and procedures (3). Knowledge of design review process (3). Knowledge of work execution in value stream (3).

Task 2 (DP025-002): Research sources of information related to export control requirements. Organize, document, store and integrate findings into the requirements hierarchy using a requirements management tool.

Quality Eng—Basic	Quality Eng—Senior	Q Manager	Supplier QE	CSoftQE	Program QE	Contract Review	MRO QE	Spec Process QE	Met Labs QE	Production Control	Regulatory Specialist
					P	P					P

Procedural Knowledge: Ability to organize export control requirements within a hierarchical framework (4).

Conceptual Knowledge: Ability to access sources of information related to export control requirements (3). Ability to follow a structured process for managing export control requirements within a requirements management tool (3).

Factual Knowledge: Knowledge of export control requirements (3). Knowledge of requirements management tools (3).

Task 3 (DP025-003): Assess the defined value stream and identify potential export control considerations. Organize, document, store and integrate findings into the requirements hierarchy using a requirements management tool.

Quality Eng—Basic	Quality Eng—Senior	Q Manager	Supplier QE	CSoftQE	Program QE	Contract Review	MRO QE	Spec Process QE	Met Labs QE	Production Control	Regulatory Specialist
S	P	S	P	S	P	S		S	S		P

Procedural Knowledge: Ability to gather information as to sources and locations of work execution throughout the value stream (3). Ability to follow a structured process for managing export control requirements within a requirements management tool (3).

Conceptual Knowledge: Ability to analyze the identified value stream and determine export control requirements based on where work is to be executed (4).

Factual Knowledge: Knowledge of export control requirements (3). Knowledge of value stream (3). Knowledge of requirements management tools (3).

Process: Create Technical Data for Suppliers

Process Purpose: In order to effectively communicate to partners/suppliers, create a master file of documents that contains all of the design information required to produce the product/service. Configuration management of applicable revisions must be assured. This may include quality (incl customer and regulatory), reliability and other related documents.

Knowledge and Skill Requirements: Ability to create a master file of documents that contains all of the design, production, and quality information required to produce the product/service (includes design and production standard work definition, design and production artifacts, regulatory requirements, and the requirement hierarchy). Ability to ensure configuration management of master file is controlled.

Task 1 (DP105-001): Create a master file of documents that contains all of the design, production, regulatory, and quality information required to produce the product/service.

Quality Eng—Basic	Quality Eng—Senior	Q Manager	Supplier QE	CSoftQE	Program QE	Contract Review	MRO QE	Spec Process QE	Met Labs QE	Production Control	Regulatory Specialist
P	P	S	S	P	P	S		S	S		S

Procedural Knowledge: Ability to follow a structured process in creating a master file of documents that contains all of the design, production, regulatory, and quality information required to produce the product/service (3).

Conceptual Knowledge: Ability to evaluate program/project requirements and create a master list of documents that contains all of the design, production, regulatory, and quality information required to produce the product/service (4). Ability to collect and populate a master file of documents that contains all of the design, production, regulatory, and quality information required to produce the product/service (3).

Factual Knowledge: Knowledge of project/program design, production, regulatory, and quality requirements (3). Knowledge of master file lists (3).

Task 2 (DP105-002): Ensure configuration management of master file/repository is controlled.

Quality Eng—Basic	Quality Eng—Senior	Q Manager	Supplier QE	CSoftQE	Program QE	Contract Review	MRO QE	Spec Process QE	Met Labs QE	Production Control	Regulatory Specialist
P	P	S	S	P	P	S		S	S		S

Procedural Knowledge: Ability to apply a structured process to assess the effective deployment of configuration management practices upon a master file (3).

Conceptual Knowledge: Ability to evaluate configuration management practices placed upon a master file to determine effective control (3).

Factual Knowledge: Knowledge of configuration management methods and processes (3). Knowledge of project/program-related quality, production and design information (3).

Process: Apply Export Control Rules (as required)

Process Purpose: Review technical data and assess if special handling processes are required to ensure compliance with export requirements.

Knowledge and Skill Requirements: Ability to review technical data and assess if special handling processes are required to ensure compliance with export control requirements. Ability to ensure suppliers/partners have adequate controls in place to ensure compliance with export control. An understanding of export control licensing and applicable regulations.

Task 1 (SS035-001): Review technical data, policies, procedures, and regulations to determine if special handling processes are required to ensure compliance with export control requirements.

Quality Eng—Basic	Quality Eng—Senior	Q Manager	Supplier QE	CSoftQE	Program QE	Contract Review	MRO QE	Spec Process QE	Met Labs QE	Production Control	Regulatory Specialist
P	P	S	P	P	P	P					P

Procedural Knowledge: Ability to apply a structured process to review technical data, policies, procedures, and regulations to determine if special handling processes are required to ensure compliance with export control requirements (3). Ability to document the special handling requirements for export control (3).

Conceptual Knowledge: Ability to review technical data, policies, procedures, to regulations and determine if special handling processes are required to ensure compliance with export control requirements at specific locations within the value stream (4). Ability to recognize the interrelatedness of the value stream and identify where export controls may need to be established at lower-levels of the value stream (4).

Factual Knowledge: Knowledge of applicable export control regulations (3). Knowledge of project/program requirements related to handling of data (3). Knowledge of the value stream structure as they related to international sources (3). Knowledge of export control plans (3).

Task 2 (SS035-002): Ensure suppliers/partners have adequate controls in place to ensure compliance with export control.

Quality Eng—Basic	Quality Eng—Senior	Q Manager	Supplier QE	CSoftQE	Program QE	Contract Review	MRO QE	Spec Process QE	Met Labs QE	Production Control	Regulatory Specialist
S	S	S	P		P	S					S

Procedural Knowledge: Ability to apply a structured process in the assessment of process and procedures within the value stream (including all levels of sub-tiers) to determine compliance to export control requirements (3).

Conceptual Knowledge: Ability to analyze the processes and procedures within the value stream to determine compliance to export control requirements (4). Ability to recognize potential export control considerations for the extended value stream (sub-tier suppliers) (4).

Factual Knowledge: Knowledge of applicable export control regulations (3). Knowledge of project/program requirements related to handling of data (3). Knowledge of the value stream structure (3). Knowledge of export control plans (3). Knowledge of audit practices (3).

IV.A.2. First Article Inspections (AS/EN/SJAC 9102)

Process: Review and Approve Development of First Article Inspection Report

Process Purpose: To ensure the organization is appropriately planning and is compliant to applicable First Article Inspection (FAI) standards (e.g., AS/EN/SJAC 9102). Ensure FAI plans addresses organization's internal requirements as well as higher-level (customer, prime) requirements. Ensure tooling is included in FAI plan. Ensure supplier activities are included in FAI plan.

Knowledge and Skill Requirements: Understanding of first article inspection standards. Ability to verify organization's FAI processes are appropriately planned and are compliant to applicable standards (e.g., AS/EN/SJAC 9102). Ability to verify FAI plans address organizational internal requirements as well as higher-level (customer, prime) requirements. Understanding of the relationship of tooling control in FAI planning. Ability to verify supplier activities are included in FAI planning.

Task 1 (PO120-001): Verify FAI processes are appropriately planned and compliant to applicable standards (e.g., AS/EN/SJAC 9102).

Quality Eng—Basic	Quality Eng—Senior	Q Manager	Supplier QE	CSoftQE	Program QE	Contract Review	MRO QE	Spec Process QE	Met Labs QE	Production Control	Regulatory Specialist
P	P				S		P				P

Procedural Knowledge: Ability to apply a structured assessment process to verify documented processes to perform First Article Inspection activities in accordance with applicable standards (3).

Conceptual Knowledge: Ability to analyze documented processes and planning to substantiate compliance to applicable standards (4).

Factual Knowledge: Knowledge of industry standards related to first article inspection processes (e.g., AS/EN/SJAC 9102) (3). Knowledge of inspection planning processes (3). Knowledge of work processes related to program/project (3). Understanding of where work processes are being performed (3). Knowledge of design and technical publications related to the organization's products (3).

Task 2 (PO120-002): Verify FAI plans address organizational internal requirements as well as higher-level (customer, prime) requirements.

Quality Eng—Basic	Quality Eng—Senior	Q Manager	Supplier QE	CSoftQE	Program QE	Contract Review	MRO QE	Spec Process QE	Met Labs QE	Production Control	Regulatory Specialist
P	P				S		P				P

Procedural Knowledge: Ability to apply a structured processes to assess organizational internal requirements as well as higher-level (customer, prime) requirements and verify that FAI plans include such requirements (3).

Conceptual Knowledge: Ability to analyze organizational internal requirements as well as higher-level (customer, prime) requirements and verify that FAI plans include such requirements (4).

Factual Knowledge: Knowledge of industry standards related to first article inspection processes (e.g., AS/EN/SJAC 9102) (3). Knowledge of inspection planning processes (3). Knowledge of work processes related to program/project (3). Understanding of where work processes are being performed (3). Understanding of the value stream structure (3).

Task 3 (PO120-003): Verify supplier activities are included in FAI planning.

Quality Eng—Basic	Quality Eng—Senior	Q Manager	Supplier QE	CSoftQE	Program QE	Contract Review	MRO QE	Spec Process QE	Met Labs QE	Production Control	Regulatory Specialist
P	P		P		S		P				P

Procedural Knowledge: Ability to apply a structured process to assess FAI planning to verify processes performed by suppliers are included (3).

Conceptual Knowledge: Ability to analyze FAI planning to verify processes performed by suppliers are included (4).

Factual Knowledge: Knowledge of industry standards related to first article inspection processes (e.g., AS/EN/SJAC 9102) (3). Knowledge of inspection planning processes (3). Knowledge of work processes related to program/project (3). Knowledge of where work is performed in the value stream (3).

Process: Ensure Deliverable Software Conformance—First Article Verification

Process Purpose: To ensure deliverable/embedded software meets design requirements and that production systems will assure conformity. Where code can be verified, ensure the review process is structured and the resulting code is maintained for future review.

Knowledge and Skill Requirements: Understanding of software quality processes (e.g., peer review). Ability to verify that deliverable/embedded software meets design requirements and that production systems will ensure conformity (e.g., review of design artifacts such as UML, verification of software performance requirements such as fault tolerance and built in test.). Ability to verify that first article inspection planning includes software considerations. Ability to review contractual documents to determine level of organizational access to source code. (Note: In cases where the organization cannot see the source code, software verification processes will be planned and arranged accordingly.) Ability to ensure personnel are capable of performing software verification activities.

Task 1 (PO125-001): Verify deliverable/embedded software meets design requirements and that production systems will assure conformity to applicable standards and program requirements.

Quality Eng—Basic	Quality Eng—Senior	Q Manager	Supplier QE	CSoftQE	Program QE	Contract Review	MRO QE	Spec Process QE	Met Labs QE	Production Control	Regulatory Specialist
P	P		S	P	S		S			S	

Procedural Knowledge: Ability to apply a structured assessment process to verify that supplier/partner has documented the verification requirements for deliverable/embedded software (3).

Conceptual Knowledge: Ability to analyze inspection planning to verify planning addresses verification requirements for deliverable/embedded software (4).

Factual Knowledge: Knowledge of industry standards related to software verification (e.g., DO-105) (3). Knowledge of inspection planning processes (3). Knowledge of work processes related to program/project (3). Knowledge of software quality assurance methods (3). Understanding of where work processes are being performed (3). Understanding of the value stream structure (3).

Task 2 (PO125-002): Verify that first article inspection planning includes software verification considerations.

Quality Eng—Basic	Quality Eng—Senior	Q Manager	Supplier QE	CSoftQE	Program QE	Contract Review	MRO QE	Spec Process QE	Met Labs QE	Production Control	Regulatory Specialist
P	P			P	S		S			S	

Procedural Knowledge: Ability to apply a structured processes to assess FAI planning to verify that FAI plans include verification requirements for software (3).

Conceptual Knowledge: Ability to analyze FAI planning to verify that FAI plans include verification requirements for software (4).

Factual Knowledge: Knowledge of industry standards related to first article inspection processes (e.g., AS/EN/SJAC 9102) (3). Knowledge of inspection planning processes (3). Knowledge of software quality assurance methods (3). Understanding of where work processes are being performed (3).

Task 3 (PO125-003): Ensure personnel are capable of performing software verification activities.

Quality Eng—Basic	Quality Eng—Senior	Q Manager	Supplier QE	CSoftQE	Program QE	Contract Review	MRO QE	Spec Process QE	Met Labs QE	Production Control	Regulatory Specialist
P	P		S	P	S		S			S	

Procedural Knowledge: Ability to apply a defined assessment process to verify that personnel are qualified to perform software verification activities (3).

Conceptual Knowledge: Ability to analyze personnel qualification and work experience records to verify that personnel are qualified of performing software verification activities (4). Ability to evaluate personnel work activities to verify that personnel are capable to perform software validation activities (5).

Factual Knowledge: Knowledge of industry standards related to first article inspection processes (e.g., AS/EN/SJAC 9102) (3). Knowledge of inspection planning processes (3). Knowledge of software quality assurance methods (3). Understanding of personnel qualification requirements related to software quality (3). Understanding of where work processes are being performed (3).

Process: Ensure NDT/Special Process Conformance—First Article Verification

Process Purpose: To ensure NDT and special processes are included in FAI plan and are verified to meet design/specification requirements. Ensure special processes have objective evidence of conformity and that NDTs are performed per requirements. Ensure personnel are qualified to perform special processes and NDT.

Knowledge and Skill Requirements: Understanding of NDT methods and applicable special process standards requirements. Ability to verify NDT and special processes are included in FAI planning. Ability to validate that defined NDT and special process verification processes defined in FAI are appropriate and adequate. Ability to verify measured results meet defined requirements. Ability to verify special processes have objective evidence of conformity and that NDTs are performed per requirements. Ability to verify personnel are qualified to perform special processes and NDT.

Task 1 (PO130-001): Verify NDT and special processes are included in FAI planning.

Quality Eng—Basic	Quality Eng—Senior	Q Manager	Supplier QE	CSoftQE	Program QE	Contract Review	MRO QE	Spec Process QE	Met Labs QE	Production Control	Regulatory Specialist
P	P	S	P		S		S	P	P	S	

Procedural Knowledge: Ability to apply a structured assessment process to verify that supplier/partner has appropriately addressed NDT verification requirements in FAI planning (3).

Ability to apply a structured assessment process to verify that the organization has appropriately addressed the control of special processes in FAI planning (3).

Conceptual Knowledge: Ability to FAI planning to verify that NDT methods are appropriately addressed (4). Ability to analyze FAI planning to verify that the control of special processes are appropriately addressed (4).

Factual Knowledge: Knowledge of industry standards related to first article inspection processes (e.g., AS/EN/SJAC 9102) (3). Knowledge of inspection planning processes (3). Knowledge of work processes related to program/project (3). Knowledge of applicable NDT methods (4). Knowledge of applicable special processes (4). Understanding of where work processes are being performed (3).

Task 2 (PO130-002): Validate that defined NDT and special process verification processes documented in FAI are appropriate and adequate.

Quality Eng—Basic	Quality Eng—Senior	Q Manager	Supplier QE	CSoftQE	Program QE	Contract Review	MRO QE	Spec Process QE	Met Labs QE	Production Control	Regulatory Specialist
S	S		P		S		S	P	P	S	

Procedural Knowledge: Ability to apply a structured assessment to validate that defined NDT and special process verification processes as documented in FAI planning are appropriately and adequately executed (3).

Conceptual Knowledge: Ability to analyze NDT and special process verification processes to ensure they are appropriately and adequately executed as documented in FAI planning (4).

Factual Knowledge: Knowledge of industry standards related to first article inspection processes (e.g., AS/EN/SJAC 9102) (3). Knowledge of inspection planning processes (3). Knowledge of work processes related to program/project (3). Knowledge of applicable NDT methods (4). Knowledge of applicable special processes (4). Understanding of where work processes are being performed (3).

Task 3 (PO130-003): Verify measured results in NDT and special processes meet defined FAI requirements.

Quality Eng—Basic	Quality Eng—Senior	Q Manager	Supplier QE	CSoftQE	Program QE	Contract Review	MRO QE	Spec Process QE	Met Labs QE	Production Control	Regulatory Specialist
S	S		S		S		S	P	P	S	

Procedural Knowledge: Ability to apply a structured assessment process to verify that measured results in NDT and special processes meet defined requirements (3).

Conceptual Knowledge: Ability to analyze measured results in NDT and special processes and determine compliance to defined requirements (4).

Factual Knowledge: Knowledge of industry standards related to first article inspection processes (e.g., AS/EN/SJAC 9102) (3). Knowledge of inspection planning processes (3). Knowledge of work processes related to program/project (3). Knowledge of applicable NDT methods (4). Knowledge of applicable special processes (4). Understanding of where work processes are being performed (3).

Task 4 (PO130-004): Ensure special processes have objective evidence of conformity and that NDTs are performed per defined FAI requirements.

Quality Eng—Basic	Quality Eng—Senior	Q Manager	Supplier QE	CSoftQE	Program QE	Contract Review	MRO QE	Spec Process QE	Met Labs QE	Production Control	Regulatory Specialist
P	P		P		S		S	P	P	S	

Procedural Knowledge: Ability to deploy a structured assessment process to verify that special processes have objective evidence of conformity and that NDTs are performed per requirements (3).

Conceptual Knowledge: Ability to analyze special processes and NDT methods, and their associated quality records, to verify that there is objective evidence of conformity per requirements (4).

Factual Knowledge: Knowledge of industry standards related to first article inspection processes (e.g., AS/EN/SJAC 9102) (3). Knowledge of inspection planning processes (3). Knowledge of work processes related to program/project (3). Knowledge of applicable NDT methods (4). Knowledge of applicable special processes (4). Understanding of where work processes are being performed (3).

Task 5 (PO130-005): Verify personnel who are performing special processes and NDT evaluations, are qualified to perform special processes and NDT.

Quality Eng—Basic	Quality Eng—Senior	Q Manager	Supplier QE	CSoftQE	Program QE	Contract Review	MRO QE	Spec Process QE	Met Labs QE	Production Control	Regulatory Specialist
S	P	S	P		S		S	P	P	S	

Procedural Knowledge: Ability to apply a defined assessment process to verify that personnel are qualified to perform special processes and NDT (3).

Conceptual Knowledge: Ability to analyze personnel qualifications and work experience records to verify that personnel who are performing special processes and NDT evaluations, are qualified to perform special processes and NDT evaluation (4). Ability to evaluate personnel work activities to verify that personnel who are performing special processes and NDT evaluations, are capable to perform special processes and NDT evaluation (5).

Factual Knowledge: Knowledge of industry standards related to first article inspection processes (e.g., AS/EN/SJAC 9102) (3). Knowledge of inspection planning processes (3). Knowledge of work processes related to program/project (3). Knowledge of applicable NDT methods (4). Knowledge of applicable special processes (4). Knowledge of personnel qualification requirements related to NDT methods and special processes (3). Understanding of where work processes are being performed (3).

Process: Ensure Hardware Conformance—First Article

Process Purpose: To ensure the geometric and physical characteristics are conformed through the FAI process. Ensure appropriate measurement devices (e.g., calibrated, and accuracy) are utilized when making inspection measurements.

Knowledge and Skill Requirements: Understanding of FAI processes and standards. Understanding of design standards. Understanding of production processes and production planning. Understanding of inspection planning. Ability to apply concepts and standards to verify the geometric and physical characteristics are conformed throughout the FAI process. Ability to verify appropriate measurement devices (e.g., calibrated, and accuracy) are utilized when making inspection measurements. Ability to ensure compliance of FAI requirements.

Task 1 (PO135-001): Verify the geometric and physical characteristics to ensure conformity to design requirements are controlled and measured in the FAI process.

Quality Eng—Basic	Quality Eng—Senior	Q Manager	Supplier QE	CSoftQE	Program QE	Contract Review	MRO QE	Spec Process QE	Met Labs QE	Production Control	Regulatory Specialist
P	P		S		S		P	S	S	P	

Procedural Knowledge: Ability to apply a defined process to verify the geometric and physical characteristics are conformed through the FAI process (3).

Conceptual Knowledge: Ability to interpret engineering specification requirements (4). Ability to analyze engineering documents and inspection planning to verify the geometric and physical characteristics are conformed through the FAI process (4).

Factual Knowledge: Knowledge of geometric tolerancing (3). Knowledge of applicable material standards (3). Knowledge of inspection planning (3). Knowledge of standards related to FAI (e.g., AS/EN/SJAC 9102)(3). Understanding of where work processes are being performed (3).

Task 2 (PO135-002): Verify the appropriate selection of measurement devices (e.g., calibrated, and accuracy) and ensure such are utilized when making inspection measurements. Verify that measurement devices are appropriately calibrated.

Quality Eng—Basic	Quality Eng—Senior	Q Manager	Supplier QE	CSoftQE	Program QE	Contract Review	MRO QE	Spec Process QE	Met Labs QE	Production Control	Regulatory Specialist
P	P		S		S		P	S	S	P	

Procedural Knowledge: Ability to apply a structured assessment methodology to verify appropriate measurement devices (e.g., calibrated, and accuracy) are utilized when making inspection measurements (3).

Conceptual Knowledge: Ability to analyze design requirements to determine accuracy requirements for measurement devices (4). Ability to interpret contract and regulatory requirements to determine unique requirements for calibration of measurement devices (4). Ability to analyze selected measurement devices to determine their adequacy to meet measurement requirements (4). Ability to audit work processes to verify that appropriate measurement devices (e.g., calibrated, and accuracy) are utilized when making inspection measurements (3).

Factual Knowledge: Understanding of design requirements (3). Understanding of regulatory requirements related to calibration (3). Understanding of standards related to measurement devices and accuracy (4). Understanding of calibration control systems (3). Understanding of auditing methods (3). Understanding of where work processes are being performed (3).

Process: Ensure Deviations are Reviewed and Approved

Process Purpose: To formally document deviations discovered in pre-production processes and to ensure appropriate review and approval is concluded. Ensure deviation authority is defined and controlled. Ensure deviations are replanned for a follow up FAI.

Knowledge and Skill Requirements: Understanding of documentation processes for product and/or process deviations. Ability to formally document deviations discovered in pre-production processes and to ensure appropriate review and approval is concluded. Ability to evaluate documented processes to verify that deviation authority is defined and controlled. Ability to verify that deviations discovered during initial FAI are replanned for a follow up FAI. Ability to evaluate FAI follow-up processes to ensure follow-up actions are effectively performed and documented. Ability to ensure FAIs are performed at the appropriate locations and sequences.

Task 1 (PO140-001): Evaluate documented processes to verify that deviation authority is defined and controlled.

Quality Eng—Basic	Quality Eng—Senior	Q Manager	Supplier QE	CSoftQE	Program QE	Contract Review	MRO QE	Spec Process QE	Met Labs QE	Production Control	Regulatory Specialist
P	P		P		P		S	S	S	S	

Procedural Knowledge: Ability to apply a structured assessment process to verify that deviation authority is defined and controlled for the pre-production and FAI processes (3).

Conceptual Knowledge: Ability to review procedures and associated quality records to verify that deviation authority is defined and controlled for the pre-production and FAI processes (3).

Factual Knowledge: Understanding of FAI process (3). Understanding of standards related to FAI (e.g., AS/EN/SJAC 9102)(3). Understanding of contract requirements related to deviations (3). Understanding of deviation approval process (3). Understanding of where work processes are being performed (3).

Task 2 (PO140-002): Document deviations discovered in pre-production processes and ensure appropriate review and approval is concluded.

Quality Eng—Basic	Quality Eng—Senior	Q Manager	Supplier QE	CSoftQE	Program QE	Contract Review	MRO QE	Spec Process QE	Met Labs QE	Production Control	Regulatory Specialist
S	S		P		S		P	S	S	P	

Procedural Knowledge: Ability to apply a structured process to document deviations discovered in pre-production processes and ensure appropriate review and approval is concluded (3).

Conceptual Knowledge: Ability to document deviations discovered in pre-production processes in terms that can be easily understood and replicated (3). Ability to coordinate with appropriate functions to ensure appropriate review and approval is concluded of deviation documentation package (3).

Factual Knowledge: Understanding of FAI process (3). Understanding of standards related to FAI (e.g., AS/EN/SJAC 9102)(3). Understanding of contract requirements related to deviations (3). Understanding of deviation approval process (3). Understanding of where work processes are being performed (3).

Task 3 (PO140-003): Verify that deviations discovered during initial FAI are replanned for a follow up FAI.

Quality Eng—Basic	Quality Eng—Senior	Q Manager	Supplier QE	CSoftQE	Program QE	Contract Review	MRO QE	Spec Process QE	Met Labs QE	Production Control	Regulatory Specialist
P	P		P		P		P	S	S	P	

Procedural Knowledge: Ability to apply a defined process to verify that deviations discovered during initial FAI are replanned for a follow up FAI (3).

Conceptual Knowledge: Ability to review applicable planning documentation to verify that deviations discovered during initial FAI are replanned for a follow up FAI (3). Ability to analyze replanned FAI activities to ensure activities accurately address the approved deviation (4).

Factual Knowledge: Understanding of FAI process (3). Understanding of standards related to FAI (e.g., AS/EN/SJAC 9102)(3). Understanding of contract/regulatory requirements related to deviations (3). Understanding of deviation approval process (3). Understanding of FAI and inspection planning (3). Understanding of where work processes are being performed (3). Understanding of production scheduling (3).

Task 4 (PO140-004): Evaluate FAI follow-up processes to ensure follow-up actions are effectively performed and documented.

Quality Eng—Basic	Quality Eng—Senior	Q Manager	Supplier QE	CSoftQE	Program QE	Contract Review	MRO QE	Spec Process QE	Met Labs QE	Production Control	Regulatory Specialist
P	P		P		P		P	S	S	P	

Procedural Knowledge: Ability to apply a defined process to verify that FAI follow-up processes to ensure follow-up actions are effectively performed and documented (3).

Conceptual Knowledge: Ability to analyze planned work activities to verify that FAI follow-up processes to ensure follow-up actions are effectively performed and documented (4).

Factual Knowledge: Understanding of FAI process (3). Understanding of standards related to FAI (e.g., AS/EN/SJAC 9102)(3). Understanding of contract/regulatory requirements related to deviations (3). Understanding of deviation approval process (3). Understanding of FAI and inspection planning (3). Understanding of where work processes are being performed (3).

Process: Approve Product Conformance—First Article Reports

Process Purpose: To ensure the entire deliverable product conforms to design and purchasing document requirements and complies with FAI requirements. Verify all lower-level/component-level FAIs are complete and accurate.

Knowledge and Skill Requirements: Understanding of FAI processes and standards. Understanding of design standards. Understanding of production processes and production planning. Understanding of inspection planning. Ability to evaluate final FAI packages to verify the entire deliverable product conforms to design and purchasing document requirements and complies with FAI requirements. Ability to verify all lower-level/component-level FAIs are complete and accurate.

Task 1 (PO145-001): Evaluate final FAI packages to verify the entire deliverable product conforms to design and purchasing document requirements and complies with FAI requirements.

Quality Eng—Basic	Quality Eng—Senior	Q Manager	Supplier QE	CSoftQE	Program QE	Contract Review	MRO QE	Spec Process QE	Met Labs QE	Production Control	Regulatory Specialist
P	P		P		P		P	S	S	P	

Procedural Knowledge: Ability to apply a defined process to review the final FAI packages to verify the entire deliverable product conforms to design and purchasing document requirements and complies with FAI requirements (3).

Conceptual Knowledge: Ability to interpret contractual, regulatory, and program requirements to ensure an accurate understanding of requirements for final FAI documentation packages (4). Ability to review final FAI packages to verify the entire deliverable product conforms to design and purchasing document requirements and complies with FAI requirements (3).

Factual Knowledge: Understanding of FAI process (3). Understanding of standards related to FAI (e.g., AS/EN/SJAC 9102)(3). Understanding of contract/regulatory requirements related to FAI packages (3). Understanding of FAI and inspection planning (3). Understanding of where work processes are being performed (3). Understanding of the value stream structure (3).

Task 2 (PO145-002): Verify all lower-level/component-level FAIs are complete and accurate.

Quality Eng—Basic	Quality Eng—Senior	Q Manager	Supplier QE	CSoftQE	Program QE	Contract Review	MRO QE	Spec Process QE	Met Labs QE	Production Control	Regulatory Specialist
P	P		P		P		P	S	S	P	

Procedural Knowledge: Ability to apply a defined process to verify all lower-level/component-level FAIs are complete and accurate (3).

Conceptual Knowledge: Ability to analyze lower-level/component-level FAI packages to verify that they are complete and accurate (4).

Factual Knowledge: Understanding of FAI process (3). Understanding of standards related to FAI (e.g., AS/EN/SJAC 9102)(3). Understanding of contract/regulatory requirements related to FAI packages (3). Understanding of FAI and inspection planning (3). Understanding of where work is performed (3). Understanding of the value stream structure (3).

Process: Review and Approve Supplier Development of First Article Inspection Report

Process Purpose: To ensure supplier's/partner's FAI is appropriately planned and compliant to applicable standards (e.g., AS/EN/SJAC 9102). Ensure FAI plan addresses supplier's/partner's internal requirements as well as higher-level (customer, prime) requirements. Ensure tooling is included in FAI plan. Ensure sub-tier supplier activities are included in FAI plan.

Knowledge and Skill Requirements: Understanding of first article inspection standards. Ability to verify that supplier's/partner's FAI processes are appropriately planned and compliant to applicable standards (e.g., AS/EN/SJAC 9102). Ability to verify FAI plans address supplier/partner internal requirements as well as higher-level (customer, prime) requirements. Understanding of the relationship of tooling control in FAI planning. Ability to verify sub-tier supplier activities are included in FAI planning.

Task 1 (PP120-001): Verify supplier's/partner's FAI processes are appropriately planned and compliant to applicable standards (e.g., AS/EN/SJAC 9102).

Quality Eng—Basic	Quality Eng—Senior	Q Manager	Supplier QE	CSoftQE	Program QE	Contract Review	MRO QE	Spec Process QE	Met Labs QE	Production Control	Regulatory Specialist
S	S		P		S		S	S	S	S	

Procedural Knowledge: Ability to apply a structured assessment process to verify that supplier/partner has documented processes to perform First Article Inspection activities in accordance with applicable standards (3).

Conceptual Knowledge: Ability to analyze supplier/partner documented processes and planning to substantiate compliance to applicable standards (4).

Factual Knowledge: Knowledge of industry standards related to first article inspection processes (e.g., AS/EN/SJAC 9102) (3). Knowledge of inspection planning processes (3). Knowledge of work processes related to program/project (3). Understanding of where work processes are being performed (3). Understanding of the value stream structure (3).

Task 2 (PP120-002): Verify FAI plans address supplier/partner internal requirements as well as higher-level (customer, prime) requirements.

Quality Eng—Basic	Quality Eng—Senior	Q Manager	Supplier QE	CSoftQE	Program QE	Contract Review	MRO QE	Spec Process QE	Met Labs QE	Production Control	Regulatory Specialist
S	S		P		S		S	S	S	S	

Procedural Knowledge: Ability to apply a structured processes to assess supplier/partner internal requirements as well as higher-level (customer, prime) requirements and verify that FAI plans include such requirements (3).

Conceptual Knowledge: Ability to analyze supplier/partner internal requirements as well as higher-level (customer, prime) requirements and verify that FAI plans include such requirements (4).

Factual Knowledge: Knowledge of industry standards related to first article inspection processes (e.g., AS/EN/SJAC 9102) (3). Knowledge of inspection planning processes (3). Knowledge of work processes related to program/project (3). Understanding of where work processes are being performed (3). Understanding of the value stream structure (3).

Task 3 (PP120-003): Verify sub-tier supplier activities are included in FAI planning.

Quality Eng—Basic	Quality Eng—Senior	Q Manager	Supplier QE	CSoftQE	Program QE	Contract Review	MRO QE	Spec Process QE	Met Labs QE	Production Control	Regulatory Specialist
S	S		P		S		S	S	S	S	

Procedural Knowledge: Ability to apply a structured process to assess FAI planning to verify processes performed by sub-tier suppliers are included (3).

Conceptual Knowledge: Ability to analyze FAI planning to verify processes performed by sub-tier suppliers are included (4).

Factual Knowledge: Knowledge of industry standards related to first article inspection processes (e.g., AS/EN/SJAC 9102) (3). Knowledge of inspection planning processes (3). Knowledge of work processes related to program/project (3). Knowledge of where work is performed in the value stream (3).

Process: Ensure Supplier Software Conformance—First Article Verification

Process Purpose: To ensure deliverable/embedded software meets design requirements and that production systems will assure conformity. Where code can be verified, ensure the review process is structured and the resulting code is maintained for future review.

Knowledge and Skill Requirements: Understanding of software quality processes (e.g., peer review). Ability to verify that deliverable/embedded software meets design requirements and that production systems will ensure conformity (e.g., review of design artifacts such as UML, verification of software performance requirements such as fault tolerance and built in test.). Ability to verify that first article inspection planning includes software considerations. Ability to review contractual documents to determine level of organizational access to source code. (Note: In cases where the organization cannot see the source code, software verification processes will be planned and arranged accordingly.) Ability to ensure personnel are capable of performing software verification activities.

Task 1 (PP125-001): Verify deliverable/embedded software meets design requirements and that production systems will assure conformity to applicable standards and program requirements, throughout the value stream.

Quality Eng—Basic	Quality Eng—Senior	Q Manager	Supplier QE	CSoftQE	Program QE	Contract Review	MRO QE	Spec Process QE	Met Labs QE	Production Control	Regulatory Specialist
S	S		P	P	S		S			S	

Procedural Knowledge: Ability to apply a structured assessment process to verify that supplier/partner has documented the verification requirements for deliverable/embedded software (3).

Conceptual Knowledge: Ability to analyze supplier/partner inspection planning to verify planning addresses verification requirements for deliverable/embedded software (4).

Factual Knowledge: Knowledge of industry standards related to software verification (e.g., DO-105) (3). Knowledge of inspection planning processes (3). Knowledge of work processes related to program/project (3). Knowledge of software quality assurance methods (3). Understanding of where work processes are being performed (3). Understanding of the value stream structure (3).

Task 2 (PP125-002): Verify that first article inspection planning includes software verification considerations, throughout the value stream.

Quality Eng—asic	Quality Eng—Senior	Q Manager	Supplier QE	CSoftQE	Program QE	Contract Review	MRO QE	Spec Process QE	Met Labs QE	Production Control	Regulatory Specialist
S	S		P	P	S		S			S	

Procedural Knowledge: Ability to apply a structured processes to assess FAI planning to verify that FAI plans include verification requirements for software (3).

Conceptual Knowledge: Ability to analyze FAI planning to verify that FAI plans include verification requirements for software (4).

Factual Knowledge: Knowledge of industry standards related to first article inspection processes (e.g., AS/EN/SJAC 9102) (3). Knowledge of inspection planning processes (3). Knowledge of software quality assurance methods (3). Understanding of where work processes are being performed (3). Understanding of the value stream structure (3).

Task 3 (PP125-003): Ensure personnel are capable of performing software verification activities, throughout the value stream.

Quality Eng—Basic	Quality Eng—Senior	Q Manager	Supplier QE	CSoftQE	Program QE	Contract Review	MRO QE	Spec Process QE	Met Labs QE	Production Control	Regulatory Specialist
			P	P	S						

Procedural Knowledge: Ability to apply a defined assessment process to verify that personnel at all levels of the supplier/partner base are qualified to perform software verification activities (3).

Conceptual Knowledge: Ability to analyze personnel qualification and work experience records to verify that personnel at all levels of the supplier/partner base are qualified of performing software verification activities (4). Ability to evaluate personnel work activities to verify that personnel at all levels of the supplier/partner base are capable to perform software validation activities (5).

Factual Knowledge: Knowledge of industry standards related to first article inspection processes (e.g., AS/EN/SJAC 9102) (3). Knowledge of inspection planning processes (3). Knowledge of software quality assurance methods (3). Understanding of personnel qualification requirements related to software quality (3). Understanding of where work processes are being performed (3). Understanding of the value stream structure (3).

Process: Ensure Supplier NDT/Special Process Conformance—First Article Verification

Process Purpose: To ensure NDT and special processes are included in FAI plan and are verified to meet design/specification requirements. Ensure special processes have objective evidence of conformity and that NDTs are performed per requirements. Ensure personnel at all levels of the supplier/partner base are qualified to perform special processes and NDT.

Knowledge and Skill Requirements: Understanding of NDT methods and applicable special process standards requirements. Ability to verify NDT and special processes are included in FAI planning. Ability to validate that defined NDT and special process verification processes defined in FAI are appropriate and adequate. Ability to verify measured results meet defined requirements. Ability to verify special processes have objective evidence of conformity and that NDTs are performed per requirements. Ability to verify personnel at all levels of the supplier/partner base are qualified to perform special processes and NDT.

Task 1 (PP130-001): Verify NDT and special processes are included in FAI planning, throughout the value stream.

Quality Eng—Basic	Quality Eng—Senior	Q Manager	Supplier QE	CSoftQE	Program QE	Contract Review	MRO QE	Spec Process QE	Met Labs QE	Production Control	Regulatory Specialist
S	S		P		S		S	P	P	S	

Procedural Knowledge: Ability to apply a structured assessment process to verify that supplier/partner has appropriately addressed NDT verification requirements in FAI planning (3). Ability to apply a structured assessment process to verify that supplier/partner has appropriately addressed the control of special processes in FAI planning (3).

Conceptual Knowledge: Ability to analyze supplier/partner FAI planning to verify that NDT methods are appropriately addressed (4). Ability to analyze supplier/partner FAI planning to verify that the control of special processes are appropriately addressed (4).

Factual Knowledge: Knowledge of industry standards related to first article inspection processes (e.g., AS/EN/SJAC 9102) (3). Knowledge of inspection planning processes (3). Knowledge of work processes related to program/project (3). Knowledge of applicable NDT methods (4). Knowledge of applicable special processes (4). Understanding of where work processes are being performed (3). Understanding of the value stream structure (3).

Task 2 (PP130-002): Validate that defined NDT and special process verification processes documented in FAI are appropriate and adequate, throughout the value stream.

Quality Eng—Basic	Quality Eng—Senior	Q Manager	Supplier QE	CSoftQE	Program QE	Contract Review	MRO QE	Spec Process QE	Met Labs QE	Production Control	Regulatory Specialist
S	S		P		S		S	P	P	S	

Procedural Knowledge: Ability to apply a structured assessment to validate that defined NDT and special process verification processes as documented in FAI planning are appropriately and adequately executed (3).

Conceptual Knowledge: Ability to analyze NDT and special process verification processes to ensure they are appropriately and adequately executed as documented in FAI planning (4).

Factual Knowledge: Knowledge of industry standards related to first article inspection processes (e.g., AS/EN/SJAC 9102) (3). Knowledge of inspection planning processes (3). Knowledge of work processes related to program/project (3). Knowledge of applicable NDT methods (4). Knowledge of applicable special processes (4). Understanding of where work processes are being performed (3). Understanding of the value stream structure (3).

Task 3 (PP130-003): Verify measured results in NDT and special processes meet defined FAI requirements, throughout the value stream.

Quality Eng—Basic	Quality Eng—Senior	Q Manager	Supplier QE	CSoftQE	Program QE	Contract Review	MRO QE	Spec Process QE	Met Labs QE	Production Control	Regulatory Specialist
S	S		P		S		S	P	P	S	

Procedural Knowledge: Ability to apply a structured assessment process to verify that measured results in NDT and special processes meet defined requirements (3).

Conceptual Knowledge: Ability to analyze measured results in NDT and special processes and determine compliance to defined requirements (4).

Factual Knowledge: Knowledge of industry standards related to first article inspection processes (e.g., AS/EN/SJAC 9102) (3). Knowledge of inspection planning processes (3). Knowledge of work processes related to program/project (3). Knowledge of applicable NDT methods (4). Knowledge of applicable special processes (4). Understanding of where work processes are being performed (3). Understanding of the value stream structure (3).

Task 4 (PP130-004): Ensure special processes have objective evidence of conformity and that NDTs are performed per defined FAI requirements throughout the value stream.

Quality Eng—Basic	Quality Eng—Senior	Q Manager	Supplier QE	CSoftQE	Program QE	Contract Review	MRO QE	Spec Process QE	Met Labs QE	Production Control	Regulatory Specialist
S	S		P		P		S	P	P	S	

Procedural Knowledge: Ability to deploy a structured assessment process to verify that special processes have objective evidence of conformity and that NDTs are performed per requirements (3).

Conceptual Knowledge: Ability to analyze special processes and NDT methods, and their associated quality records, to verify that there is objective evidence of conformity per requirements (4).

Factual Knowledge: Knowledge of industry standards related to first article inspection processes (e.g., AS/EN/SJAC 9102) (3). Knowledge of inspection planning processes (3). Knowledge of work processes related to program/project (3). Knowledge of applicable NDT methods (4). Knowledge of applicable special processes (4). Understanding of where work processes are being performed (3). Understanding of the value stream structure (3).

Task 5 (PP130-005): Verify personnel at all levels of the supplier/partner base, who are performing special processes and NDT evaluations, are qualified to perform special processes and NDT.

Quality Eng—Basic	Quality Eng—Senior	Q Manager	Supplier QE	CSoftQE	Program QE	Contract Review	MRO QE	Spec Process QE	Met Labs QE	Production Control	Regulatory Specialist
S	S		P		S			P	P		

Procedural Knowledge: Ability to apply a defined assessment process to verify that personnel at all levels of the supplier/partner base are qualified to perform special processes and NDT (3).

Conceptual Knowledge: Ability to analyze personnel qualifications and work experience records to verify that personnel at all levels of the supplier/partner base, who are performing special processes and NDT evaluations, are qualified to perform special processes and NDT evaluation (4). Ability to evaluate personnel work activities to verify that personnel at all levels of the supplier/partner base, who are performing special processes and NDT evaluations, are capable to perform special processes and NDT evaluation (5).

Factual Knowledge: Knowledge of industry standards related to first article inspection processes (e.g., AS/EN/SJAC 9102) (3). Knowledge of inspection planning processes (3). Knowledge of work processes related to program/project (3). Knowledge of applicable NDT methods (4). Knowledge of applicable special processes (4). Knowledge of personnel qualification requirements related to NDT methods and special processes (3). Understanding of where work processes are being performed (3). Understanding of the value stream structure (3).

Process: Ensure Supplier Hardware Conformance—First Article Verification

Process Purpose: To ensure the geometric and physical characteristics are conformed through the FAI process. Ensure appropriate measurement devices (e.g., calibrated, and accuracy) are utilized when making inspection measurements.

Knowledge and Skill Requirements: Understanding of FAI processes and standards. Understanding of design standards. Understanding of production processes and production planning. Understanding of inspection planning. Ability to apply concepts and standards to verify the geometric and physical characteristics are conformed throughout the FAI process. Ability to verify appropriate measurement devices (e.g., calibrated, and accuracy) are utilized when making inspection measurements. Ability to ensure compliance of FAI requirements throughout the value stream.

Task 1 (PP135-001): Verify the geometric and physical characteristics to ensure conformity to design requirements are controlled and measured in the FAI process throughout the value stream.

Quality Eng—Basic	Quality Eng—Senior	Q Manager	Supplier QE	CSoftQE	Program QE	Contract Review	MRO QE	Spec Process QE	Met Labs QE	Production Control	Regulatory Specialist
S	S		P		S			S	S		

Procedural Knowledge: Ability to apply a defined process to verify the geometric and physical characteristics are conformed through the FAI process (3).

Conceptual Knowledge: Ability to interpret engineering specification requirements (4). Ability to analyze engineering documents and inspection planning to verify the geometric and physical characteristics are conformed through the FAI process (4).

Factual Knowledge: Knowledge of geometric tolerancing (3). Knowledge of applicable material standards (3). Knowledge of inspection planning (3). Knowledge of standards related to FAI (e.g., AS/EN/SJAC 9102)(3). Understanding of where work processes are being performed (3). Understanding of the value stream structure (3).

Task 2 (PP135-002): Verify the supplier's/partner's appropriate selection of measurement devices (e.g., calibrated, and accuracy) and ensure such are utilized when making inspection measurements, throughout the value stream. Verify that measurement devices are appropriately calibrated.

Quality Eng—Basic	Quality Eng—Senior	Q Manager	Supplier QE	CSoftQE	Program QE	Contract Review	MRO QE	Spec Process QE	Met Labs QE	Production Control	Regulatory Specialist
S	S		P		S				S		

Procedural Knowledge: Ability to apply a structured assessment methodology to verify appropriate measurement devices (e.g., calibrated, and accuracy) are utilized when making inspection measurements (3).

Conceptual Knowledge: Ability to analyze design requirements to determine accuracy requirements for measurement devices (4). Ability to interpret contract and regulatory requirements to determine unique requirements for calibration of measurement devices (4). Ability to analyze selected measurement devices to determine their adequacy to meet measurement requirements (4). Ability to audit work processes to verify that appropriate measurement devices (e.g., calibrated, and accuracy) are utilized when making inspection measurements (3).

Factual Knowledge: Understanding of design requirements (3). Understanding of regulatory requirements related to calibration (3). Understanding of standards related to measurement devices and accuracy (4). Understanding of calibration control systems (3). Understanding of auditing methods (3). Understanding of where work processes are being performed (3). Understanding of the value stream structure (3).

Process: Ensure Deviations Are Reviewed and Approved—in the Value Stream

Process Purpose: To formally document deviations discovered in pre-production processes and to ensure appropriate review and approval is concluded. Ensure deviation authority is defined and controlled. Ensure deviations are replanned for a follow up FAI.

Knowledge and Skill Requirements: Understanding of documentation processes for product and/or process deviations. Ability to formally document deviations discovered in pre-production processes and to ensure appropriate review and approval is concluded. Ability to evaluate documented processes to verify that deviation authority is defined and controlled. Ability to verify that deviations discovered during initial FAI are replanned for a follow up FAI. Ability to evaluate FAI follow-up processes to ensure follow-up actions are effectively performed and documented. Ability to ensure FAIs are performed at the appropriate locations and sequences within the value stream.

Task 1 (PP140-001): Evaluate documented processes to verify that deviation authority is defined and controlled, throughout the value stream.

Quality Eng—Basic	Quality Eng—Senior	Q Manager	Supplier QE	CSoftQE	Program QE	Contract Review	MRO QE	Spec Process QE	Met Labs QE	Production Control	Regulatory Specialist
S	S		P		S			S	S		

Procedural Knowledge: Ability to apply a structured assessment process to verify that deviation authority is defined and controlled for the pre-production and FAI processes (3).

Conceptual Knowledge: Ability to review procedures and associated quality records to verify that deviation authority is defined and controlled for the pre-production and FAI processes (3).

Factual Knowledge: Understanding of FAI process (3). Understanding of standards related to FAI (e.g., AS/EN/SJAC 9102)(3). Understanding of contract requirements related to deviations (3). Understanding of deviation approval process (3). Understanding of where work processes are being performed (3). Understanding of the value stream structure (3).

Task 2 (PP140-002): Document deviations discovered in pre-production processes and ensure appropriate review and approval is concluded, throughout the value stream.

Quality Eng—Basic	Quality Eng—Senior	Q Manager	Supplier QE	CSoftQE	Program QE	Contract Review	MRO QE	Spec Process QE	Met Labs QE	Production Control	Regulatory Specialist
S	S		P		S			S	S		

Procedural Knowledge: Ability to apply a structured process to document deviations discovered in pre-production processes and ensure appropriate review and approval is concluded (3).

Conceptual Knowledge: Ability to document deviations discovered in pre-production processes in terms that can be easily understood and replicated (3). Ability to coordinate with appropriate functions to ensure appropriate review and approval is concluded of deviation documentation package (3).

Factual Knowledge: Understanding of FAI process (3). Understanding of standards related to FAI (e.g., AS/EN/SJAC 9102)(3). Understanding of contract requirements related to deviations (3). Understanding of deviation approval process (3). Understanding of where work processes are being performed (3). Understanding of the value stream structure (3).

Task 3 (PP140-003): Verify that deviations discovered during initial FAI are replanned for a follow up FAI, throughout the value stream.

Quality Eng—Basic	Quality Eng—Senior	Q Manager	Supplier QE	CSoftQE	Program QE	Contract Review	MRO QE	Spec Process QE	Met Labs QE	Production Control	Regulatory Specialist
S	S		P		P			S	S		

Procedural Knowledge: Ability to apply a defined process to verify that deviations discovered during initial FAI are replanned for a follow up FAI (3).

Conceptual Knowledge: Ability to review applicable planning documentation to verify that deviations discovered during initial FAI are replanned for a follow up FAI (3). Ability to analyze replanned FAI activities to ensure activities accurately address the approved deviation (4).

Factual Knowledge: Understanding of FAI process (3). Understanding of standards related to FAI (e.g., AS/EN/SJAC 9102)(3). Understanding of contract/regulatory requirements related to deviations (3). Understanding of deviation approval process (3). Understanding of FAI and inspection planning (3). Understanding of where work processes are being performed (3). Understanding of the value stream structure (3).

Task 4 (PP140-004): Evaluate FAI follow-up processes to ensure follow-up actions are effectively performed and documented, throughout the value stream.

Quality Eng—Basic	Quality Eng—Senior	Q Manager	Supplier QE	CSoftQE	Program QE	Contract Review	MRO QE	Spec Process QE	Met Labs QE	Production Control	Regulatory Specialist
S	S		P		P			S	S		

Procedural Knowledge: Ability to apply a defined process to verify that FAI follow-up processes to ensure follow-up actions are effectively performed and documented (3).

Conceptual Knowledge: Ability to analyze planned work activities to verify that FAI follow-up processes to ensure follow-up actions are effectively performed and documented (4).

Factual Knowledge: Understanding of FAI process (3). Understanding of standards related to FAI (e.g., AS/EN/SJAC 9102)(3). Understanding of contract/regulatory requirements related to deviations (3). Understanding of deviation approval process (3). Understanding of FAI and inspection planning (3). Understanding of where work processes are being performed (3). Understanding of the value stream structure (3).

Process: Approve Product Conformance—First Article Reports (Value Stream)

Process Purpose: To ensure the entire deliverable product conforms to design and purchasing document requirements and complies with FAI requirements. Verify all lower-level/component-level FAIs are complete and accurate.

Knowledge and Skill Requirements: Understanding of FAI processes and standards. Understanding of design standards. Understanding of production processes and production planning. Understanding of inspection planning. Ability to evaluate final FAI packages to verify the entire deliverable product conforms to design and purchasing document requirements and complies with FAI requirements. Ability to verify all lower-level/component-level FAIs are complete and accurate.

Task 1 (PP145-001): Evaluate final supplier/partner FAI packages to verify the entire deliverable product conforms to design and purchasing document requirements and complies with FAI requirements.

Quality Eng—Basic	Quality Eng—Senior	Q Manager	Supplier QE	CSoftQE	Program QE	Contract Review	MRO QE	Spec Process QE	Met Labs QE	Production Control	Regulatory Specialist
S	S		P		P			S	S		

Procedural Knowledge: Ability to apply a defined process to review the final FAI packages to verify the entire deliverable product conforms to design and purchasing document requirements and complies with FAI requirements (3).

Conceptual Knowledge: Ability to interpret contractual, regulatory, and program requirements to ensure an accurate understanding of requirements for final FAI documentation packages (4). Ability to review final FAI packages to verify the entire deliverable product conforms to design and purchasing document requirements and complies with FAI requirements (3).

Factual Knowledge: Understanding of FAI process (3). Understanding of standards related to FAI (e.g., AS/EN/SJAC 9102)(3). Understanding of contract/regulatory requirements related to FAI packages (3). Understanding of FAI and inspection planning (3). Understanding of where work processes are being performed (3). Understanding of the value stream structure (3).

Task 2 (PP145-002): Verify all supplier/partner lower-level/component-level FAIs are complete and accurate.

Quality Eng—Basic	Quality Eng—Senior	Q Manager	Supplier QE	CSoftQE	Program QE	Contract Review	MRO QE	Spec Process QE	Met Labs QE	Production Control	Regulatory Specialist
S	S		P		P			S	S		

Procedural Knowledge: Ability to apply a defined process to verify all lower-level/component-level FAIs are complete and accurate (3).

Conceptual Knowledge: Ability to analyze lower-level/component-level FAI packages to verify that they are complete and accurate (4).

Factual Knowledge: Understanding of FAI process (3). Understanding of standards related to FAI (e.g., AS/EN/SJAC 9102)(3). Understanding of contract/regulatory requirements related to FAI packages (3). Understanding of FAI and inspection planning (3). Understanding of where work is performed (3).

IV.A.3. Process/Manufacturing/Servicing/Quality Planning

Process: Verify Production/ Servicing Work Instructions

Process Purpose: Review production/servicing work instructions and ensure the process of completing work is accurately defined. Work instruction review will require review of design and process documents and assurance of correlating revisions (configuration management).

Knowledge and Skills Requirements: Ability to evaluate production/servicing work instructions and verify that the process of completing work is accurately defined. Ability to assess production/servicing work instruction creation processes and verify that the definition of the scope of work captures appropriate verification steps, and are revision controlled. Ability to verify that a configuration management process is in place at all levels of production/servicing.

Task 1 (PO040-001): Assess production/servicing work instructions and verify that required work is accurately and appropriately defined.

Quality Eng—Basic	Quality Eng—Senior	Q Manager	Supplier QE	CSoftQE	Program QE	Contract Review	MRO QE	Spec Process QE	Met Labs QE	Production Control	Regulatory Specialist
P	P				S			P	P		

Procedural Knowledge: Ability to apply a structured process to determine the compliance of work instruction management (3).

Conceptual Knowledge: Ability to review production/servicing work instructions and determine the accuracy of work scope definition (4). Ability to review production/servicing work instructions and assess the influence of human factors in communication (4). Ability to assess production/servicing work instruction processes to ensure compliance to stated requirements (4). Ability to assess production/servicing processes and determine if verification processes are appropriately planned (4).

Factual Knowledge: Understanding of work instruction processes (3). Understanding of effective communication practices (3). Knowledge of assessment practices related to work instruction controls (3). Knowledge of configuration management systems (3).

Task 2 (PO040-003): Verify that a configuration management process is deployed in work instructions at all levels of the organization.

Quality Eng—Basic	Quality Eng—Senior	Q Manager	Supplier QE	CSoftQE	Program QE	Contract Review	MRO QE	Spec Process QE	Met Labs QE	Production Control	Regulatory Specialist
P	P	S			P			S	S		

Procedural Knowledge: Ability to apply a structured process to determine the appropriate application of configuration revision in work instructions throughout the organization (3).

Conceptual Knowledge: Ability to review and determine compliance and appropriate application of configuration revision in work instructions throughout the organization (4).

Factual Knowledge: Knowledge of configuration management practices (3). Knowledge of organization design definition processes (3). Knowledge of project/program requirements (3). Knowledge of work instruction control processes (3).

Process: Review Product/Servicing Process Planning

Process Purpose: Review documented process planning to ensure processes are adequately defined and controlled. Process planning may include statistical process control plans, process control sheets, and other means of process definition. Ensure process planning is controlled by configuration management processes.

Knowledge and Skill Requirements: Ability to review documented process planning to ensure processes are appropriately documented. Ability to understand and evaluate control methods to ensure adequacy and capability of process planning may include statistical process control plans, process control sheets, and other means of process definition. Ability to evaluate defined organizational processes to ensure process planning is controlled by configuration management processes.

Task 1 (PO041-001): Evaluate control methods to ensure adequacy and capability of process planning that may include statistical process control plans, process control sheets, and other means of process definition.

Quality Eng—Basic	Quality Eng—Senior	Q Manager	Supplier QE	CSoftQE	Program QE	Contract Review	MRO QE	Spec Process QE	Met Labs QE	Production Control	Regulatory Specialist
P	P	S			P		S	S	S	S	

Procedural Knowledge: Ability to apply a structured process to assess the compliance of process planning processes (3).

Conceptual Knowledge: Ability to review process planning processes to determine the compliance and adequacy of process planning processes (4). Ability to analyze the effectiveness of established process control methods (5).

Factual Knowledge: Understanding of process control methods (4). Knowledge of process planning methods (3). Understanding of assessment practices related to process planning (3). Understanding of applicable technical processes (4).

Task 2 (PO041-002): Evaluate defined organizational processes to ensure process planning is controlled by configuration management processes.

Quality Eng—Basic	Quality Eng—Senior	Q Manager	Supplier QE	CSoftQE	Program QE	Contract Review	MRO QE	Spec Process QE	Met Labs QE	Production Control	Regulatory Specialist
P	P	P			P		S	S	S	S	

Procedural Knowledge: Ability to apply a structured process to determine the compliance of organizational processes that ensure appropriate configuration revisions are flowed to all applicable process planning documents (3).

Conceptual Knowledge: Ability to analyze, for compliance and adequacy, organizational processes that ensure appropriate configuration revisions are flowed to all applicable process planning documents (4).

Factual Knowledge: Understanding of process control methods (4). Knowledge of process planning methods (3). Understanding of assessment practices related to process planning (3). Understanding of applicable technical processes (4). Understanding of configuration management processes (4).

Process: Ensure In-Process Inspection/Operator Certification

Process Purpose: To ensure inspections are being performed at appropriate points in production/servicing and the inspections are being performed by qualified persons. Where operators are performing inspection functions, ensure a comprehensive operator certification program is in place. If required, ensure organization has approval to perform operator certification. (E.g., Approval may be required by primes and regulatory authorities.) Where special processes are being performed, persons responsible for inspection/verification shall be appropriately qualified and may require special certification (e.g., welding, NDT, chemical processes.)

Knowledge and Skill Requirements: Ability to determine the appropriate points of inspection in the production/servicing work processes. Ability to verify inspections are being performed at appropriate points in production/servicing and the inspections are being performed by qualified persons. Ability to validate the effectiveness of inspection planning. Ability to assess the effectiveness of an operator certification process. Ability to verify that operator certification programs have the appropriate approvals needed to deploy.

Task 1 (PO085-001): Verify inspections are being performed at appropriate points in production/servicing and the inspections are being performed by qualified persons.

Quality Eng—Basic	Quality Eng—Senior	Q Manager	Supplier QE	CSoftQE	Program QE	Contract Review	MRO QE	Spec Process QE	Met Labs QE	Production Control	Regulatory Specialist
S	P	P			S		P	S	S	P	

Procedural Knowledge: Ability to apply a structured process to review production/servicing and inspection planning to verify that inspections are being performed at appropriate points in production/servicing (3). Ability to apply a structured process to verify that planned inspections are being performed by qualified persons (3). Ability to apply a structured process to review personnel training and certifications records to verify that personnel have adequate qualifications to perform assigned inspection functions (3).

Conceptual Knowledge: Ability to understand the influence an extended value stream has on inspection functions (5). Ability to understand where inspection activities should be performed in production/servicing processes (4). Ability to review production/servicing and inspection planning to verify that inspections are being performed at appropriate points in production/servicing (4). Ability to verify that planned inspections are being performed by qualified persons (4). Ability to review personnel training and certifications records to verify that personnel have adequate qualifications to perform assigned inspection functions (4). Ability to interview personnel assigned inspection responsibilities to determine appropriate understanding (4). Ability to review quality data to determine the effectiveness of personnel assigned inspection responsibilities (4).

Factual Knowledge: Knowledge of work instruction processes (3). Understanding of quality records (3). Understanding of inspection training requirements (3). Knowledge of personnel training records (3). Knowledge of auditing methods (3). Knowledge of certification requirements related to product verification (3). Knowledge of contractual and regulatory requirements related to inspection personnel (3). Knowledge of where work is being performed (3).

Task 2 (PO085-002): Validate the effectiveness of inspection planning, through evaluation of quality records and product review.

Quality Eng—Basic	Quality Eng—Senior	Q Manager	Supplier QE	CSoftQE	Program QE	Contract Review	MRO QE	Spec Process QE	Met Labs QE	Production Control	Regulatory Specialist
P	P	S			P		P	S	S	P	

Procedural Knowledge: Ability to apply a structured process (proactively and forensically) to verify the effectiveness of inspection planning (3). Ability to apply a structured process to obtain quality data (3).

Conceptual Knowledge: Ability to recognize and understand what data is needed for effectiveness review (4). Ability to evaluate quality data and determine the effectiveness of inspection planning processes (5).

Factual Knowledge: Knowledge of work instruction processes (3). Understanding of quality records (3). Knowledge of personnel training records (3). Knowledge of auditing methods (3). Knowledge of certification requirements related to product verification (3). Knowledge of contractual and regulatory requirements related to inspection personnel (3). Knowledge of where work is being performed (3). Knowledge of effectiveness measurement practices (3).

Task 3 (PO085-003): Assess the effectiveness of operator certification processes.

Quality Eng—Basic	Quality Eng—Senior	Q Manager	Supplier QE	CSoftQE	Program QE	Contract Review	MRO QE	Spec Process QE	Met Labs QE	Production Control	Regulatory Specialist
P	P	S			S		P			P	

Procedural Knowledge: Ability to apply a structured assessment process to measure the effectiveness of operator certification processes (3).

Conceptual Knowledge: Ability to analyze operator certification processes to establish the effectiveness of the process to assure product and process quality (4).

Factual Knowledge: Knowledge of operator certification processes (3). Knowledge of regulatory requirements related to operator certification processes (3). Knowledge of contractual requirements related to operator certification processes (3). Knowledge of product and process verification methods (3).

Task 4 (PO085-004): Verify that operator certification programs have required approvals.

Quality Eng—Basic	Quality Eng—Senior	Q Manager	Supplier QE	CSoftQE	Program QE	Contract Review	MRO QE	Spec Process QE	Met Labs QE	Production Control	Regulatory Specialist
P	P	P			P	S	S			S	P

Procedural Knowledge: Ability to apply a structured process to verify that operator certification programs have required approvals (3).

Conceptual Knowledge: Ability to review and assess that operator certification programs have required approvals (3).

Factual Knowledge: Knowledge of operator certification processes (3). Knowledge of regulatory requirements related to operator certification processes (3). Knowledge of contractual requirements related to operator certification processes (3).

Process: Ensure In-Process Inspection/Machine Probing are Adequate

Process Purpose: To ensure in-process inspections and machine probing are appropriately planned, sequenced, implemented and controlled. May require process auditing and documentation review. Ensure inspection personnel are appropriately qualified to perform inspection and are available when inspection is required. Ensure the in-line probing equipment has adequate capability to perform the required inspections. Ensure the use of machine probing in incorporated into organizational procedures and work instructions and utilization is documented in the first article inspection (FAI) reports, as appropriate.

Knowledge and Skill Requirements: Ability to assess production and inspection processes and verify that in-process inspections and machine probing are appropriately planned, sequenced, implemented and controlled. Ability to plan and perform auditing and documentation review of in-process inspection activities. Ability to assess and determine that inspection personnel are appropriately qualified to perform inspection. Ability to verify that inspection personnel are available when inspection is required. Ability to verify in-line probing equipment has adequate capability to perform the required inspections. Ability to validate the use of machine probing into procedures and work instructions and utilization is documented in the FAI reports.

Task 1 (PO105-001): Plan and perform auditing and documentation review of in-process inspection activities throughout the organization.

Quality Eng—Basic	Quality Eng—Senior	Q Manager	Supplier QE	CSoftQE	Program QE	Contract Review	MRO QE	Spec Process QE	Met Labs QE	Production Control	Regulatory Specialist
P	P				S		P	S	S	P	

Procedural Knowledge: Ability to apply a structured process to adequately plan an audit of production and inspection processes to verify that in-process inspections and machine probing are appropriately planned, sequenced, implemented and controlled (3). Ability to apply a structured process in the review of documentation related to in-process inspection activities to determine compliance to requirements (3). Ability to apply a structured and defined process to assess work activities for compliance to requirements (3).

Conceptual Knowledge: Ability to review and evaluate internal and external standards and requirements to adequately plan an audit of production and inspection processes to verify that in-process inspections and machine probing are appropriately planned, sequenced, implemented and controlled (4). Ability to analyze documentation related to in-process inspection activities to determine compliance to requirements (3). Ability to analyze established work activities for compliance to requirements (4).

Factual Knowledge: Understanding of inspection methods (3). Understanding of in-line probing methods (3). Understanding of inspection technologies (4). Understanding of inspection planning methods (3). Understanding of product work-flow (3). Understanding of documentation requirements (3). Understanding of auditing planning methods (3). Understanding of where work is being performed (3). Knowledge of product realization requirements from internal and external standards and requirements (e.g., AS/EN/JISQ 9100.)(3).

Task 2 (PO105-002): Audit production and inspection processes and verify that in-process inspections and machine probing are appropriately planned, sequenced, implemented and controlled throughout the organization.

Quality Eng—Basic	Quality Eng—Senior	Q Manager	Supplier QE	CSoftQE	Program QE	Contract Review	MRO QE	Spec Process QE	Met Labs QE	Production Control	Regulatory Specialist
P	P				S		P	P	S	P	

Procedural Knowledge: Ability to apply structured assessment methods to verify that in-process inspections and machine probing are appropriately planned, sequenced, implemented and controlled (3).

Conceptual Knowledge: Ability to analyze production/servicing and inspection work-flow to identify where inspection activities are to be performed (4). Ability to analyze production/servicing and inspection processes to determine if in-process inspections and machine probing are appropriately planned, sequenced, implemented and controlled (4).

Factual Knowledge: Understanding of inspection methods (3). Understanding of in-line probing methods (3). Understanding of inspection technologies (4). Understanding of inspection planning methods (4). Understanding of product work-flow (3). Understanding of documentation requirements (3). Understanding of audit practices (3). Understanding of where work is being performed (3). Knowledge of product realization requirements from internal and external standards and requirements (e.g., AS/EN/JISQ 9100.)(3).

Task 3 (PO105-003): Assess and determine that inspection personnel are appropriately qualified to perform inspections.

Quality Eng—Basic	Quality Eng—Senior	Q Manager	Supplier QE	CSoftQE	Program QE	Contract Review	MRO QE	Spec Process QE	Met Labs QE	Production Control	Regulatory Specialist
P	P				S		P	P	S	P	

Procedural Knowledge: Ability to apply a structured process to identify inspection personnel qualification (3). Ability to apply a structured process to assess inspection personnel qualifications (3).

Conceptual Knowledge: Ability to analyze inspection activities and determine qualification requirements for inspection personnel (4). Ability to assess inspection personnel qualification records (incl. past work experience) to determine compliance to inspection qualification requirements (includes certification) (4). Ability to evaluate the capabilities of inspection personnel to determine adequacy (4).

Factual Knowledge: Understanding of inspection methods (3). Understanding of project/program inspection requirements (3). Understanding of technical processes (3). Understanding of personnel capability assessments (3). Understanding of applicable inspection technologies (3). Knowledge of applicable industry standards related to inspection/evaluation (3).

Task 4 (PO105-004): Verify that probing inspection personnel are available when inspection is required.

Quality Eng—Basic	Quality Eng—Senior	Q Manager	Supplier QE	CSoftQE	Program QE	Contract Review	MRO QE	Spec Process QE	Met Labs QE	Production Control	Regulatory Specialist
S	S	P			P		S	S		S	

Procedural Knowledge: Ability to apply a structured assessment process to establish adequacy of inspection personnel availability (3).

Conceptual Knowledge: Ability to evaluate work flow, schedule, and availability of inspection personnel to determine adequate availability of qualified personnel (5).

Factual Knowledge: Understanding of inspection work processes (3). Understanding of work flow (3). Understanding of production/servicing schedules (3). Knowledge of inspection capacity planning (3). Knowledge of where work is performed (3).

Task 5 (PO105-005): Verify in-line probing equipment has adequate capability to perform the required inspections.

Quality Eng—Basic	Quality Eng—Senior	Q Manager	Supplier QE	CSoftQE	Program QE	Contract Review	MRO QE	Spec Process QE	Met Labs QE	Production Control	Regulatory Specialist
P	P	S			S		P	S	S	P	

Procedural Knowledge: Ability to apply a structured assessment process to verify capability of in-line probing processes (4).

Conceptual Knowledge: Ability to analyze the capability of in-line probing processes (4). Ability to evaluate the capability of in-line probing against design requirements (5).

Factual Knowledge: Understanding of in-line probing technologies (4). Understanding of design requirements (3). Understanding of probing equipment capability (4).

Task 6 (PO105-006): Validate the use of machine probing is appropriately incorporated into procedures and work instructions. Verify that utilization of machine probing is documented in the FAI reports.

Quality Eng—Basic	Quality Eng—Senior	Q Manager	Supplier QE	CSoftQE	Program QE	Contract Review	MRO QE	Spec Process QE	Met Labs QE	Production Control	Regulatory Specialist
P	P	S			S		P			P	

Procedural Knowledge: Ability to apply a structured assessment process to determine if suppliers have integrated the application of in-line probing into organizational procedures (3). Ability to apply a structured assessment process to determine if suppliers have integrated the application of in-line probing into work instructions and quality records (3). Ability to apply a structured assessment process to determine if in-line probing has been appropriately integrated into FAI records (3).

Conceptual Knowledge: Ability to assess organizational procedures to determine if the application of in-line probing has been appropriately integrated, as applicable (3). Ability to analyze organizational work instructions and quality records to determine if the application of in-line probing into work instructions and quality records has been appropriately integrated, as appropriate (4). Ability to analyze FAI records to determine if the application of in-line probing has been appropriately integrated, as applicable (4).

Factual Knowledge: Knowledge of in-line probing methods (3). Understanding of applicable production/servicing processes (3). Understanding of requirements flow into procedures (3). Understanding of work instruction processes (3). Understanding of FAI planning processes (3). Understanding of FAI documentation processes (3). Understanding of industry standards related to FAI (e.g., AS/EN/JISQ 9102) (3).

Process: Review Supplier Process Planning

Process Purpose: Review documented process planning at suppliers/partners to ensure processes are adequately defined and controlled. Process planning may include statistical process control plans, process control sheets, and other means of process definition. Ensure process planning is controlled by configuration management processes.

Knowledge and Skill Requirements: Ability to review documented process planning at suppliers/partners to ensure processes are documented. Ability to understand and evaluate control methods to ensure adequacy and capability of process planning may include statistical process control plans, process control sheets, and other means of process definition. Ability to evaluate supplier defined processes to ensure process planning is controlled by configuration management processes.

Task 1 (PP035-001): Review documented process planning at suppliers/partners to ensure processes are documented. (Note: Program QE support role includes the assurance of access to secured locations.)

Quality Eng—Basic	Quality Eng—Senior	Q Manager	Supplier QE	CSoftQE	Program QE	Contract Review	MRO QE	Spec Process QE	Met Labs QE	Production Control	Regulatory Specialist
			P		S			S	S		

Procedural Knowledge: Ability to apply a structured process in assessing supplier/partner work definition documentation to determine compliance to applicable requirements (3).

Conceptual Knowledge: Ability to evaluate supplier/partner work definition documentation to determine effectiveness and adequacy of work directions (5). Ability to assess supplier/partner processes for work definition documentation control to ensure adequacy of control, to include changes (4).

Factual Knowledge: Knowledge of work instruction processes (3). Knowledge of documentation change processes (3). Knowledge of applicable technical processes (3). Knowledge of configuration management processes (3). Understanding of where technical standards reside (3).

Task 2 (PP035-002): Review documented process planning at suppliers/partners to ensure processes are adequately defined and controlled. Process planning may include statistical process control plans, process control sheets, and other means of process definition. Ensure process planning is controlled by configuration management processes. (Note: Program QE support role includes the assurance of access to secured locations.)

Quality Eng—Basic	Quality Eng—Senior	Q Manager	Supplier QE	CSoftQE	Program QE	Contract Review	MRO QE	Spec Process QE	Met Labs QE	Production Control	Regulatory Specialist
			P		S			S	S		

Procedural Knowledge: Evaluate control methods to ensure adequacy and capability of process planning may include statistical process control plans, process control sheets, and other means of process definition.

Conceptual Knowledge: Ability to apply a structured process to assess the compliance of process planning processes (3).

Factual Knowledge: Ability to review process planning processes to determine the compliance and adequacy of process planning processes (4). Ability to analyze the effectiveness of established process control methods (5).

Task 3 (PP035-003): Evaluate supplier/partner defined processes to ensure process planning is controlled by configuration management processes.

Quality Eng—Basic	Quality Eng—Senior	Q Manager	Supplier QE	CSoftQE	Program QE	Contract Review	MRO QE	Spec Process QE	Met Labs QE	Production Control	Regulatory Specialist
			P		S			S	S		

Procedural Knowledge: Ability to apply a structured process to determine the compliance of supplier/partner processes that ensure appropriate configuration revisions are flowed to all applicable process planning documents (3).

Conceptual Knowledge: Ability to analyze, for compliance and adequacy, supplier/partner processes that ensure appropriate configuration revisions are flowed to all applicable process planning documents (4).

Factual Knowledge: Understanding of process control methods (4). Knowledge of process planning methods (3). Understanding of assessment practices related to process planning (3). Understanding of applicable technical processes (4). Understanding of configuration management processes (4).

Process: Verify Supplier Work Instructions

Process Purpose: Review supplier/partner work instructions and ensure the process of completing work is accurately defined. Work instruction review will require review of design and process documents and assurance of correlating revisions (configuration management).

Knowledge and Skill Requirements: Ability to assess supplier/partner work instructions and verify that the process of completing work is accurately defined. Ability to assess sub-tier suppliers work instruction creation processes and verify that they define the scope of work, capture verification steps, and are revision controlled. Ability to verify that a configuration management process is in place at all levels of the value stream.

Task 1 (PP040-001): Assess supplier/partner work instructions and verify that the process of completing work is accurately and appropriately defined. (Note: Program QE support role includes the assurance of access to secured locations.)

Quality Eng—Basic	Quality Eng—Senior	Q Manager	Supplier QE	CSoftQE	Program QE	Contract Review	MRO QE	Spec Process QE	Met Labs QE	Production Control	Regulatory Specialist
			P		S			S	S		

Procedural Knowledge: Ability to apply a structured process to determine the compliance of work instruction management (3).

Conceptual Knowledge: Ability to review supplier/partner work instructions and determine the accuracy of work scope definition (4). Ability to review supplier/partner work instructions and assess the influence of human factors in communication (4). Ability to assess supplier/partner work instruction processes to ensure compliance to stated requirements (4). Ability to assess production/servicing processes and determine if verification processes are appropriately planned (4).

Factual Knowledge: Understanding of work instruction processes (3). Understanding of effective communication practices (3). Knowledge of assessment practices related to work instruction controls (3).

Task 2 (PP040-002): Assess sub-tier suppliers work instruction creation processes and verify that they define the scope of work, capture verification steps, and are revision controlled.

Quality Eng—Basic	Quality Eng—Senior	Q Manager	Supplier QE	CSoftQE	Program QE	Contract Review	MRO QE	Spec Process QE	Met Labs QE	Production Control	Regulatory Specialist
			P		S			S	S		

Procedural Knowledge: Ability to apply a structured process to determine the compliance of work instruction management within the value stream (3).

Conceptual Knowledge: Ability to review sub-tier supplier work instructions and determine the accuracy of work scope definition (4). Ability to review sub-tier supplier work instructions and assess the influence of human factors in communication (4). Ability to assess sub-tier supplier work instruction processes to ensure compliance to stated requirements (4). Ability to assess production/servicing processes and determine if verification processes are appropriately planned (4).

Factual Knowledge: Understanding of work instruction processes (3). Understanding of effective communication practices (3). Knowledge of assessment practices related to work instruction controls (3). Knowledge of value stream structure (3).

Task 3 (PP040-003): Verify that a configuration management process is in place at all levels of the value stream.

Quality Eng—Basic	Quality Eng—Senior	Q Manager	Supplier QE	CSoftQE	Program QE	Contract Review	MRO QE	Spec Process QE	Met Labs QE	Production Control	Regulatory Specialist
S	S	S	P	S	P			S	S		

Procedural Knowledge: Ability to apply a structured process to determine the appropriate application of configuration revision in work instructions throughout the value stream (3).

Conceptual Knowledge: Ability to review and determine compliance and appropriate application of configuration revision in work instructions throughout the value stream (4).

Factual Knowledge: Knowledge of configuration management practices (3). Knowledge of project/program requirements (3). Knowledge of work instruction control processes (3). Knowledge of purchasing documents control processes (3).

Process: Ensure Supplier In-Process Inspection/Operator Certification

Process Purpose: To ensure inspections are being performed at appropriate points in production/servicing and the inspections are being performed by qualified persons. Where operators are performing inspection functions, ensure a comprehensive operator certification program is in place. If required, ensure supplier/partner has approval to perform operator certification. (E.g., Approval may be required by primes, major suppliers, and regulatory authorities.) Where special processes are being performed, persons responsible for inspection/verification shall be appropriately qualified and may require special certification (e.g., welding, NDT, chemical processes.)

Knowledge and Skill Requirements: Ability to determine the appropriate points of inspection in the production/servicing work processes. Ability to verify inspections are being performed at appropriate points in production/servicing and the inspections are being performed by qualified persons. Ability to validate the effectiveness of inspection planning. Ability to assess the effectiveness of an operator certification process. Ability to verify that operator certification programs have the appropriate approvals needed to deploy.

Task 1 (PP085-001): Verify inspections are being performed at appropriate points in production/servicing (in the value stream) and the inspections are being performed by qualified persons.

Quality Eng—Basic	Quality Eng—Senior	Q Manager	Supplier QE	CSoftQE	Program QE	Contract Review	MRO QE	Spec Process QE	Met Labs QE	Production Control	Regulatory Specialist
		S	P		P			S	S		

Procedural Knowledge: Ability to apply a structured process to review production/servicing and inspection planning to verify that inspections are being performed at appropriate points in production/servicing (3). Ability to apply a structured process to verify that planned inspections are being performed by qualified persons (3). Ability to apply a structured process to review personnel training and certifications records to verify that personnel have adequate qualifications to perform assigned inspection functions (3).

Conceptual Knowledge: Ability to understand the influence an extended value stream has on inspection functions (5). Ability to understand where inspection activities should be performed in production/servicing processes (4). Ability to review production/servicing and inspection planning to verify that inspections are being performed at appropriate points in production/servicing (4). Ability to verify that planned inspections are being performed by qualified persons (4). Ability to review personnel training and certifications records to verify that personnel have adequate qualifications to perform assigned inspection functions (4). Ability to interview personnel assigned inspection responsibilities to determine appropriate understanding (4). Ability to review quality data to determine the effectiveness of personnel assigned inspection responsibilities (4).

Factual Knowledge: Knowledge of work instruction processes (3). Understanding of quality records (3). Understanding of inspection training requirements (3). Knowledge of personnel training records (3). Knowledge of auditing methods (3). Knowledge of certification requirements related to product verification (3). Knowledge of contractual and regulatory requirements related to inspection personnel (3). Knowledge of value stream structure (3).

Task 2 (PP085-002): Validate the effectiveness of inspection planning (in the value stream), through evaluation of quality records and product review.

Quality Eng—Basic	Quality Eng—Senior	Q Manager	Supplier QE	CSoftQE	Program QE	Contract Review	MRO QE	Spec Process QE	Met Labs QE	Production Control	Regulatory Specialist
			P		S			S	S		

Procedural Knowledge: Ability to apply a structured process (proactively and forensically) to verify the effectiveness of inspection planning (3). Ability to apply a structured process to obtain quality data (3).

Conceptual Knowledge: Ability to recognize and understand what data is needed for effectiveness review (4). Ability to evaluate quality data and determine the effectiveness of inspection planning processes (5).

Factual Knowledge: Knowledge of work instruction processes (3). Understanding of quality records (3). Knowledge of personnel training records (3). Knowledge of auditing methods (3). Knowledge of certification requirements related to product verification (3). Knowledge of contractual and regulatory requirements related to inspection personnel (3). Knowledge of value stream structure (3). Knowledge of effectiveness measurement practices (3).

Task 3 (PP085-003): Assess the effectiveness of operator certification (in the value stream) processes.

Quality Eng—Basic	Quality Eng—Senior	Q Manager	Supplier QE	CSoftQE	Program QE	Contract Review	MRO QE	Spec Process QE	Met Labs QE	Production Control	Regulatory Specialist
S	S		P								

Procedural Knowledge: Ability to apply a structured assessment process to measure the effectiveness of operator certification processes (3).

Conceptual Knowledge: Ability to analyze operator certification processes to establish the effectiveness of the process to assure product and process quality (4).

Factual Knowledge: Knowledge of operator certification processes (3). Knowledge of regulatory requirements related to operator certification processes (3). Knowledge of contractual requirements related to operator certification processes (3). Knowledge of product and process verification methods (3).

Task 4 (PP085-004): Verify that operator certification programs (in the value stream) have required approvals.

Quality Eng—Basic	Quality Eng—Senior	Q Manager	Supplier QE	CSoftQE	Program QE	Contract Review	MRO QE	Spec Process QE	Met Labs QE	Production Control	Regulatory Specialist
S	S	S	P		P	S					S

Procedural Knowledge: Ability to apply a structured process to verify that operator certification programs have required approvals (3).

Conceptual Knowledge: Ability to review and assess that operator certification programs have required approvals (3).

Factual Knowledge: Knowledge of operator certification processes (3). Knowledge of regulatory requirements related to operator certification processes (3). Knowledge of contractual requirements related to operator certification processes (3).

Process: Ensure Process Verification (Process Tryout)—Value Stream

Process Purpose: To ensure processes perform as designed a process verification process is conducted. Ensure processes are adequately defined and operating in a conforming fashion. As processes are submitted for production, ensure a documented process tryout is performed.

Knowledge and Skill Requirements: Ability to verify organizational and supplier/partner documented processes to ensure process verifications (process try-outs) are conducted as required. Ability to assess the adequacy of process verification processes. Ability to verify that process verification processes are appropriately documented.

Task 1 (PP090-001): Verify supplier/partner documented processes to ensure risk factors identified in process try-out are addressed and process verifications are conducted as required.

Quality Eng—Basic	Quality Eng—Senior	Q Manager	Supplier QE	CSoftQE	Program QE	Contract Review	MRO QE	Spec Process QE	Met Labs QE	Production Control	Regulatory Specialist
	S		P		P			S	S		

Procedural Knowledge: Ability to apply a structured process to identify the risk factors associated with the work process (3). Ability to apply a structured assessment process to verify that supplier/partner has documented processes to perform process try-out activities, as required (3).

Conceptual Knowledge: Ability to assess supplier/partner documented processes to verify compliance to defined requirements for process try-outs (3). Ability to recognize, through work analysis, potential risk factors associated with related work processes (4).

Factual Knowledge: Knowledge of manufacturing/servicing processes related to the project/program (3). Understanding of process try-out practices (3). Understanding of contract requirements related to process try-outs (3). Knowledge of risk factors related to related work processes (4).

Task 2 (PP090-002): Assess the adequacy of process verification processes throughout the value stream.

Quality Eng—Basic	Quality Eng—Senior	Q Manager	Supplier QE	CSoftQE	Program QE	Contract Review	MRO QE	Spec Process QE	Met Labs QE	Production Control	Regulatory Specialist
			P		P			S	S		

Procedural Knowledge: Ability to apply a structured assessment process to determine the adequacy of process verification controls (3).

Conceptual Knowledge: Ability to analyze work flow processes to identify where process verification activities should occur within the value stream (4). Ability to analyze established process verification controls to determine adequacy to measure and verify process capability (4).

Factual Knowledge: Knowledge of manufacturing/servicing processes related to the project/program (3). Understanding of process try-out practices (3). Knowledge of process capability measurement (3). Knowledge of the value stream structure (3). Knowledge of where work is being performed (3).

Task 3 (PP090-003): Verify that process verification activities have been accomplished and are appropriately documented throughout the value stream.

Quality Eng—Basic	Quality Eng—Senior	Q Manager	Supplier QE	CSoftQE	Program QE	Contract Review	MRO QE	Spec Process QE	Met Labs QE	Production Control	Regulatory Specialist
		S	P		P			S	S		

Procedural Knowledge: Ability to apply a structured process to audit the execution and documentation of process verification activities (3).

Conceptual Knowledge: Ability to analyze the execution and documentation of process verification activities to determine compliance to process procedures (4).

Factual Knowledge: Knowledge of manufacturing/servicing processes related to the project/program (3). Understanding of process try-out practices (3). Knowledge of process capability measurement (3). Knowledge of quality records requirements (3). Knowledge of management practices (3). Knowledge of the value stream (3). Knowledge of where work is being performed (3).

Process: Ensure In-Process Inspection/Machine Probing Are Adequate—Value Stream

Process Purpose: To ensure in-process inspections and machine probing are appropriately planned, sequenced, implemented and controlled. May require process auditing and documentation review. Ensure inspection personnel are appropriately qualified to perform inspection and are available when inspection is required. Ensure the in-line probing equipment has adequate capability to perform the required inspections. Ensure suppliers incorporate the use of machine probing into their procedures and work instructions and utilization is documented in the first article inspection (FAI) reports.

Knowledge and Skill Requirements: Ability to assess production and inspection processes and verify that in-process inspections and machine probing are appropriately planned, sequenced, implemented, and controlled. Ability to plan and perform auditing and documentation review of in-process inspection activities. Ability to assess and determine that inspection personnel are appropriately qualified to perform inspection. Ability to verify that inspection personnel are available when inspection is required. Ability to verify in-line probing equipment has adequate capability to perform the required inspections. Ability to validate that suppliers incorporate the use of machine probing into their procedures and work instructions and utilization is documented in the FAI reports.

Task 1 (PP105-001): Plan and perform auditing and documentation review of in-process inspection activities throughout the value stream.

Quality Eng—Basic	Quality Eng—Senior	Q Manager	Supplier QE	CSoftQE	Program QE	Contract Review	MRO QE	Spec Process QE	Met Labs QE	Production Control	Regulatory Specialist
			P		S			S	S		

Procedural Knowledge: Ability to apply a structured process to adequately plan an audit of production and inspection processes to verify that in-process inspections and machine probing are appropriately planned, sequenced, implemented and controlled (3). Ability to apply a structured process in the review of documentation related to in-process inspection activities to determine compliance to requirements (3). Ability to apply a structured and defined process to assess work activities for compliance to requirements (3).

Conceptual Knowledge: Ability to review and evaluate internal and external standards and requirements to adequately plan an audit of production and inspection processes to verify that in-process inspections and machine probing are appropriately planned, sequenced, implemented and controlled (4). Ability to analyze documentation related to in-process inspection activities to determine compliance to requirements (3). Ability to analyze established work activities for compliance to requirements (4).

Factual Knowledge: Understanding of inspection methods (3). Understanding of in-line probing methods (3). Understanding of inspection technologies (4). Understanding of inspection planning methods (3). Understanding of product work-flow (3). Understanding of documentation requirements (3). Understanding of auditing planning methods (3). Understanding of the value stream structure (3). Understanding of where work is being performed (3). Knowledge of product realization requirements from internal and external standards and requirements (e.g., AS/EN/JISQ 9100.)(3).

Task 2 (PP105-002): Audit production and inspection processes and verify that in-process inspections and machine probing are appropriately planned, sequenced, implemented and controlled throughout the value stream.

Quality Eng—Basic	Quality Eng—Senior	Q Manager	Supplier QE	CSoftQE	Program QE	Contract Review	MRO QE	Spec Process QE	Met Labs QE	Production Control	Regulatory Specialist
			P		S			S	S		

Procedural Knowledge: Ability to apply structured assessment methods to verify that in-process inspections and machine probing are appropriately planned, sequenced, implemented and controlled (3).

Conceptual Knowledge: Ability to analyze production/servicing and inspection work-flow to identify where inspection activities are to be performed (4). Ability to analyze production/servicing and inspection processes to determine if in-process inspections and machine probing are appropriately planned, sequenced, implemented and controlled (4).

Factual Knowledge: Understanding of inspection methods (3). Understanding of in-line probing methods (3). Understanding of inspection technologies (4). Understanding of inspection planning methods (4). Understanding of product work-flow (3). Understanding of documentation requirements (3). Understanding of audit practices (3). Understanding of the value stream structure (3). Understanding of where work is being performed (3). Knowledge of product realization requirements from internal and external standards and requirements (e.g., AS/EN/JISQ 9100.)(3).

Task 3 (PP105-003): Assess and determine that supplier/partner inspection personnel are appropriately qualified to perform inspections.

Quality Eng—Basic	Quality Eng—Senior	Q Manager	Supplier QE	CSoftQE	Program QE	Contract Review	MRO QE	Spec Process QE	Met Labs QE	Production Control	Regulatory Specialist
			P					S	S		

Procedural Knowledge: Ability to apply a structured process to identify inspection personnel qualification (3). Ability to apply a structured process to assess inspection personnel qualifications (3).

Conceptual Knowledge: Ability to analyze inspection activities and determine qualification requirements for inspection personnel (4). Ability to assess inspection personnel qualification records (incl. past work experience) to determine compliance to inspection qualification requirements (includes certification) (4). Ability to evaluate the capabilities of inspection personnel to determine adequacy (4).

Factual Knowledge: Understanding of inspection methods (3). Understanding of project/program inspection requirements (3). Understanding of technical processes (3). Understanding of personnel capability assessments (3). Understanding of applicable inspection technologies (3). Knowledge of applicable industry standards related to inspection/evaluation (3).

Task 4 (PP105-004): Verify that supplier/partner inspection personnel are available when inspection is required.

Quality Eng—Basic	Quality Eng—Senior	Q Manager	Supplier QE	CSoftQE	Program QE	Contract Review	MRO QE	Spec Process QE	Met Labs QE	Production Control	Regulatory Specialist
		S	P		S			S	S		

Procedural Knowledge: Ability to apply a structured assessment process to establish adequacy of inspection personnel availability (3).

Conceptual Knowledge: Ability to evaluate work flow, schedule, and availability of inspection personnel to determine adequate availability of qualified personnel (5).

Factual Knowledge: Understanding of inspection work processes (3). Understanding of work flow (3). Understanding of production/servicing schedules (3). Knowledge of inspection capacity planning (3). Knowledge of value stream structure (3). Knowledge of where work is performed (3).

Task 5 (PP105-005): Verify supplier/partner in-line probing equipment has adequate capability to perform the required inspections.

Quality Eng—Basic	Quality Eng—Senior	Q Manager	Supplier QE	CSoftQE	Program QE	Contract Review	MRO QE	Spec Process QE	Met Labs QE	Production Control	Regulatory Specialist
	S		P		S			S	S		

Procedural Knowledge: Ability to apply a structured assessment process to verify capability of in-line probing processes (4).

Conceptual Knowledge: Ability to analyze the capability of in-line probing processes (4). Ability to evaluate the capability of in-line probing against design requirements (5).

Factual Knowledge: Understanding of in-line probing technologies (4). Understanding of design requirements (3). Understanding of probing equipment capability (4).

Task 6 (PP105-006): Validate that suppliers incorporate the use of machine probing into their procedures and work instructions. Verify that utilization of machine probing is documented in the FAI reports.

Quality Eng—Basic	Quality Eng—Senior	Q Manager	Supplier QE	CSoftQE	Program QE	Contract Review	MRO QE	Spec Process QE	Met Labs QE	Production Control	Regulatory Specialist
			P		S			S	S		

Procedural Knowledge: Ability to apply a structured assessment process to determine if suppliers have integrated the application of in-line probing into organizational procedures (3). Ability to apply a structured assessment process to determine if suppliers have integrated the application of in-line probing into work instructions and quality records (3). Ability to apply a structured assessment process to determine if suppliers have integrated the application of in-line probing into FAI records (3).

Conceptual Knowledge: Ability to assess supplier procedures to determine if suppliers have integrated the application of in-line probing, as applicable (3). Ability to analyze suppliers work instructions and quality records to determine if suppliers have integrated the application of in-line probing into work instructions and quality records, as appropriate (4). Ability to analyze supplier FAI records to determine if suppliers have integrated the application of in-line probing, as applicable (4).

Factual Knowledge: Knowledge of in-line probing methods (3). Understanding of applicable production/servicing processes (3). Understanding of requirements flow into procedures (3). Understanding of work instruction processes (3). Understanding of FAI planning processes (3). Understanding of FAI documentation processes (3). Understanding of industry standards related to FAI (e.g., AS/EN/JISQ 9102) (3).

IV.A.3.a.Mandatory Inspections

Process: Establish Delegated Inspection Process

Process Purpose: To provide for organizational personnel to perform delegated governmental/customer inspection functions. Delegation may include FAA, company/organization, prime, and other entities. Ensure delegation process is well defined, means for delegation process oversight provided, and training for inspectors is available.

Knowledge and Skill Requirements: Understanding of regulatory requirements for inspection and design delegations. Understanding of customer requirements for inspection and design delegations. Ability to assess organizational procedures related to delegations functions to ensure compliance to external and internal requirements. Ability to evaluate the capabilities and qualifications of candidate personnel to perform delegated inspection functions. Ability to evaluate delegated functions to ensure compliance to applicable requirements. Ability to ensure delegated functions are controlled.

Task 1 (PO150-001)(PP150-001): Assess organizational procedures related to delegations functions to ensure compliance with external and internal requirements.

Quality Eng—Basic	Quality Eng—Senior	Q Manager	Supplier QE	CSoftQE	Program QE	Contract Review	MRO QE	Spec Process QE	Met Labs QE	Production Control	Regulatory Specialist
P	P	S	P		P	P					P

Procedural Knowledge: Ability to apply a structured process in review of documented procedures, to verify that the application of verification delegation functions is accurately defined and complies with external and internal requirements (3).

Conceptual Knowledge: Ability to review applicable contractual and regulatory requirements to identify requirements related to verification delegations (3). Ability to documented procedures, to verify that the application of verification delegation functions is accurately defined and complies with external and internal requirements (3).

Factual Knowledge: Understanding of contractual and regulatory requirements related to verification delegation activities (3). Understanding of inspection delegation processes (3).

Task 2 (PO150-002)(PP150-002): Evaluate the capabilities and qualifications of candidate personnel to perform delegated inspection functions (at individual and/or organizational level).

Quality Eng—Basic	Quality Eng—Senior	Q Manager	Supplier QE	CSoftQE	Program QE	Contract Review	MRO QE	Spec Process QE	Met Labs QE	Production Control	Regulatory Specialist
P	P		P		S			P			P

Procedural Knowledge: Ability to apply a structured process to verify candidate personnel performing delegated inspection functions possess requisite capabilities (3).

Conceptual Knowledge: Ability to analyze supplier/partner personnel work experience to verify they possess requisite capabilities to perform the delegated inspection activities (4). Ability to analyze personnel training and education records to verify they possess requisite capabilities to perform the delegated inspection activities (4). Ability to evaluate personnel work activities to verify they possess requisite capabilities to perform the delegated inspection activities (5).

Factual Knowledge: Understanding of contractual and regulatory requirements related to verification delegation activities (3). Understanding of inspection delegation processes (3). Understanding of personnel qualification requirements related to verification activities (3). Understanding of where work is performed (3).

Task 3 (PO150-003)(PP-150-03): Evaluate actual delegated functions to ensure compliance to applicable requirements.

Quality Eng—Basic	Quality Eng—Senior	Q Manager	Supplier QE	CSoftQE	Program QE	Contract Review	MRO QE	Spec Process QE	Met Labs QE	Production Control	Regulatory Specialist
P	P		P		S			P			P

Procedural Knowledge: Ability to apply a structured assessment process to verify deployed delegated verification functions are in compliance with applicable requirements (3).

Conceptual Knowledge: Ability to analyze deployed delegated verification functions, and verify compliance with applicable requirements (4).

Factual Knowledge: Understanding of contractual and regulatory requirements related to verification delegation activities (3). Understanding of inspection delegation processes (3). Understanding of personnel qualification requirements related to verification activities (3). Understanding of where work is performed (3).

Task 4 (PO150-004)(PP-150-004): Perform performance analysis of delegated activities to determine if additional qualification review and process inspection is required (e.g. remedial activities).

Quality Eng—Basic	Quality Eng—Senior	Q Manager	Supplier QE	CSoftQE	Program QE	Contract Review	MRO QE	Spec Process QE	Met Labs QE	Production Control	Regulatory Specialist
P	P		P		S			P			P

Procedural Knowledge: Ability to apply a structured approach to conduct performance analysis of delegated activities to determine if additional qualification review and process inspection is required (e.g. remedial activities) (4).

Conceptual Knowledge: Ability to analyze work product, quality records, and product/process performance of delegated activities to determine if additional qualification review and process inspection is required (e.g. remedial activities) (4).

Factual Knowledge: Understanding of contractual and regulatory requirements related to verification delegation activities (3). Understanding of inspection delegation processes (3). Understanding of personnel qualification requirements related to verification activities (3). Understanding of where work is performed (3).

Process: Perform Supplier Designee Management

Process Purpose: To ensure persons who have been delegated inspection authority are capable to execute inspection responsibilities and are in compliance with applicable requirements.

Knowledge and Skill Requirements: Understanding of regulations and processes for delegations of inspection functions. Ability to ensure persons who have been delegated inspection authority are capable to execute inspection responsibilities and are in compliance with applicable requirements.

Task (PM035-001): Ensure persons who have been delegated inspection authority are capable to execute inspection responsibilities and are in compliance with applicable requirements.

Quality Eng—Basic	Quality Eng—Senior	Q Manager	Supplier QE	CSoftQE	Program QE	Contract Review	MRO QE	Spec Process QE	Met Labs QE	Production Control	Regulatory Specialist
P	P	P	P	P	S		P	P		P	

Procedural Knowledge: Ability to apply a structured process to verify supplier/partner personnel performing delegated inspection functions possess requisite capabilities (3). Ability to apply a structured approach to ensure persons who have been delegated inspection authority are executing inspection responsibilities in compliance with applicable requirements (3).

Conceptual Knowledge: Ability to analyze supplier/partner personnel work experience to verify they possess requisite capabilities to perform the delegated inspection activities (4). Ability to analyze supplier/partner personnel training and education records to verify they possess requisite capabilities to perform the delegated inspection activities (4). Ability to analyze supplier/partner personnel work activities to verify they possess requisite capabilities to perform the delegated inspection activities (4). Ability to evaluate supplier/partner personnel work activities to verify they are executing delegated responsibilities in compliance with applicable requirements (5).

Factual Knowledge: Understanding of contractual and regulatory requirements related to verification delegation activities (3). Understanding of inspection delegation processes (3). Understanding of personnel qualification requirements related to verification activities (3). Understanding of where work is performed (3).

Process: Provide Operator/Inspector Training/Supervision Support

Process Purpose: To ensure personnel have accurate understanding of requirements. To ensure personnel have access to organizational decision-making processes. To provide instruction to operators and/or inspectors in the methods of product/service verification.

Knowledge and Skill Requirements: Understanding of skill development processes. Understanding of various industry personnel, process, and system certifications. Understanding of how skills relate to task. Ability to verify personnel have accurate understanding of requirements. Ability to verify that personnel have access to organizational decision-making processes. Ability to provide instruction to operators and/or inspectors in the methods of product/service verification. Ability to verify that effective training management systems have been deployed throughout the value stream.

Task 1 (PMO60-001): Provide instruction to operators and/or inspectors in the methods of product/service verification, throughout the value stream.

Quality Eng—Basic	Quality Eng—Senior	Q Manager	Supplier QE	CSoftQE	Program QE	Contract Review	MRO QE	Spec Process QE	Met Labs QE	Production Control	Regulatory Specialist
P	P	P	P	P	S		P	P	P	P	S

Procedural Knowledge: Ability to apply defined processes in providing operators and/or inspectors with interpretative data (3). Ability to apply defined processes for accessing organizational personnel in providing interpretative information (3).

Conceptual Knowledge: Ability to understand and evaluate unique supplier/partner data (5). Ability to determine appropriate support requirements for interpretation and route to appropriate organization members for evaluation (4). Ability to provide formal instruction to suppliers/partners on matters relating to product quality, in accordance with applicable requirements (4).

Factual Knowledge: Understanding of contractual, regulatory, design, and quality requirements as they relate to program/project requirements (3). Knowledge of technical support resources within the organization (3). Knowledge of cultural differences as they relate to technical communication (3). Understanding of program requirements (3). Understanding of data processing methods (3). Understanding of formal instruction techniques (3).

Task 2 (PMO60-002): Verify that effective training management systems have been deployed throughout the value stream.

Quality Eng—Basic	Quality Eng—Senior	Q Manager	Supplier QE	CSoftQE	Program QE	Contract Review	MRO QE	Spec Process QE	Met Labs QE	Production Control	Regulatory Specialist
S	P	P	P		S						

Procedural Knowledge: Ability to apply a structured approach to verify that effective training management systems have been deployed throughout the value stream (3).

Conceptual Knowledge: Ability to analyze training management systems throughout the value stream to ensure effectiveness and compliance to requirements (4).

Factual Knowledge: Understanding of regulatory requirements related to training management systems (3). Understanding of contractual requirements related to training management systems (3). Understanding of project/program requirements related to training management systems (3). Understanding of effective training management systems and their application to the various production/servicing processes (3). Understanding of applicable work processes (3). Understanding of the value stream structure and the locations of work execution (3).

Process: Ensure Self Release Program

Process Purpose: To provide a means of delegating inspection authority to suppliers/partners and create appropriate quality and shipping documentation. Ensure self-release programs at suppliers/partners are approved, personnel understand requirements, documentation is maintained, and product is conforming.

Knowledge and Skill Requirements: Understanding of regulatory requirements related to self-release. Understanding of organizational procedures related to self-release processes. Ability to create appropriate forms related to self-release. Ability to verify that supplier/partner processes for self-release meet applicable regulatory and organizational requirements. Ability to verify that self-release programs at suppliers/partners are approved, as required. Ability to verify that personnel performing or managing self-release programs understand requirements. Ability to verify that documentation related to self-release processes are maintained, and product is conforming.

Task 1 (PM075-001): Verify that supplier/partner processes for self-release meet applicable regulatory and organizational requirements and appropriate approvals are granted prior to implementation.

Quality Eng—Basic	Quality Eng—Senior	Q Manager	Supplier QE	CSoftQE	Program QE	Contract Review	MRO QE	Spec Process QE	Met Labs QE	Production Control	Regulatory Specialist
S	P	P	P		P	P					P

Procedural Knowledge: Ability to apply a structured approach to verify that supplier/partner processes for self-release meet applicable regulatory and organizational requirements and appropriate approvals are granted prior to implementation (3).

Conceptual Knowledge: Ability to analyze supplier/partner self-release processes to verify they meet applicable regulatory and organizational requirements and appropriate approvals are granted prior to implementation (4).

Factual Knowledge: Understanding of requirements management system (3)Understanding of regulatory requirements related to self-release processes (3). Understanding of contractual and project/program requirements related to self-release processes (3). Understanding of self-release processes (3). Understanding of applicable work processes subject to self-release controls (3). Understanding of value stream structure (3). Knowledge of product/service verification processes (3). Knowledge of control of nonconforming material processes (3).

Task 2 (PM075-002): Verify that personnel performing or managing self-release programs understand requirements.

Quality Eng—Basic	Quality Eng—Senior	Q Manager	Supplier QE	CSoftQE	Program QE	Contract Review	MRO QE	Spec Process QE	Met Labs QE	Production Control	Regulatory Specialist
P	P	S	P		P		P			P	P

Procedural Knowledge: Ability to apply a structured assessment to verify that personnel performing or managing self-release programs understand requirements (3).

Conceptual Knowledge: Ability interview and perform analysis of work product and quality records to verify that personnel performing or managing self-release programs understand requirements (4).

Factual Knowledge: Understanding of requirements management system (3). Understanding of regulatory requirements related to self-release processes (3). Understanding of contractual and project/program requirements related to self-release processes (3). Understanding of self-release processes (3). Understanding of applicable work processes subject to self-release controls (3). Understanding of value stream structure (3). Knowledge of product/service verification processes (3). Knowledge of control of nonconforming material processes (3).

Task 3 (PM075-003): Verify that documentation related to self-release processes are maintained, and product is conforming.

Quality Eng—Basic	Quality Eng—Senior	Q Manager	Supplier QE	CSoftQE	Program QE	Contract Review	MRO QE	Spec Process QE	Met Labs QE	Production Control	Regulatory Specialist
P	P	S	P		P						P

Procedural Knowledge: Ability to apply a structured assessment process to verify that documentation related to self-release processes are maintained, and product is conforming (3).

Conceptual Knowledge: Ability to review established self-release systems and perform analysis on quality records to verify that documentation related to self-release processes are maintained, and product is conforming (4).

Factual Knowledge: Understanding of requirements management system (3). Understanding of regulatory requirements related to self-release processes (3). Understanding of contractual and project/program requirements related to self-release processes (3). Understanding of self-release processes (3). Understanding of applicable work processes subject to self-release controls (3). Understanding of value stream structure (3). Knowledge of product/service verification processes (3). Knowledge of control of nonconforming material processes (3).

Process: Monitor Operator Certification Controls

Process Purpose: When required, to provide a means of delegating inspection authority to personnel other than quality inspectors. Ensure the integrity of the certification process and the capability of delegated inspections.

Knowledge and Skill Requirements: Understanding of regulatory requirements related to inspection processes and authorizations. Ability to validate the effectiveness of operator certification processes. Ability to assess the competency of persons delegated inspection authorization. Ability to verify that delegation records are maintained, complete, and accurate. Ability to review the process oversight processes associated with delegated inspection authorization to verify the process is in compliance and capable.

Task 1 (PM225-001): Review the operator certification oversight processes to verify the processes are in compliance and capable. Validate the effectiveness of operator certification processes.

Quality Eng—Basic	Quality Eng—Senior	Q Manager	Supplier QE	CSoftQE	Program QE	Contract Review	MRO QE	Spec Process QE	Met Labs QE	Production Control	Regulatory Specialist
P	P	P	P		S		P	S	S	P	

Procedural Knowledge: Ability to apply a defined and structured process to assess operator certification oversight processes to verify processes are in compliance and capable (3). Ability to apply a defined and structured process to assess the effectiveness of operator certification processes (3).

Conceptual Knowledge: Ability to analyze procedures, quality records, work product, and personnel activities to verify that operator certification oversight processes are in compliance with applicable requirements and capable (4). Ability to analyze procedures, quality records, work product, personnel activities, and product/service performance to determine the effectiveness of operator certification processes (4).

Factual Knowledge: Understanding of regulatory requirements related to inspection processes and authorizations (3). Understanding of contractual and project/program requirements related to inspection processes and authorizations (3). Understanding of requirements management tools (3). Understanding of effective operator certification processes (3). Understanding of production/service planning processes (3). Understanding of inspection/product verification activities related to the product/service (3). Understanding of audit practices (3). Understanding of value stream structure and where work is performed (3).

Task 2 (PM225-002): Assess the competency of persons with delegated inspection authorization (operator certification).

Quality Eng—Basic	Quality Eng—Senior	Q Manager	Supplier QE	CSoftQE	Program QE	Contract Review	MRO QE	Spec Process QE	Met Labs QE	Production Control	Regulatory Specialist
S	P	P	P		S		P	S	S	P	

Procedural Knowledge: Ability to apply a structured approach to verify the competency of persons with delegated inspection authorization (operator certification) to perform the assigned responsibilities (3).

Conceptual Knowledge: Ability to analyze procedures, quality records, personnel records, work product, personnel activities, and product/service performance to assess the competency of persons with delegated inspection authorization (operator certification) to perform the assigned responsibilities (4).

Factual Knowledge: Understanding of regulatory requirements related to inspection processes and authorizations (3). Understanding of contractual and project/program requirements related to inspection processes and authorizations (3). Understanding of requirements management tools (3). Understanding of effective operator certification processes (3). Understanding of production/service planning processes (3). Understanding of inspection/product verification activities related to the product/service (3). Understanding of skill requirements related to product verification (3). Understanding of audit practices (3). Understanding of personnel training records (3). Understanding of value stream structure and where work is being performed (3).

Task 3 (PM225-003): Verify that operator certification delegation records are maintained, complete, and accurate.

Quality Eng—Basic	Quality Eng—Senior	Q Manager	Supplier QE	CSoftQE	Program QE	Contract Review	MRO QE	Spec Process QE	Met Labs QE	Production Control	Regulatory Specialist
P	P	P	P		S		P	S	S	P	

Procedural Knowledge: Ability to apply a defined and structured assessment process to verify that operator certification delegation records are maintained, complete, and accurate (3).

Conceptual Knowledge: Ability to assess quality records to verify that operator certification delegation records are maintained, complete, and accurate (3).

Factual Knowledge: Understanding of regulatory requirements related to inspection processes and authorizations (3). Understanding of contractual and project/program requirements related to inspection processes and authorizations (3). Understanding of requirements management tools (3). Understanding of effective operator certification processes (3). Understanding of production/service planning processes (3). Understanding of inspection/product verification activities related to the product/service (3). Understanding of audit practices (3). Understanding of value stream structure and where work is being performed (3).

IV.B.1. Material Identification, Status, and Traceability

Process: Ensure Traceability Controls

Process Purpose: To ensure materials, components, and products are controlled in such a manner that traceability is assured to all specified requirements (Note: Traceability requirements may originate from various sources.). Ensure drawings, process documents, work instructions, and quality records accurately reflect applicable traceability requirements and that traceability is maintained to the level of product definition that requirements specify. Ensure product identification processes are appropriately defined.

Knowledge and Skill Requirements: Ability to verify that the organization has processes in place to ensure materials, components, and products are controlled in such a manner that traceability is assured to specified requirements. Ability to review and verify that drawings, process documents, work instructions, and quality records accurately reflect applicable traceability requirements and that traceability is maintained to the level of product definition that requirements specify. Ability to ensure traceability controls are effectively deployed throughout the organization and into the value stream.

Task 1 (PO075-001): Verify the organization has documented processes in place to ensure materials, components, and products are controlled in such a manner that traceability is assured to specified requirements.

Quality Eng—Basic	Quality Eng—Senior	Q Manager	Supplier QE	CSoftQE	Program QE	Contract Review	MRO QE	Spec Process QE	Met Labs QE	Production Control	Regulatory Specialist
P	P	S			P	S					S

Procedural Knowledge: Ability to apply a structured process to assess and verify the organization has documented processes in place to ensure materials, components, and products are controlled in such a manner that traceability is assured to specified requirements (3).

Conceptual Knowledge: Ability to assess and verify the organization has documented processes in place to ensure materials, components, and products are controlled in such a manner that traceability is assured to specified requirements (4).

Factual Knowledge: Knowledge of quality management system auditing methods (3). Knowledge of traceability requirements (4). Knowledge of applicable technology (3). Knowledge of applicable product/service (3). Knowledge of applicable materials (4). Knowledge of inspection methods (3). Understanding of traceability practices (4). Knowledge of requirements management tools (3).

Task 2 (PO075-002): Review and verify that drawings, process documents, work instructions, and quality records accurately reflect applicable traceability requirements and that traceability is maintained to the level of product definition that requirements specify. Verify that all applicable traceability requirements are reflected in the appropriate drawings, process documents, work instructions, and quality records, as applicable. (Note: This task requires verification of traceability in artifacts in both directions.)

Quality Eng—Basic	Quality Eng—Senior	Q Manager	Supplier QE	CSoftQE	Program QE	Contract Review	MRO QE	Spec Process QE	Met Labs QE	Production Control	Regulatory Specialist
P	P	S			S		S	S	S	S	

Procedural Knowledge: Ability to apply a structured process to assess and verify the organization has identified traceability requirements and have implemented the documented control of such in drawings, process documents, work instructions, and quality records (3).

Conceptual Knowledge: Ability to assess and verify the organization has documented traceability requirements into the appropriate level of documentation (4).

Factual Knowledge: Knowledge of quality management system auditing methods (3). Knowledge of traceability requirements (4). Understanding of traceability practices (4). Understanding requirements management tools (3). Understanding of project/program requirements related to traceability (3).

Task 3 (PO075-003): Ensure traceability controls are effectively deployed into the organization's work processes (Product review).

Quality Eng—Basic	Quality Eng—Senior	Q Manager	Supplier QE	CSoftQE	Program QE	Contract Review	MRO QE	Spec Process QE	Met Labs QE	Production Control	Regulatory Specialist
P	P	S			P		P	S	S	P	

Procedural Knowledge: Ability to apply a structured process to assess the compliance to traceability requirements throughout the organization (3).

Conceptual Knowledge: Ability to recognize where traceability requirements are to be implemented (4). Ability to understand the influences of work processes on traceability (5). Ability to assess compliance to traceability requirements throughout the organization's work processes (4).

Factual Knowledge: Knowledge of quality management system auditing methods (3). Knowledge of traceability requirements (4). Understanding of traceability practices (4). Understanding requirements management tools (3). Understanding of project/program requirements related to traceability (3).

Task 4 (PP075-001): Verify that suppliers/partners have documented processes in place to ensure materials, components, and products are controlled in such a manner that traceability is assured to specified requirements.

Quality Eng—Basic	Quality Eng—Senior	Q Manager	Supplier QE	CSoftQE	Program QE	Contract Review	MRO QE	Spec Process QE	Met Labs QE	Production Control	Regulatory Specialist
		S	P		S			S	S		

Procedural Knowledge: Ability to apply a structured process to assess and verify that suppliers/partners have documented processes in place to ensure materials, components, and products are controlled in such a manner that traceability is assured to specified requirements (3).

Conceptual Knowledge: Ability to assess and verify that suppliers/partners have documented processes in place to ensure materials, components, and products are controlled in such a manner that traceability is assured to specified requirements (4).

Factual Knowledge: Knowledge of quality management system auditing methods (3). Knowledge of value stream structure (3). Knowledge of applicable technology (3). Knowledge of applicable product/service (3). Knowledge of applicable materials (4). Knowledge of inspection methods (3). Understanding of traceability practices (4). Knowledge of requirements management tools (3).

Task 5 (PP075-002): Review and verify that drawings, process documents, work instructions, and quality records accurately reflect applicable traceability requirements and that traceability is maintained to the level of product definition that requirements specify. Verify that all applicable traceability requirements are reflected in the appropriate drawings, process documents, work instructions, and quality records, as applicable. (Note: This task requires verification of traceability in artifacts in both directions.)

Quality Eng—Basic	Quality Eng—Senior	Q Manager	Supplier QE	CSoftQE	Program QE	Contract Review	MRO QE	Spec Process QE	Met Labs QE	Production Control	Regulatory Specialist
		S	P		S			S	S		

Procedural Knowledge: Ability to apply a structured process to assess and verify that suppliers/partners have identified traceability requirements and have implemented the documented control of such in drawings, process documents, work instructions, and quality records (3).

Conceptual Knowledge: Ability to assess and verify that suppliers/partners have documented traceability requirements into the appropriate level of documentation (4).

Factual Knowledge: Knowledge of quality management system auditing methods (3). Knowledge of value stream structure (3). Understanding of traceability practices (4). Understanding requirements management tools (3). Understanding of project/program requirements related to traceability (3).

Task 6 (PP075-003): Ensure traceability controls are effectively deployed into the value stream (Product review).

Quality Eng—Basic	Quality Eng—Senior	Q Manager	Supplier QE	CSoftQE	Program QE	Contract Review	MRO QE	Spec Process QE	Met Labs QE	Production Control	Regulatory Specialist
		S	P		S			S	S		

Procedural Knowledge: Ability to apply a structured process to assess the compliance to traceability requirements throughout the value stream (3).

Conceptual Knowledge: Ability to recognize where traceability requirements are to be implemented (4). Ability to understand the influences of extended value stream traceability (5). Ability to assess compliance to traceability requirements throughout the value stream (4).

Factual Knowledge: Knowledge of quality management system auditing methods (3). Knowledge of value stream structure (3). Understanding of traceability practices (4). Understanding requirements management tools (3). Understanding of project/program requirements related to traceability (3).

Process: Monitor Parts Release Kitting Verification

Process Purpose: To ensure parts assembled into kits are effectively identified, documented, and controlled. Parts making up the kit must be identified and controlled with the appropriate revision status.

Knowledge and Skill Requirements: Understanding of material control processes. Understanding of foreign object debris prevention. Understanding of packaging and handling methods. Ability to verify that kitting is conforming to stated requirements and that the configuration of kitted parts is correct.

Task (PM230-001): Verify that kitting is conforming to stated requirements and that the configuration of kitted parts is correct.

Quality Eng—Basic	Quality Eng—Senior	Q Manager	Supplier QE	CSoftQE	Program QE	Contract Review	MRO QE	Spec Process QE	Met Labs QE	Production Control	Regulatory Specialist
S	P	S	P		S		P			P	

Procedural Knowledge: Ability to apply a defined and structured assessment process to verify that kitting is conforming to stated requirements and that the configuration of kitted parts is correct (3).

Conceptual Knowledge: Ability to analyze project/program requirements and production/servicing planning to identify kitting requirements (4). Ability to analyze kitting activities to verify that such practices are conforming to stated requirements and that the configuration of kitted parts is correct (4).

Factual Knowledge: Understanding of requirements management tools (3). Understanding of material control processes (3). Understanding of foreign object debris prevention (3). Understanding of packaging and handling methods (3). Understanding of production/service planning processes (3). Understanding of configuration control processes (3). Understanding of value stream structure and where work is being performed (3). Understanding of audit practices (3).

Process: Monitor Stamp Control

Process Purpose: To ensure stamps and signatures utilized to perform acceptance of work processes are effectively controlled and traceable to the individual accepting work. (Includes all media, including electronic.)

Knowledge and Skill Requirements: Understanding of stamp control methods. Understanding of regulatory and industry standard requirements related to stamp control. Ability to verify that stamps and signatures utilized to perform acceptance of work processes are effectively controlled and traceable to the individual accepting work. (Includes all types of media, including electronic.)

Task (PM235-001): Verify that stamps and signatures utilized to perform acceptance of work processes are effectively controlled and traceable to the individual accepting work. (Includes all media, including electronic.)

Quality Eng—Basic	Quality Eng—Senior	Q Manager	Supplier QE	CSoftQE	Program QE	Contract Review	MRO QE	Spec Process QE	Met Labs QE	Production Control	Regulatory Specialist
S	P	P	P		S		P	P	P	P	S

Procedural Knowledge: Ability to apply a defined and structured assessment process to verify that stamps and signatures utilized to perform acceptance of work processes are effectively controlled and traceable to the individual accepting work. (Includes all media, including electronic.) (3).

Conceptual Knowledge: Ability to analyze quality records, procedures, and work product to verify that stamps and signatures utilized to perform acceptance of work processes are effectively controlled and traceable to the individual accepting work. (Includes all media, including electronic.) (4).

Factual Knowledge: Understanding of stamp control methods (3). Understanding of regulatory and industry standard requirements related to stamp control (3). Understanding of production/service planning practices (3). Understanding of inspection planning practices (3). Understanding of inspection/product verification activities related to the product/service (3). Understanding of audit practices (3). Understanding of controls for electronic media (3). Understanding of the value stream structure and where work in being performed (3). Understanding of sampling methods related to compliance to organizational procedures (3).

IV.B.4. Material Review Board (MRB)

Process: Document Product Nonconformance

Process Purpose: To ensure product and process nonconformities are appropriately documented.

Knowledge and Skill Requirements: Understanding of regulatory requirements related to the documentation of product and/or process nonconformities. Understanding of industry standard and/or customer requirements related to the documentation of product and/or process nonconformities. Understanding of material review board processes. Understanding of contract requirements. Ability to document product and/or process nonconformities on appropriate material review/quality records. Ability to verify appropriate approvals are obtained. Ability to verify that the nonconformance document is provided to appropriate MRB personnel.

Task 1 (PF070-001): Document product and/or process nonconformities on appropriate material review/quality records.

Quality Eng—Basic	Quality Eng—Senior	Q Manager	Supplier QE	CSoftQE	Program QE	Contract Review	MRO QE	Spec Process QE	Met Labs QE	Production Control	Regulatory Specialist
P	P	P	P	P	S	S	P	P	P	P	S

Procedural Knowledge: Ability to apply a defined and structured process to document product and/or process nonconformities on appropriate material review/quality records (3).

Conceptual Knowledge: Ability to analyze engineering requirements and identify variations thereof in product and process (4). Ability to analyze nonconformities and accurately describe the encountered conditions on a material/quality review record (4).

Factual Knowledge: Understanding of regulatory requirements related to the documentation of product and/or process nonconformities (3). Understanding of industry standard and/or customer requirements related to the documentation of product and/or process nonconformities (3). Understanding of material review board processes (3). Understanding of contract requirements related to nonconformities (3). Knowledge of inspection planning (3). Knowledge of production/service planning (3). Knowledge of requirements management tools (3).

Task 2 (PF070-002): Verify appropriate approvals are obtained for nonconformance records.

Quality Eng—Basic	Quality Eng—Senior	Q Manager	Supplier QE	CSoftQE	Program QE	Contract Review	MRO QE	Spec Process QE	Met Labs QE	Production Control	Regulatory Specialist
P	P	P	P	P	P	S	P	P	P	P	S

Procedural Knowledge: Ability to apply a structured and defined process to verify appropriate approvals are obtained for nonconformance records (3).

Conceptual Knowledge: Ability to review nonconformity records and verify that appropriate approvals have been obtained (3).

Factual Knowledge: Understanding of regulatory requirements related to the documentation of product and/or process nonconformities (3). Understanding of industry standard and/or customer requirements related to the documentation of product and/or process nonconformities (3). Understanding of material review board processes (3). Understanding of contract requirements related to nonconformities (3). Knowledge of inspection planning (3). Knowledge of production/service planning (3). Understanding of requirements management tools (3).

Process: Manage Quality Review (MRB Activities)—Product

Process Purpose: To ensure appropriate segregation, review, disposition, corrective work, and approvals are made in the event of a product nonconformance. Ensure appropriate MRB authorizations are provided. Ensure authorized (e.g., MRB) personnel have approved the corrective work activities.

Knowledge and Skill Requirements: Understanding of regulatory requirements related to the documentation of product and/or process nonconformities. Understanding of industry standard and/or customer requirements related to the documentation of product and/or process nonconformities. Understanding of material review board processes. Understanding of contract requirements. Ability to validate the appropriate segregation, review, disposition, corrective work, and approvals are made in the event of a product/process nonconformance. Ability to verify that authorized (e.g., MRB) personnel have approved the corrective work activities.

Task 1 (PF075-001): Validate the appropriate segregation, review, disposition, corrective work, and approvals are made in the event of a product/process nonconformance.

Quality Eng—Basic	Quality Eng—Senior	Q Manager	Supplier QE	CSoftQE	Program QE	Contract Review	MRO QE	Spec Process QE	Met Labs QE	Production Control	Regulatory Specialist
P	P	P	P	P	P	S	P	P	S	P	S

Procedural Knowledge: Ability to apply a defined and structured process in validating that appropriate segregation, review, disposition, corrective work, and approvals have been made in product/process nonconformance events (3).

Conceptual Knowledge: Ability analyze material review process to validate that appropriate segregation, review, disposition, corrective work, and approvals have been made in product/process nonconformance events (4).

Factual Knowledge: Understanding of regulatory requirements related to the documentation of product and/or process nonconformities (3). Understanding of industry standard and/or customer requirements related to the documentation of product and/or process nonconformities (3). Understanding of material review board processes (3). Understanding of contract requirements related to nonconformities (3). Knowledge of inspection planning (3). Knowledge of production/service planning (3). Understanding of material control processes (3) Knowledge of requirement management tools (3).

Task 2 (PF075-002): Verify that authorized (e.g., MRB) personnel have approved the corrective work activities. (Includes the verification that MRB authorization has been provided by relevant customer/government.)

Quality Eng—Basic	Quality Eng—Senior	Q Manager	Supplier QE	CSoftQE	Program QE	Contract Review	MRO QE	Spec Process QE	Met Labs QE	Production Control	Regulatory Specialist
P	P	P	P	P	P	S	P	P	P	P	S

Procedural Knowledge: Ability to apply a defined and structured process to verify that personnel are authorized (e.g., MRB) personnel have approved the corrective work activities (3).

Conceptual Knowledge: Ability to analyze corrective work activities to verify that personnel are authorized (e.g., MRB) personnel and have approved disposition and applied work (4).

Factual Knowledge: Understanding of regulatory requirements related to the documentation of product and/or process nonconformities (3). Understanding of industry standard and/or customer requirements related to the documentation of product and/or process nonconformities (3). Understanding of material review board processes (3). Understanding of contract requirements related to nonconformities (3). Knowledge of inspection planning (3). Knowledge of production/service planning (3). Understanding of material control processes (3) Knowledge of requirements management tool (3).

Process: Define Material Review Board (MRB) Requirements

Process Purpose: Define requirements for material review processes for creation of nonconforming product/services. Ensure the process is documented in a controlled procedure and identifies the responsibilities of the entities involved in the design, procurement, and production of the product/service.

Knowledge and Skill Requirements: Ability to identify requirements for material review processes of nonconforming product/services at suppliers/partners. Ability to ensure MRB processes at suppliers/partners are documented in a controlled procedure and the responsibilities of the entities involved in the design, procurement, and production of the product/service are identified.

Task 1 (SS015-001): Identify requirements for material review processes of nonconforming product/services are created at suppliers/partners.

Quality Eng—Basic	Quality Eng—Senior	Q Manager	Supplier QE	CSoftQE	Program QE	Contract Review	MRO QE	Spec Process QE	Met Labs QE	Production Control	Regulatory Specialist
S	P	S	P		P	P					P

Procedural Knowledge: Ability to apply a structured process to review requirements and identify specific requirements related to the control of nonconforming material created in the value stream (3).

Conceptual Knowledge: Ability to review requirements and identify specific requirements related to the control of nonconforming material created in the value stream (3). Ability to elicit undefined requirements related to the control of nonconforming material created in the value stream (4).

Factual Knowledge: Knowledge of MRB processes (3). Knowledge of applicable regulatory requirements related to the control of nonconforming material (3). Understanding of requirements flow-down documents (3). Knowledge of requirements management tool (3).

Task 2 (SS015-002): Ensure MRB processes at suppliers/partners are documented in a controlled procedure and the responsibilities of the entities involved in the design, procurement, and production of the product/service are identified. Ensure contractual documents at suppliers/partners provide requirements to all points where value is being created (including applicable sub-tier suppliers).

Quality Eng—Basic	Quality Eng—Senior	Q Manager	Supplier QE	CSoftQE	Program QE	Contract Review	MRO QE	Spec Process QE	Met Labs QE	Production Control	Regulatory Specialist
S	P	S	P		P			S	S		

Procedural Knowledge: Ability to apply a structured process to assess the compliance of controlled procedures to ensure the positive control of nonconforming materials within the value stream (3). Ability to apply a structured process to assess contractual documents at suppliers/partners to ensure requirement flow to all points where value is being created (3).

Conceptual Knowledge: Ability to review and assess the compliance and adequacy of controlled procedures to ensure the positive control of nonconforming materials within the value stream (4).

Factual Knowledge: Knowledge of MRB processes (3). Knowledge of applicable regulatory requirements related to the control of nonconforming material (3). Understanding of requirements flow-down documents (3). Knowledge of requirements management tool (3).

IV.B.4.a. Escape Containment

Process: Perform Escapes (nonconformance to field) Management and Field Notification/Recall Process

Process Purpose: To ensure nonconforming materials that have left the quality system are positively controlled in notification/recall processes and/or field management dispositions.

Knowledge and Skill Requirements: Understanding of regulatory requirements related to field escapes. Understanding of organizational MRB processes. Ability to verify that suppliers/partners have established appropriate field recall/notification processes. Ability to support investigation processes to determine how well nonconforming material populations are positively controlled. Ability to apply audit processes to determine capability of MRB processes in product recall. Understanding of industry and government notification systems related to nonconforming or suspected nonconforming products/materials (e.g., GIDEP).

Task 1 (PF105-001): Verify appropriate field recall/notification processes.

Quality Eng—Basic	Quality Eng—Senior	Q Manager	Supplier QE	CSoftQE	Program QE	Contract Review	MRO QE	Spec Process QE	Met Labs QE	Production Control	Regulatory Specialist
P	P	P	P	P	P	S	P	P	P	P	S

Procedural Knowledge: Ability to apply a structured approach to verify appropriate field recall/notification processes have been established (3).

Conceptual Knowledge: Ability to analyze processes, procedures, and past work product to verify appropriate field recall/notification processes have been established (4).

Factual Knowledge: Understanding of regulatory requirements related to the documentation of product and/or process nonconformities (3). Understanding of industry standards and/or customer requirements related to the documentation of product and/or process nonconformities (3). Understanding of contract requirements related to product recall processes (3). Understanding of recall process urgency (3). Understanding of customer base (3). Understanding of basic auditing practices (3).

Task 2 (PF105-002): Support investigation processes to determine how well nonconforming material populations are positively controlled.

Quality Eng—Basic	Quality Eng—Senior	Q Manager	Supplier QE	CSoftQE	Program QE	Contract Review	MRO QE	Spec Process QE	Met Labs QE	Production Control	Regulatory Specialist
P	P	P	P	P	P	S	P	P	P	P	S

Procedural Knowledge: Ability to adhere to structured process in support of investigation processes to determine how well nonconforming material populations are positively controlled (3).

Conceptual Knowledge: Ability to evaluate investigation needs and provide support of investigation processes to determine how well nonconforming material populations are positively controlled (5).

Factual Knowledge: Understanding of regulatory requirements related to the documentation of product and/or process nonconformities (3). Understanding of industry standards and/or customer requirements related to the documentation of product and/or process nonconformities (3). Understanding of contract requirements related to product recall processes (3). Understanding of recall process urgency (3). Understanding of customer base (3). Understanding of organizational authority lines (3). Understanding of value stream and where work is performed (3). Understanding of requirement management tools (3). Understanding of basic audit practices (3).

Process: Manage Reportable Nonconformance, Defect, and Malfunction (e.g., FAA 14 CFR Part 21.3) Reporting Process

Process Purpose: To ensure discrepancies defined in regulatory requirements (e.g., 14 CFR Part 21.3) are appropriately reported within specified time requirements.

Knowledge and Skill Requirements: Understanding of regulatory requirements and advisory materials (e.g., FAA Advisory Circular 00-58) related to the notification of reportable defects and failures. Ability to verify that discrepancies defined in regulations (e.g., 14 CFR Part 21.3) are appropriately reported within the time requirements. Ability to verify that suppliers, partners, and sub-tier suppliers have processes in place to report potential issues to the organization.

Task 1 (PF115-001): Verify that the organization, suppliers, partners, and sub-tier suppliers have processes in place to report potential product and process issues to the organization, customers, and regulatory authorities, as appropriate.

Quality Eng—Basic	Quality Eng—Senior	Q Manager	Supplier QE	CSoftQE	Program QE	Contract Review	MRO QE	Spec Process QE	Met Labs QE	Production Control	Regulatory Specialist
P	P	P	P	P	P	S	S	S	S	S	P

Procedural Knowledge: Ability to apply a defined and structured assessment process to verify the organization, suppliers, partners, and sub-tier suppliers have processes in place to report potential product and process issues to the organization, customers, and to regulatory authorities, as appropriate (3).

Conceptual Knowledge: Ability to verify the organization, suppliers, partners, and sub-tier suppliers have processes in place to report potential product and process issues to the organization, customers, and regulatory authorities, as appropriate (4).

Factual Knowledge: Understanding of regulatory requirements and advisory materials related to the notification of reportable defects and failures (3). Understanding of organizational MRB processes (3). Understanding of reporting communication methods (3). Understanding of organizational legal responsibilities regarding reporting activities (4). Understanding of audit practices (3). Understanding of the value stream structure and where work is performed (3).

Task 2 (PF115-002): Verify that discrepancies defined in regulatory requirements (e.g., CFR 14 Part 21.3) are appropriately reported to the organization, customers, and applicable regulatory authorities within defined time requirements.

Quality Eng—Basic	Quality Eng—Senior	Q Manager	Supplier QE	CSoftQE	Program QE	Contract Review	MRO QE	Spec Process QE	Met Labs QE	Production Control	Regulatory Specialist
P	P	P	P	P	P	S	S	S	S	S	P

Procedural Knowledge: Ability to apply a structured approach to ensure that discrepancies defined in regulatory requirements (e.g., CFR 14 Part 21.3) are appropriately reported to the organization, customers, and applicable regulatory authorities within defined time requirements (3).

Conceptual Knowledge: Ability to analyze organizational and value stream processes to ensure that discrepancies defined in regulatory requirements (e.g., CFR 14 Part 21.3) are appropriately reported to the organization and applicable regulatory authorities within defined time requirements (4). Ability to evaluate individual issues to determine if reporting is required as defined in regulatory requirements (e.g., CFR 14 Part 21.3) and are appropriately reported to the organization and applicable regulatory authorities within defined time requirements (5).

Factual Knowledge: Understanding of regulatory requirements and advisory materials related to the notification of reportable defects and failures (4). Understanding of organizational MRB processes (3). Understanding of reporting communication methods (3). Understanding of organizational legal responsibilities regarding reporting activities (4). Understanding of the value stream structure and where work is performed (3).

IV.B.4.e. Positive Control of Scrap

Process: Monitor Control of Scrap

Process Purpose: To ensure scrap materials are positively controlled.

Knowledge and Skill Requirements: Understanding of regulatory and industry standard requirements related to the control of scrap materials and product. Understanding of organizational MRB processes. Ability to substantiate that scrap materials are positively controlled. Ability to verify that documentation related to the control and disposition of scrap materials is maintained as quality records. Understanding of material disposal requirements. Understanding of audit practices.

Task 1 (PF110-001): Substantiate that scrap materials are positively controlled.

Quality Eng—Basic	Quality Eng—Senior	Q Manager	Supplier QE	CSoftQE	Program QE	Contract Review	MRO QE	Spec Process QE	Met Labs QE	Production Control	Regulatory Specialist
P	P	S	P		P	S	P	S	S	P	P

Procedural Knowledge: Ability to apply a structured process to ensure processes positively control the final disposition of scrap materials (3). Ability to apply a structured process to ensure scrap materials are positively controlled through the extended value stream (3).

Conceptual Knowledge: Ability to analyze processes positively control the final disposition of scrap materials (4). Ability to analyze processes throughout the value stream to ensure scrap materials are positively controlled (4). Ability to assess work product throughout the value stream to ensure scrap materials are positively controlled (3). Ability to assess and ensure scrap materials are appropriately disposed of i/a/w requirements (4).

Factual Knowledge: Understanding of regulatory and industry standards requirements related to the control of scrap materials (3). Understanding of organizational MRB processes (3). Understanding of suspected unapproved parts (3). Understanding of audit practices (3). Understanding of the value stream structure and where work is performed (3). Understanding of audit practices (3). Understanding of disposal requirements for applicable materials (3). Understanding of requirements management tools (3).

Task 2 (PF110-002): Verify that documentation related to the control and disposition of scrap materials is maintained as quality records.

Quality Eng—Basic	Quality Eng—Senior	Q Manager	Supplier QE	CSoftQE	Program QE	Contract Review	MRO QE	Spec Process QE	Met Labs QE	Production Control	Regulatory Specialist
P	P	S	P		S	S	P	S	S	P	P

Procedural Knowledge: Ability to apply a defined and structured assessment process to verify that documentation related to the control and disposition of scrap materials is maintained as quality records (3).

Conceptual Knowledge: Ability to assess quality records throughout the value stream to verify that documentation related to the control and disposition of scrap materials is maintained as quality records (3).

Factual Knowledge: Understanding of regulatory and industry standards requirements related to the control of scrap materials (3). Understanding of organizational MRB processes (3). Understanding of suspected unapproved parts (3). Understanding of audit practices (3). Understanding of the value stream structure and where work is performed (3).

IV.C. Acceptance Sampling (Non-SPC)

Process: Ensure Sampling Inspection Processes Are Appropriate and Controlled

Process Purpose: To ensure the sampling plans established and controlled in compliance with appropriate sampling standards, and that such plans are appropriately utilized. Ensure personnel utilizing sampling plans understand sampling methods.

Knowledge and Skill Requirements: Basic understanding of statistics. Understanding of inspection techniques. Understanding of key characteristics. Ability to verify that established sampling plans are controlled and in compliance with appropriate sampling standards. Ability to verify that approved sampling plans are appropriately utilized. Ability to verify that personnel utilizing sampling plans understand sampling methods. Ability to verify that records related to sampling activities are maintained. Ability to verify that supplier/partner personnel understand the use of sampling plans. Ability to verify that when sampling plans are utilized, the requirements are flowed down to all levels of the supplier base. Ability to verify, when required, that sampling plans have appropriate approvals prior to implementation.

Task 1 (PM070-001): Verify that established sampling plans are controlled and in compliance with appropriate sampling standards.

Quality Eng—Basic	Quality Eng—Senior	Q Manager	Supplier QE	CSoftQE	Program QE	Contract Review	MRO QE	Spec Process QE	Met Labs QE	Production Control	Regulatory Specialist
P	P	S	P		S	S	P			P	S

Procedural Knowledge: Ability to apply a defined and structured review and assessment process to verify that established sampling plans are controlled and in compliance with appropriate sampling standards (3).

Conceptual Knowledge: Ability to analyze sampling plans to verify that they are in compliance with appropriate sampling standards (4). Ability to assess established control processes to verify that sampling plans are controlled (3).

Factual Knowledge: Understanding of sampling plans (3). Understanding of regulatory requirements related to sampling plans (3). Understanding of contractual and project/program requirements related to sampling plans (3). Understanding of industry standards related to sampling plans (3).

Task 2 (PM070-002): Verify that approved sampling plans are appropriately utilized, throughout the value stream. Verify that records related to sampling activities are appropriately controlled and maintained, throughout the value stream.

Quality Eng—Basic	Quality Eng—Senior	Q Manager	Supplier QE	CSoftQE	Program QE	Contract Review	MRO QE	Spec Process QE	Met Labs QE	Production Control	Regulatory Specialist
P	P	S	P		S	S	P			P	S

Procedural Knowledge: Ability to apply defined and structured assessment processes to verify that approved sampling plans are appropriately utilized, through the value stream (3). Verify that records related to sampling activities are appropriately controlled and maintained (3).

Conceptual Knowledge: Ability to audit where sampling plans are utilized to verify that they are utilized in accordance to defined requirements (3). Ability to audit sampling plans to ensure they are appropriately approved. (3). Ability to audit where sampling plans are utilized and that records related to sampling activities are appropriately controlled and maintained (3).

Factual Knowledge: Understanding of requirements management system (3) Understanding of sampling plans (3). Understanding of regulatory requirements related to sampling plans (3). Understanding of contractual and project/program requirements related to sampling plans (3). Understanding of industry standards related to sampling plans (3).

Task 3 (PM070-003): Verify that personnel utilizing sampling plans understand approved sampling methods, throughout the value stream.

Quality Eng—Basic	Quality Eng—Senior	Q Manager	Supplier QE	CSoftQE	Program QE	Contract Review	MRO QE	Spec Process QE	Met Labs QE	Production Control	Regulatory Specialist
S	P	S	P		S		P			P	

Procedural Knowledge: Ability to apply a structured approach to verify that personnel utilizing sampling plans understand approved sampling methods (3).

Conceptual Knowledge: Ability to assess personnel utilizing sampling plans to verify that they correctly understand the appropriate applications of approved sampling methods (4). Ability to review personnel training records to determine adequacy of qualifications for sampling responsibilities (3).

Factual Knowledge: Understanding of requirements management system (3). Understanding of sampling plans (3). Understanding of regulatory requirements related to sampling plans (3). Understanding of contractual and project/program requirements related to sampling plans (3). Understanding of industry standards related to sampling plans (3). Understanding of the value stream structure and the locations of work execution (3). Understanding of requisite skills related to sampling processes (3).

IV.D. Measurement and Test

Process: Perform Source Inspection

Process Purpose: To assure product conformity at source of production/servicing. Plan, schedule, and perform product inspection activities and document results in appropriate quality records. Ensure personnel performing source inspection have the appropriate skills and experience to understand and access product/service conformity.

Knowledge and Skill Requirements: Understanding of production and inspection planning processes. Understanding of MRP processes. Ability to evaluate that personnel performing source inspection have the appropriate skills and experience to understand and assess product/service conformity. Ability to plan, schedule, and perform product inspection activities and document results in appropriate quality records. Through inspection and test methods, the ability to verify product conformity at source of production/servicing. Ability to document and verify that inspection results are documented and appropriately maintained.

Task 1 (PF045-001): Evaluate personnel performing source inspection have the appropriate skills and experience to understand and assess product/service conformity.

Quality Eng—Basic	Quality Eng—Senior	Q Manager	Supplier QE	CSoftQE	Program QE	Contract Review	MRO QE	Spec Process QE	Met Labs QE	Production Control	Regulatory Specialist
S	P	S	P		S		S	P	P	S	S

Procedural Knowledge: Ability to apply a structured assessment process to determine if personnel performing source inspection have the appropriate skills and experience to understand and assess product/service conformity (3).

Conceptual Knowledge: Ability to analyze inspection processes and determine skill requirements for inspection personnel (4). Ability to analyze inspection personnel training records to determine if assigned personnel have adequate training (4). Ability to analyze quality records of previous work product to determine the effectiveness of inspection personnel (4).

Factual Knowledge: Understanding of skill requirements related to inspection personnel (3). Understanding of physical requirements for personnel performing inspection related activities (3). Understanding of personnel training records processes (3). Knowledge of training materials and courses for inspection personnel (3). Understanding of where work processes are being performed (3). Understanding of the value stream structure (3). Understanding of quality records requirements for source inspection (3). Knowledge of requirements management tools (3).

Task 2 (PF045-002): Plan and schedule product inspection activities at source.

Quality Eng—Basic	Quality Eng—Senior	Q Manager	Supplier QE	CSoftQE	Program QE	Contract Review	MRO QE	Spec Process QE	Met Labs QE	Production Control	Regulatory Specialist
S	S	S	P		S		S	P	P	S	

Procedural Knowledge: Ability to apply a structured process to plan and schedule product inspection activities in the value stream (3).

Conceptual Knowledge: Ability to analyze production schedules to determine when and what quality source inspection requirements apply (4). Ability to review inspection requirements and plan and schedule source inspection activities (3).

Factual Knowledge: Understanding of production and inspection planning processes (3). Understanding of MRP processes (3). Understanding of applicable inspection processes (3). Understanding of where work processes are being performed (3). Understanding of the value stream structure (3).

Task 3 (PF045-003): Through inspection and test methods, ensure product conformity at source of production/servicing; and document results in appropriate quality records.

Quality Eng—Basic	Quality Eng—Senior	Q Manager	Supplier QE	CSoftQE	Program QE	Contract Review	MRO QE	Spec Process QE	Met Labs QE	Production Control	Regulatory Specialist
S	S	S	P		S		S	P	P	S	

Procedural Knowledge: Ability to apply a structured approach to apply inspection and test methods to verify product conformity at source of production/servicing; and ensure results are documented in appropriate quality records (3).

Conceptual Knowledge: Ability to understand and apply inspection methods to ensure products/services conform to requirements (4). Ability to ensure inspection results are documented in appropriate quality records (3).

Factual Knowledge: Understanding of inspection planning processes (3). Understanding of regulatory requirements related to product verification (2). Understanding of contractual and project/program requirements related to product verification (2). Understanding of inspection methods (3). Understanding of audit practices (3). Understanding of the value stream structure and where work is performed (3). Understanding of requirements management tools (3).

Process: Perform Product Conformity Audits

Process Purpose: To ensure the established production/servicing processes continuously produce conforming product. Plan, schedule, perform, and document product conformity audits. Utilize risk based methods as appropriate.

Knowledge and Skill Requirements: Understanding of and ability to apply various inspection and test methods. Ability to plan, schedule, perform, and document product conformity audits. Ability to apply risk based methods as appropriate.

Task 1 (PF055-001): Analyze production/servicing performance, determine audit needs and plan and schedule product conformity audits.

Quality Eng—Basic	Quality Eng—Senior	Q Manager	Supplier QE	CSoftQE	Program QE	Contract Review	MRO QE	Spec Process QE	Met Labs QE	Production Control	Regulatory Specialist
P	P	P	P		P		P	P	P	P	S

Procedural Knowledge: Ability to apply a structured process to plan and schedule product conformity audits (3).

Conceptual Knowledge: Ability to analyze the production/servicing performance to determine the need to perform oversight evaluations (4). Ability to assess governmental, contractual, and/or project/program requirements to determine product conformity verification requirements (4).

Factual Knowledge: Understanding of applicable inspection methods (3). Understanding of regulatory, contractual and/or project/program requirements related to product conformity assessments at suppliers (3). Understanding of where work processes are being performed (3). Understanding of the value stream structure (3). Understanding of requirements management tools (3).

Task 2 (PF055-002): Apply risk based methods to product conformity inspections, as appropriate.

Quality Eng—Basic	Quality Eng—Senior	Q Manager	Supplier QE	CSoftQE	Program QE	Contract Review	MRO QE	Spec Process QE	Met Labs QE	Production Control	Regulatory Specialist
P	P	P	P		P		P	P	P	P	S

Procedural Knowledge: Ability to apply a structured approach to analyzing production/servicing performance and product characteristics in application of risk based oversight methods. (3)

Conceptual Knowledge: Ability to analyze design and production requirements to determine the appropriateness of risk based oversight methods. (4) Apply to analyze production/servicing performance to determine the appropriateness of risk based oversight methods. (4) Ability to analyze product service information to identify additional potential risk factors in the value stream (4).

Factual Knowledge: Understanding of product design requirements related to special processes (3). Understanding of quality performance measurements (3). Understanding of various NDT methods (3). Understanding of special process controls (3). Understanding of regulatory, contractual and/or project/program requirements related to evaluations of special process suppliers (3). Understanding of risk measurement processes (3). Understanding of where work processes are being performed (3). Understanding of the value stream structure (3). Understanding of requirements management tools (3).

Task 3 (PF055-003): Perform periodic product conformity evaluations.

Quality Eng—Basic	Quality Eng—Senior	Q Manager	Supplier QE	CSoftQE	Program QE	Contract Review	MRO QE	Spec Process QE	Met Labs QE	Production Control	Regulatory Specialist
P	P	S	P		P		P	P	P	P	S

Procedural Knowledge: Ability to apply a structured process in the evaluation of product to verify conformity to requirements (3).

Conceptual Knowledge: Ability to inspect product, work instructions, personnel qualifications, and quality records to determine conformance to design requirements and compliance to system requirements (4).

Factual Knowledge: Understanding of applicable inspection methods (3). Understanding of applicable special process controls (3). Understanding of regulatory, contractual and/or project/program requirements related to product conformity (3). Understanding of product conformity evaluation practices (3). Understanding of where work processes are being performed (3). Understanding of the value stream structure (3). Understanding of personnel training records (3). Understanding of requirements management tools (3).

Process: Monitor Inspection Control Processes

Process Purpose: To ensure inspection processes are appropriately planned and performed throughout production/servicing processes by qualified persons. Evaluate effectiveness of inspection/testing processes, ensure authorized personnel perform inspection/test functions, and ensure appropriate documentation.

Knowledge and Skill Requirements: Understanding of inspection methods. Understanding of regulatory requirements related to inspection processes and authorizations. Ability to verify that inspection processes are appropriately planned and performed throughout production/servicing processes by qualified persons. Ability to evaluate effectiveness of inspection/testing processes, ensure authorized personnel perform inspection/test functions, and ensure appropriate documentation is maintained. Ability to assess certification requirements for work processes. Ability to assess individual records to determine if persons performing work processes have requisite certifications.

Task 1 (PM220-001): Verify that inspection processes are appropriately planned and performed throughout production/servicing processes by qualified persons.

Quality Eng—Basic	Quality Eng—Senior	Q Manager	Supplier QE	CSoftQE	Program QE	Contract Review	MRO QE	Spec Process QE	Met Labs QE	Production Control	Regulatory Specialist
P	P	S	P	P	P		P	P	S	P	S

Procedural Knowledge: Ability to apply a defined and structured assessment process to verify that inspection processes are appropriately planned and performed throughout production/servicing processes by qualified persons (3).

Conceptual Knowledge: Ability to analyze quality records and work product to verify that inspection processes are appropriately planned and performed throughout production/servicing processes by qualified persons (4).

Factual Knowledge: Understanding of regulatory requirements related to personnel qualifications (3). Understanding of contractual and project/program requirements related to personnel qualifications (3). Understanding of requirements management tools (3). Understanding of technical work processes related to the project/program (3). Understanding of industry standards related to applied technologies of the project/program (3). Understanding of personnel skill qualification and certification practices (3). Understanding of personnel training records (3). Understanding of value stream structure and where work is being performed (3).

Task 2 (PM220-002): Evaluate effectiveness of inspection/testing processes and ensure appropriate documentation is maintained.

Quality Eng—Basic	Quality Eng—Senior	Q Manager	Supplier QE	CSoftQE	Program QE	Contract Review	MRO QE	Spec Process QE	Met Labs QE	Production Control	Regulatory Specialist
P	P	P	P	P	P		P	P	S	P	S

Procedural Knowledge: Ability to apply a structured approach to determine the effectiveness of inspection/testing processes and ensure appropriate documentation is maintained (3).

Conceptual Knowledge: Ability to analyze quality records, work product, supplier/partner performance, and product/service performance to determine the effectiveness of inspection/testing processes and ensure appropriate documentation is maintained (4).

Factual Knowledge: Understanding of technical work processes related to the project/program (3). Understanding of industry standards related to applied technologies of the project/program (3). Understanding of value stream structure and where work is being performed (3). Understanding of performance measures (3). Understanding of audit practices (3).

IV.D.2. Destructive and Nondestructive Testing, Inspection, and Evaluation

Process: Ensure Non-Destructive Test Sequences are Appropriately Applied—within the Organization

Process Purpose: Review work instructions, process planning, and quality records to ensure nondestructive testing is being performed at the appropriate points in production/servicing. The NDT process shall be documented in appropriate records and define the standards of reference and the limits of acceptance. Ensure the documentation of NDT results are accurately defined in applicable work instructions and quality records.

Knowledge and Skill Requirements: Ability to evaluate work instructions, process planning, and quality records to ensure nondestructive testing is being performed at the appropriate points in production/servicing. Ability to evaluate NDT process requirements to determine if appropriate records are established and standards of reference and the limits of acceptance are defined. Ability to evaluate certification requirements for NDT inspectors and validate that persons assigned to associated NDT responsibilities are appropriately certified. Ability to assess inspection records to validate that the documentation of NDT results are accurately defined in applicable work instructions and quality records.

Task 1 (PO045-001)(PP045-001): Evaluate work instructions, process planning, and quality records to ensure nondestructive testing is being performed at the appropriate points in production/servicing.

Quality Eng—Basic	Quality Eng—Senior	Q Manager	Supplier QE	CSoftQE	Program QE	Contract Review	MRO QE	Spec Process QE	Met Labs QE	Production Control	Regulatory Specialist
S	P		P		S		S	P	S	S	

Procedural Knowledge: Ability to apply a structured process to identify relevant documents for review (3). Ability to apply a structured process in review of work instructions, process planning, and quality records to determine the adequacy of such instructions to describe applicable nondestructive testing (3). Ability to apply a structured process to determine that applicable nondestructive evaluation methods are planned at the appropriate points in production/servicing in applicable work instructions, process planning, and quality records (3).

Conceptual Knowledge: Ability to review production/servicing processes and identify relevant documents for review (4). Ability to review work instructions, process planning, and quality records to determine the adequacy of such instructions to describe applicable nondestructive testing (4). Ability to review and determine that applicable nondestructive evaluation methods are planned at the appropriate points in production/servicing in applicable work instructions, process planning, and quality records (4).

Factual Knowledge: Knowledge of applicable nondestructive evaluation methods (4). Knowledge of process planning methods (3). Knowledge of inspection planning processes (3). Knowledge of industry standards related to applicable nondestructive evaluation methods (4). Knowledge of applicable technologies (3). Knowledge of where work is performed (3).

Task 2 (PO045-002)(PP045-002): Evaluate NDT process requirements to determine if appropriate records are established and standards of reference and the limits of acceptance are defined.

Quality Eng—Basic	Quality Eng—Senior	Q Manager	Supplier QE	CSoftQE	Program QE	Contract Review	MRO QE	Spec Process QE	Met Labs QE	Production Control	Regulatory Specialist
P	P		P		S		S	P	S	S	

Procedural Knowledge: Ability to apply a structured process in review of work instructions, process planning, and quality records to determine the adequacy of such instructions to describe applicable nondestructive testing (3). Ability to apply a structured process to determine that applicable nondestructive evaluation methods are described by standards of reference and limits of acceptance (3).

Conceptual Knowledge: Ability to analyze applicable nondestructive evaluation methods and determine that such are adequately described by standards of reference and limits of acceptance (4).

Factual Knowledge: Knowledge of applicable nondestructive evaluation methods (4). Knowledge of process planning methods (3). Knowledge of inspection planning processes (3). Knowledge of industry standards related to applicable nondestructive evaluation methods (4). Knowledge of applicable technologies (3). Knowledge of where work is performed (3).

Task 3 (PO045-003)(PP045-003): Evaluate certification requirements for NDT inspectors and validate that persons assigned to associated NDT responsibilities are appropriately certified.

Quality Eng—Basic	Quality Eng—Senior	Q Manager	Supplier QE	CSoftQE	Program QE	Contract Review	MRO QE	Spec Process QE	Met Labs QE	Production Control	Regulatory Specialist
S	P		P		S		S	P	S	S	

Procedural Knowledge: Ability to apply a structured process to review applicable nondestructive evaluation methods and identify personnel certification requirements (3). Ability to apply a structured process to review training, education, and certification records of personnel certified in nondestructive evaluation methods to determine compliance to applicable requirements (3).

Conceptual Knowledge: Ability to review applicable nondestructive evaluation methods and identify personnel certification requirements (4). Ability to analyze training, education, and certification records of personnel certified in nondestructive evaluation methods to determine compliance to applicable requirements (4).

Factual Knowledge: Knowledge of applicable nondestructive evaluation methods (4). Knowledge of process planning methods (3). Knowledge of inspection planning processes (3). Knowledge of industry standards related to applicable nondestructive evaluation methods (4). Knowledge of personnel certification processes in nondestructive evaluation methods (3). Knowledge of applicable technologies (3). Knowledge of where work is performed (3).

Task 4 (PO045-004)(PP045-004): Assess inspection records to validate that the documentation of NDT results are accurately defined in applicable work instructions and quality records.

Quality Eng—Basic	Quality Eng—Senior	Q Manager	Supplier QE	CSoftQE	Program QE	Contract Review	MRO QE	Spec Process QE	Met Labs QE	Production Control	Regulatory Specialist
P	P		P		S		S	P	S	S	

Procedural Knowledge: Ability to apply a structure process in review of quality records related to nondestructive evaluation methods to verify that the results of evaluation methods are adequately recorded (3).

Conceptual Knowledge: Ability to review quality records related to nondestructive evaluation methods to verify that the results of evaluation methods are adequately recorded (4).

Factual Knowledge: Knowledge of applicable nondestructive evaluation methods (4). Knowledge of process planning methods (3). Knowledge of inspection planning processes (3). Knowledge of industry standards related to applicable nondestructive evaluation methods (4). Knowledge of personnel certification processes in nondestructive evaluation methods (3). Knowledge of applicable technologies (3). Knowledge of where work is performed (3).

Process: Ensure Non-Destructive Test Sequences Are Appropriately Applied—within the Value Stream

Process Purpose: Review work instructions, process planning, and quality records to ensure nondestructive testing is being performed at the appropriate points in production/servicing. The NDT process shall be documented in appropriate records and define the standards of reference and the limits of acceptance. Ensure the documentation of NDT results are accurately defined in applicable work instructions and quality records.

Knowledge and Skill Requirements: Ability to evaluate work instructions, process planning, and quality records to ensure nondestructive testing is being performed at the appropriate points in production/servicing. Ability to evaluate NDT process requirements to determine if appropriate records are established and standards of reference and the limits of acceptance are defined. Ability to evaluate certification requirements for NDT inspectors and validate that persons assigned to associated NDT responsibilities are appropriately certified. Ability to assess inspection records to validate that the documentation of NDT results are accurately defined in applicable work instructions and quality records.

Task 1 (PP045-001): Evaluate work instructions, process planning, and quality records to ensure nondestructive testing is being performed at the appropriate points in production/servicing.

Quality Eng—Basic	Quality Eng—Senior	Q Manager	Supplier QE	CSoftQE	Program QE	Contract Review	MRO QE	Spec Process QE	Met Labs QE	Production Control	Regulatory Specialist
S	S	S	P		S			P	S		

Procedural Knowledge: Ability to apply a structured process to identify relevant documents for review (3). Ability to apply a structured process in review of work instructions, process planning, and quality records to determine the adequacy of such instructions to describe applicable nondestructive testing (3). Ability to apply a structured process to determine that applicable nondestructive evaluation methods are planned at the appropriate points in production/servicing in applicable work instructions, process planning, and quality records (3).

Conceptual Knowledge: Ability to review production/servicing processes and identify relevant documents for review (4). Ability to review work instructions, process planning, and quality records to determine the adequacy of such instructions to describe applicable nondestructive testing (4). Ability to review and determine that applicable nondestructive evaluation methods are planned at the appropriate points in production/servicing in applicable work instructions, process planning, and quality records (4).

Factual Knowledge: Knowledge of applicable nondestructive evaluation methods (4). Knowledge of process planning methods (3). Knowledge of inspection planning processes (3). Knowledge of industry standards related to applicable nondestructive evaluation methods (4). Knowledge of applicable technologies (3). Knowledge of where work is performed (3).

Task 2 (PP045-002): Evaluate NDT process requirements to determine if appropriate records are established and standards of reference and the limits of acceptance are defined.

Quality Eng—Basic	Quality Eng—Senior	Q Manager	Supplier QE	CSoftQE	Program QE	Contract Review	MRO QE	Spec Process QE	Met Labs QE	Production Control	Regulatory Specialist
S	S		P		S			P	S		

Procedural Knowledge: Ability to apply a structured process in review of work instructions, process planning, and quality records to determine the adequacy of such instructions to describe applicable nondestructive testing (3). Ability to apply a structured process to determine that applicable nondestructive evaluation methods are described by standards of reference and limits of acceptance (3).

Conceptual Knowledge: Ability to analyze applicable nondestructive evaluation methods and determine that such are adequately described by standards of reference and limits of acceptance (4).

Factual Knowledge: Knowledge of applicable nondestructive evaluation methods (4). Knowledge of process planning methods (3). Knowledge of inspection planning processes (3). Knowledge of industry standards related to applicable nondestructive evaluation methods (4). Knowledge of applicable technologies (3). Knowledge of where work is performed (3).

Task 3 (PP045-003): Evaluate certification requirements for NDT inspectors and validate that persons assigned to associated NDT responsibilities are appropriately certified.

Quality Eng—Basic	Quality Eng—Senior	Q Manager	Supplier QE	CSoftQE	Program QE	Contract Review	MRO QE	Spec Process QE	Met Labs QE	Production Control	Regulatory Specialist
S	S		P		S			P	S		

Procedural Knowledge: Ability to apply a structured process to review applicable nondestructive evaluation methods and identify personnel certification requirements (3). Ability to apply a structured process to review training, education, and certification records of personnel certified in nondestructive evaluation methods to determine compliance to applicable requirements (3).

Conceptual Knowledge: Ability to review applicable nondestructive evaluation methods and identify personnel certification requirements (4). Ability to analyze training, education, and certification records of personnel certified in nondestructive evaluation methods to determine compliance to applicable requirements (4).

Factual Knowledge: Knowledge of applicable nondestructive evaluation methods (4). Knowledge of process planning methods (3). Knowledge of inspection planning processes (3). Knowledge of industry standards related to applicable nondestructive evaluation methods (4). Knowledge of personnel certification processes in nondestructive evaluation methods (3). Knowledge of applicable technologies (3). Knowledge of where work is performed (3).

Task 4 (PP045-004): Assess inspection records to validate that the documentation of NDT results are accurately reflected in applicable work instructions and quality records.

Quality Eng—Basic	Quality Eng—Senior	Q Manager	Supplier QE	CSoftQE	Program QE	Contract Review	MRO QE	Spec Process QE	Met Labs QE	Production Control	Regulatory Specialist
S	P		P		S			P	S		

Procedural Knowledge: Ability to apply a structure process in review of quality records related to nondestructive evaluation methods to verify that the results of evaluation methods are adequately recorded (3).

Conceptual Knowledge: Ability to review quality records related to nondestructive evaluation methods to verify that the results of evaluation methods are adequately recorded (4).

Factual Knowledge: Knowledge of applicable nondestructive evaluation methods (4). Knowledge of process planning methods (3). Knowledge of inspection planning processes (3). Knowledge of industry standards related to applicable nondestructive evaluation methods (4). Knowledge of personnel certification processes in nondestructive evaluation methods (3). Knowledge of applicable technologies (3). Knowledge of where work is performed (3).

Process: Ensure Hardware Conformance—NDT Inspection

Process Purpose: To ensure physical product qualities are verified by appropriate nondestructive evaluation techniques. Ensure supplier/partner NDT processes are performed to requirements, equipment is appropriately maintained (to include calibration and shelf-life), and personnel performing NDT hold appropriate qualifications.

Knowledge and Skill Requirements: Verify that physical product qualities are verified by appropriate nondestructive evaluation techniques and approved persons. (Verification of NDT processes on product.)

Task 1 (PF035-001): Verify that physical product qualities are verified by appropriate nondestructive evaluation techniques and approved persons. (Verification of NDT processes on product.)

Quality Eng—Basic	Quality Eng—Senior	Q Manager	Supplier QE	CSoftQE	Program QE	Contract Review	MRO QE	Spec Process QE	Met Labs QE	Production Control	Regulatory Specialist
P	P	S	P		S		P	P	S	P	

Procedural Knowledge: Ability to apply a structured approach to verify that physical product qualities are verified by appropriate nondestructive evaluation techniques and approved persons. (3)

Conceptual Knowledge: Ability to analyze that physical product qualities are being verified by appropriate nondestructive evaluation techniques and approved persons. (4) Ability to evaluate NDT practices to ensure they are in compliance with applicable requirements (e.g., contractual, regulatory, industry standards) (5).

Factual Knowledge: Knowledge of inspection planning processes (3). Knowledge of work processes related to program/project (3). Knowledge of applicable NDT methods (4). Knowledge of NDT personnel qualification requirements (3). Knowledge of NDT methods development and approval processes (3). Knowledge of applicable special processes (4). Understanding of where work processes are being performed (3). Understanding of the value stream structure (3). Knowledge of industry standards related to NDT practices (3). Knowledge of contractual and/or regulatory requirements related to NDT practices (3). Knowledge of requirement management tools (3).

Task 2 (PF035-002): Verify that NDT processes are performed to requirements, equipment is appropriately maintained (to include calibration and shelf-life), and personnel performing NDT hold appropriate qualifications. (Assurance of NDT processes.)

Quality Eng—Basic	Quality Eng—Senior	Q Manager	Supplier QE	CSoftQE	Program QE	Contract Review	MRO QE	Spec Process QE	Met Labs QE	Production Control	Regulatory Specialist
P	P	S	P		S		P	P	S	P	

Procedural Knowledge: Ability to apply a structured approach to verify that NDT processes are performed to requirements, equipment is appropriately maintained (to include calibration and shelf-life), and personnel performing NDT hold appropriate qualifications (3).

Conceptual Knowledge: Ability to evaluate the processes, procedures, equipment, personnel, and work product to determine if NDT processes are being performed to requirements, equipment is appropriately maintained (to include calibration and shelf-life), and personnel performing NDT hold appropriate qualifications (5).

Factual Knowledge: Knowledge of inspection planning processes (3). Knowledge of work processes related to program/project (3). Knowledge of applicable NDT methods (4). Knowledge of NDT personnel qualification requirements (3). Knowledge of NDT methods development and approval processes (3). Knowledge of applicable special processes (4). Understanding of where work processes are being performed (3). Understanding of the value stream structure (3). Understanding of requirements management tools (3).

Process: Perform Special Process and/or NDT Evaluations—within the Value Stream

Process Purpose: When special processes and/or nondestructive testing is to be performed within the supplier/partner base, such shall be evaluated to the applicable specifications and standards to ensure capability and compliance. These evaluations will include review of people, process and equipment.

Knowledge and Skill Requirements: Ability to define the scope of evaluation to determine the compliance, capability, and capacity of supplier's/partner's controls for special processes and/or nondestructive testing. Understanding of applicable industry standards for special processes and nondestructive testing. Understanding of regulatory requirements related to special processes and nondestructive testing. Ability to perform evaluation of supplier/partner special processes and nondestructive testing to determine compliance, capability, and capacity. Ability to identify and document risks associated with noncompliance, capability concerns, and capacity constraints found during evaluation at supplier/partner facilities. Ability to document evaluation results.

Task 1 (SS105-001): In review of quality requirements, define the scope of evaluation (plan the audit) to determine the compliance, capability, and capacity of supplier's/partner's controls for special processes and/or nondestructive testing.

Quality Eng—Basic	Quality Eng—Senior	Q Manager	Supplier QE	CSoftQE	Program QE	Contract Review	MRO QE	Spec Process QE	Met Labs QE	Production Control	Regulatory Specialist
S	P	S	P		S			P	P		

Procedural Knowledge: Ability to apply a structured process to review quality requirements and identify audit/evaluation scope criteria to determine the compliance, capability, and capacity of supplier's/partner's controls for special processes and/or nondestructive testing (3). Ability to apply a structured process to document audit/evaluation scope (3).

Conceptual Knowledge: Ability to analyze quality requirements and identify audit/evaluation scope criteria to determine the compliance, capability, and capacity of supplier's/partner's controls for special processes and/or nondestructive testing (4). Ability effectively plan and document an audit/evaluation (4).

Factual Knowledge: Understanding of special processes (4). Understanding of non-destructive evaluation processes (4). Understanding of audit scope planning processes (3). Understanding of audit practices (3). Knowledge of quality requirements (3). Knowledge of requirements management tools (3).

Task 2 (SS105-002): Perform evaluation of supplier/partner special processes and nondestructive testing to determine compliance, capability, and capacity. Document evaluation results.

Quality Eng—Basic	Quality Eng—Senior	Q Manager	Supplier QE	CSoftQE	Program QE	Contract Review	MRO QE	Spec Process QE	Met Labs QE	Production Control	Regulatory Specialist
S	S	S	P		S			P	P		

Procedural Knowledge: Ability to apply a structured process to perform an evaluation of supplier/partner special processes and nondestructive testing to determine compliance, capability, and capacity (3). Ability to apply a structured process to document results of an evaluation (3).

Conceptual Knowledge: Ability to analyze supplier/partner special processes and nondestructive testing to determine compliance, capability, and capacity to applicable specifications and requirements (4). Ability to effectively document results of an evaluation (3).

Factual Knowledge: Understanding of special processes (4). Understanding of non-destructive evaluation processes (4). Understanding of audit scope planning processes (3). Understanding of audit practices (3). Knowledge of quality requirements (3).

Task 3 (SS105-003): Identify and document risks associated with noncompliance, capability concerns, and capacity constraints found during evaluation at supplier/partner facilities.

Quality Eng—Basic	Quality Eng—Senior	Q Manager	Supplier QE	CSoftQE	Program QE	Contract Review	MRO QE	Spec Process QE	Met Labs QE	Production Control	Regulatory Specialist
S	P	S	P		P			P	P		

Procedural Knowledge: Ability to apply a structured process to identify and document risks associated with noncompliance, capability concerns, and capacity constraints found during evaluation at supplier/partner facilities (3).

Conceptual Knowledge: Ability to analyze risks associated with noncompliance, capability concerns, and capacity constraints found during evaluation at supplier/partner facilities and establish risk measurement to findings (4).

Factual Knowledge: Understanding of risk assessment and measurement processes (4). Understanding of special processes (4). Understanding of non-destructive processes (4). Understanding of audit scope planning processes (3). Understanding of audit practices (3). Knowledge of quality requirements (3).

IV.D.3. Measuring Techniques

Process: Ensure First Piece Setup

Process Purpose: To ensure the first article produced for each production run in recurring production processes is appropriately set-up so that it will yield conforming characteristics and meets all stated requirements.

Knowledge and Skill Requirements: Ability to perform geometry math. Ability to lay a part out for inspection. Understanding of applicable production methods. Ability to utilize inspection gauging. Ability to perform inspection techniques to verify that the first article produced for each production run in recurring production processes is appropriately set-up so that it will yield conforming characteristics and meets all stated requirements (not AS/EN/SJAC 9102 FAI).

Task (PF010-001): Perform inspection techniques to verify that the first article produced for each production run in recurring production processes is appropriately set-up so that it will yield conforming characteristics and meets all stated requirements (not AS/EN/SJAC 9102 FAI).

Quality Eng—Basic	Quality Eng—Senior	Q Manager	Supplier QE	CSoftQE	Program QE	Contract Review	MRO QE	Spec Process QE	Met Labs QE	Production Control	Regulatory Specialist
S	P	S	S		S		P			P	

Procedural Knowledge: Ability to apply a structured process to perform inspection techniques to verify that the first production piece for each production/service run in recurring production/service processes is appropriately set-up so that it will yield conforming characteristics and meets all stated requirements (not AS/EN/SJAC 9102 FAI) (3).

Conceptual Knowledge: Ability to analyze design and production/servicing requirements to determine and plan a lay-out inspection (4). Ability to apply inspection methods to verify that the first production piece for each production/service run in recurring production/service processes is appropriately set-up so that it will yield conforming characteristics and meets all stated requirements (4).

Factual Knowledge: Understanding of geometry math. Understanding of lay out part inspection practices (3). Understanding of applicable production methods (3). Understanding of proper inspection gauging utilization (3). Understanding of design media (3). Understanding of production/servicing planning (3). Understanding of inspection planning (3). Understanding of requirements management tool (3).

Process: Ensure Hidden Dimension Verification Is Performed (as required)

Process Purpose: To ensure all required verification characteristics are positively conformed to design and engineering data sets. Ensure required inspections and tests on product characteristics are performed that cannot be verified at a later point. Ensure quality records accurately reflect inspection of hidden characteristics.

Knowledge and Skill Requirements: Understanding of applicable production methods. Ability to utilize inspection gauging. Ability to verify that inspection planning accurately reflects inspection requirements at appropriate points of inspection. Ability to apply inspection techniques to verify all required verification characteristics are positively conformed to design and engineering data sets at the points where verification can be made. Ability to verify that quality records accurately reflect inspection of hidden characteristics.

Task (PF015-001): Apply inspection techniques to verify all required verification characteristics are positively conformed to design and engineering data sets at the points where verification can be made. Verify that quality records accurately reflect inspection of hidden characteristics.

Quality Eng—Basic	Quality Eng—Senior	Q Manager	Supplier QE	CSoftQE	Program QE	Contract Review	MRO QE	Spec Process QE	Met Labs QE	Production Control	Regulatory Specialist
P	P	S	P		S		P	S	S	P	

Procedural Knowledge: Ability to apply a structured approach to inspection techniques in verifying that all required verification characteristics are positively conformed to design and engineering data sets at the points where verification can be made (3). Ability to apply a structured approach to verify that quality records accurately reflect inspection of hidden characteristics (3).

Conceptual Knowledge: Ability to apply inspection techniques to verify that all required verification characteristics are positively conformed to design and engineering data sets at the points where verification can be made (3). Ability to verify that quality records accurately reflect inspection of hidden characteristics (3).

Factual Knowledge: Understanding of geometry math. Understanding of practices related to measurement/verification of hidden dimensions (3). Understanding of applicable production methods (3). Understanding of proper inspection gauging utilization (3). Understanding of design media (3). Understanding of production/servicing planning (3). Understanding of inspection planning (3). Understanding of quality records (3).

Process: Ensure Test and Inspection Using Measurement Software, Equipment etc. Is Performed (as required)—within the Value Stream

Process Purpose: Ensure inspections (NDT and visual) and verification (e.g., testing, software model runs) of product characteristics at appropriate points and with appropriate techniques. Ensure process, people, and equipment are appropriate and controlled. Results of inspection and test activities must be defined and documented.

Knowledge and Skill Requirements: Understanding of production and inspection planning processes. Understanding of special process controls. Understanding of NDT methods. Understanding of operator/technician qualification requirements. Ability to verify that supplier's/partner's perform inspections (NDT and visual) and verification (e.g., testing, software model runs) of product characteristics at appropriate points and with appropriate techniques. Ability to evaluate that process, people, and equipment are appropriate and controlled throughout the supplier/partner base. Ability to verify that results of inspection and test activities are defined and documented.

Task 1 (PF020-002): Evaluate that process, people, and equipment are appropriate to effectively perform product/service verification activities and are controlled throughout the supplier/partner base. Ability to verify that results of inspection and test activities are defined and documented.

Quality Eng—Basic	Quality Eng—Senior	Q Manager	Supplier QE	CSoftQE	Program QE	Contract Review	MRO QE	Spec Process QE	Met Labs QE	Production Control	Regulatory Specialist
S	P	S	S		S			S	S		

Procedural Knowledge: Ability to apply a structured approach to ensure that process, people, and equipment are appropriate to effectively perform product/service verification activities and are controlled throughout the supplier/partner base (3). Ability to apply a structured approach to verify that results of inspection and test activities are defined and documented (3).

Conceptual Knowledge: Ability to analyze the capability and capacity of process, people, and equipment throughout the value stream to determine that adequate capability and capacity is present (4). Ability to analyze the work product of process, people, and equipment throughout the value stream to determine that product/service verification activities are effective (4). Ability to review quality records to verify that results of inspection and test activities are defined and documented (3).

Factual Knowledge: Understanding of production and inspection planning processes (3). Understanding of special process controls (3). Understanding of NDT methods (3). Understanding of operator/technician qualification requirements (3). Understanding of inspection methods (3). Knowledge of applicable inspection equipment (3). Understanding of audit practices (3). Understanding of the value stream structure and where work is performed (3).

Task 2 (PF020-003): Verify that supplier's/partner's perform inspections (NDT and visual) and verification (e.g., testing, software model runs) of product characteristics at appropriate points and with appropriate techniques.

Quality Eng—Basic	Quality Eng—Senior	Q Manager	Supplier QE	CSoftQE	Program QE	Contract Review	MRO QE	Spec Process QE	Met Labs QE	Production Control	Regulatory Specialist
S	S	S	P		P			P	P		

Procedural Knowledge: Ability to apply a defined and structured assessment process to verify that supplier's/partner's perform inspections (NDT and visual) and verification (e.g., testing, software model runs) of product characteristics at appropriate points and with appropriate techniques (3).

Conceptual Knowledge: Ability to analyze inspection planning processes to verify that supplier's/partner's perform inspections (NDT and visual) and verification (e.g., testing, software model runs) of product characteristics at appropriate points and with appropriate techniques (4).

Factual Knowledge: Understanding of production and inspection planning processes (3). Understanding of special process controls (3). Understanding of NDT methods (3). Understanding of operator/technician qualification requirements (3). Understanding of inspection methods (3). Understanding of audit practices (3). Understanding of the value stream structure and where work is performed (3).

Task 3 (PF020-003): Verify that supplier's/partner's perform inspections (NDT and visual) and verification (e.g., testing, software model runs) of product characteristics at appropriate points and with appropriate techniques.

Quality Eng—Basic	Quality Eng—Senior	Q Manager	Supplier QE	CSoftQE	Program QE	Contract Review	MRO QE	Spec Process QE	Met Labs QE	Production Control	Regulatory Specialist
S	S	S	P		P			P	P		

Procedural Knowledge: Ability to apply a defined and structured assessment process to verify that supplier's/partner's perform inspections (NDT and visual) and verification (e.g., testing, software model runs) of product characteristics at appropriate points and with appropriate techniques (3).

Conceptual Knowledge: Ability to analyze inspection planning processes to verify that supplier's/partner's perform inspections (NDT and visual) and verification (e.g., testing, software model runs) of product characteristics at appropriate points and with appropriate techniques (4).

Factual Knowledge: Understanding of production and inspection planning processes (3). Understanding of special process controls (3). Understanding of NDT methods (3). Understanding of operator/technician qualification requirements (3). Understanding of inspection methods (3). Understanding of audit practices (3). Understanding of the value stream structure and where work is performed (3).

Process: Ensure Test and Inspection Using Measurement Software, Equipment, etc. Is Performed (as required)—within the Organization

Process Purpose: Ensure inspections (NDT and visual) and verification (e.g., testing, software model runs) of product characteristics at appropriate points and with appropriate techniques. Ensure process, people, and equipment are appropriate and controlled. Results of inspection and test activities must be defined and documented.

Knowledge and Skill Requirements: Understanding of production and inspection planning processes. Understanding of special process controls. Understanding of NDT methods. Understanding of operator/technician qualification requirements. Ability to verify the organization's inspections (NDT and visual) and verification (e.g., testing, software model runs) of product characteristics at appropriate points and with appropriate techniques. Ability to evaluate that process, people, and equipment are appropriate and controlled throughout the organization. Ability to verify that results of inspection and test activities are defined and documented.

Task 1 (PF021-001): Evaluate that process, people, and equipment are appropriate to effectively perform product/service verification activities and are controlled throughout the organization. Ability to verify that results of inspection and test activities are defined and documented.

Quality Eng—Basic	Quality Eng—Senior	Q Manager	Supplier QE	CSoftQE	Program QE	Contract Review	MRO QE	Spec Process QE	Met Labs QE	Production Control	Regulatory Specialist
P	P	P	S		P		P	P	P	P	

Procedural Knowledge: Ability to apply a structured approach to ensure that process, people, and equipment are appropriate to effectively perform product/service verification activities and are controlled throughout the organization (3). Ability to apply a structured approach to verify that results of inspection and test activities are defined and documented (3).

Conceptual Knowledge: Ability to analyze the capability and capacity of process, people, and equipment throughout the value stream to determine that adequate capability and capacity is present (4). Ability to analyze the work product of process, people, and equipment throughout the organization to determine that product/service verification activities are effective (4). Ability to review quality records to verify that results of inspection and test activities are defined and documented (3).

Factual Knowledge: Understanding of production and inspection planning processes (3). Understanding of special process controls (3). Understanding of NDT methods (3). Understanding of operator/technician qualification requirements (3). Understanding of inspection methods (3). Knowledge of applicable inspection equipment (3). Understanding of audit practices (3). Understanding of where work is performed (3).

Task 2 (PF021-002): Verify the organization performs inspections (NDT and visual) and verification (e.g., testing, software model runs) of product characteristics at appropriate points and with appropriate techniques.

Quality Eng—Basic	Quality Eng—Senior	Q Manager	Supplier QE	CSoftQE	Program QE	Contract Review	MRO QE	Spec Process QE	Met Labs QE	Production Control	Regulatory Specialist
P	P	P	S		P		P	P	P	P	

Procedural Knowledge: Ability to apply a defined and structured assessment process to verify the organization performs inspections (NDT and visual) and verification (e.g., testing, software model runs) of product characteristics at appropriate points and with appropriate techniques (3).

Conceptual Knowledge: Ability to analyze inspection planning processes to verify the organization performs inspections (NDT and visual) and verification (e.g., testing, software model runs) of product characteristics at appropriate points and with appropriate techniques (4).

Factual Knowledge: Understanding of production and inspection planning processes (3). Understanding of special process controls (3). Understanding of NDT methods (3). Understanding of operator/technician qualification requirements (3). Understanding of inspection methods (3). Understanding of audit practices (3). Understanding of where work is performed (3).

Process: Ensure Hardware Conformance—Dimensional Inspection—CMM/Layout/Bench

Process Purpose: Ensure dimensional inspections are performed utilizing appropriate equipment and gages, utilizing appropriate inspection techniques, utilizing qualified personnel, and are appropriately documented.

Knowledge and Skill Requirements: Understanding of inspection planning processes. Ability to utilize inspection measurement devices. Understanding of metrology control requirements. Ability to verify that dimensional inspections are utilizing appropriate equipment and gages, utilizing appropriate inspection techniques, utilizing qualified personnel, and are appropriately documented.

Task (PF025-001): Verify dimensional inspections utilize appropriate equipment and gages, utilizing appropriate inspection techniques, utilizing qualified personnel, and are appropriately documented. Ensure all inspection requirements (contractual and regulatory) are appropriately applied.

Quality Eng—Basic	Quality Eng—Senior	Q Manager	Supplier QE	CSoftQE	Program QE	Contract Review	MRO QE	Spec Process QE	Met Labs QE	Production Control	Regulatory Specialist
P	P	P	P		P	S	P	P	P	P	S

Procedural Knowledge: Ability to apply a defined and structured assessment process to verify dimensional inspections are utilizing appropriate equipment and gages, utilizing appropriate inspection techniques, utilizing qualified personnel, are appropriately documented, and are effective (3). Ability to apply a structured approach to verify and ensure that all inspection requirements (contractual and regulatory) are appropriately applied (3).

Conceptual Knowledge: Ability to analyze process, procedures, and work product throughout the organization and value stream to verify that dimensional inspections are utilizing appropriate equipment and gages, utilizing appropriate inspection techniques, utilizing qualified personnel, and are appropriately documented, and are in compliance with applicable requirements (4).

Factual Knowledge: Understanding of inspection planning processes (3). Understanding of inspection equipment utilization (3). Understanding of metrology control requirements (3). Understanding of regulatory requirements related to product verification (3). Understanding of contractual and project/program requirements related to product verification (3). Understanding of personnel quality function requirements related to product verification (3). Understanding of inspection methods (3). Understanding of audit practices (3). Understanding of the value stream structure and where work is performed (3). Understanding of requirements management tools (3).

Process: Ensure Hardware Conformance—Visual Inspection

Process Purpose: To ensure physical product qualities are verified by appropriate visual means. Ensure visual inspection processes are appropriately defined and personnel performing the visual inspection process are qualified to the scope of their responsibilities.

Knowledge and Skill Requirements: Understanding of production and inspection planning processes. Ability to verify that physical product qualities were verified by appropriate visual means. Ability to verify that visual inspection processes are appropriately defined and personnel performing the visual inspection process are qualified to the scope of their responsibilities.

Task 1 (PF030-001): Verify, where appropriate, that physical product qualities were verified by appropriate visual means. Ensure visual inspection activities are performed in compliance with applicable requirements (e.g., contractual, industry standards, regulatory).

Quality Eng—Basic	Quality Eng—Senior	Q Manager	Supplier QE	CSoftQE	Program QE	Contract Review	MRO QE	Spec Process QE	Met Labs QE	Production Control	Regulatory Specialist
P	P	P	P		P		P	S	S	P	

Procedural Knowledge: Ability to apply a defined and structured assessment process to verify that physical product qualities were verified by appropriate visual means, where appropriate (3).

Conceptual Knowledge: Ability to review applicable engineering drawings and inspection planning to identify visual inspection requirements (3). Ability to assess, where appropriate, visual inspection activities to verify that physical product qualities were verified by appropriate visual means (3).

Factual Knowledge: Understanding of inspection planning processes (3). Understanding of regulatory requirements related to product verification (2). Understanding of contractual and project/program requirements related to product verification (2). Understanding of personnel quality function requirements related to product verification (2). Understanding of inspection methods (3). Understanding of audit practices (3). Understanding of the value stream structure and where work is performed (2). Understanding of requirements management tools (3).

Task 2 (PF030-002): Verify that visual inspection processes are appropriately defined and personnel performing the visual inspection process are qualified to the scope of their responsibilities.

Quality Eng—Basic	Quality Eng—Senior	Q Manager	Supplier QE	CSoftQE	Program QE	Contract Review	MRO QE	Spec Process QE	Met Labs QE	Production Control	Regulatory Specialist
P	P	P	P		P		P	S	S	P	

Procedural Knowledge: Ability to apply a structured review process on applicable engineering drawings and inspection planning to identify visual inspection requirements (3). Ability to apply a structured approach in review of visual inspection requirements to identify personnel qualification requirements (3).

Conceptual Knowledge: Ability to review applicable engineering drawings and inspection planning to identify visual inspection requirements (3). Ability to review visual inspection requirements to identify personnel qualification requirements (3). Ability to analyze quality records, processes, work product, and personnel to determine if visual inspection processes are effectively controlled and deployed (4).

Factual Knowledge: Understanding of inspection planning processes (3). Understanding of regulatory requirements related to product verification (3). Understanding of contractual and project/program requirements related to product verification (3). Understanding of personnel quality function requirements related to product verification (3). Understanding of inspection methods (3). Understanding of audit practices (3). Understanding of the value stream structure and where work is performed (3). Understanding of requirement management tools (3).

Process: Ensure Internal Receiving Inspection Controls Are Adequate

Process Purpose: To ensure receiving inspection processes that receive supplier/partner products and services are capable of performing required verification processes. Inspection requirements must meet the requirements of the customer, design, and production complexity. Receiving inspection processes may include physical inspection, documentation review, and other means of verifying conformity of design and contract requirements.

Knowledge and Skill Requirements: Understanding of regulatory requirements related to inspection processes. Understanding of inspection skill requirements. Understanding of inspection methods and utilization of inspection equipment. Ability to verify that receiving inspection processes that receive supplier/partner products and services are capable of performing required verification processes. Ability to verify that inspection requirements meet the requirements of the customer, design, and production complexity. Ability to verify that receiving inspection documentation is maintained as quality records.

Task 1 (PF090-001): Verify that the process of receiving inspection is capable of meeting the requirements of the customer, design, and production/servicing complexity (includes assessment of receiving inspection planning).

Quality Eng—Basic	Quality Eng—Senior	Q Manager	Supplier QE	CSoftQE	Program QE	Contract Review	MRO QE	Spec Process QE	Met Labs QE	Production Control	Regulatory Specialist
P	P	P	S		P		P	S	S	P	

Procedural Knowledge: Ability to apply a structured process to verify that receiving inspection requirements meet the requirements of the customer, design, and production complexity (3).

Conceptual Knowledge: Ability to analyze processes, procedures, work product, and performance measures to verify that receiving inspection requirements meet the requirements of the customer, design, and production complexity (4).

Factual Knowledge: Understanding of regulatory requirements related to receiving inspection processes (3). Understanding of contractual and project/program requirements related to receiving inspection processes (3). Understanding of receiving inspection skill requirements (3). Understanding of inspection methods and utilization of inspection equipment (3). Understanding of products and/or services received by the organization (3).

Task 2 (PF090-002): Verify that receiving inspection processes, personnel, and facilities that receive supplier/partner products and services are capable of performing required verification processes (Capability shall be aligned with customer, design, and production/servicing complexity.)

Quality Eng—Basic	Quality Eng—Senior	Q Manager	Supplier QE	CSoftQE	Program QE	Contract Review	MRO QE	Spec Process QE	Met Labs QE	Production Control	Regulatory Specialist
P	P	P	S		P		P	S	S	P	

Procedural Knowledge: Ability to apply a structured approach to verify that receiving inspection processes, personnel, and facilities that receive supplier/partner products and services are capable of performing required verification processes (3).

Conceptual Knowledge: Ability to analyze receiving inspection processes, personnel, and facilities that receive supplier/partner products and services to ensure they are capable of performing required verification processes, to include the personnel accepting the products/services (4).

Factual Knowledge: Understanding of regulatory requirements related to receiving inspection processes (3). Understanding of contractual and project/program requirements related to receiving inspection processes (3). Understanding of receiving inspection skill requirements (3). Understanding of inspection methods and utilization of inspection equipment (3). Understanding of products and/or services received by the organization (3). Understanding of personnel training records (3). Understanding of facilities requirements for receiving inspection (e.g., lighting, cleanliness) (3). Understanding of material handling requirements related to receiving inspection (e.g., ESD) (3). Understanding of requirements management tools (3).

Task 3 (PF090-003): Verify that receiving inspection documentation is maintained as quality records.

Quality Eng—Basic	Quality Eng—Senior	Q Manager	Supplier QE	CSoftQE	Program QE	Contract Review	MRO QE	Spec Process QE	Met Labs QE	Production Control	Regulatory Specialist
P	P	P	S		P		P	S	S	P	

Procedural Knowledge: Ability to apply a defined and structured assessment process to verify that receiving inspection documentation is maintained as quality records (3).

Conceptual Knowledge: Ability to assess processes, work product, and quality records to verify that receiving inspection documentation is maintained as quality records (3).

Factual Knowledge: Understanding of regulatory requirements related to receiving inspection processes (including retention) (3). Understanding of contractual and project/program requirements related to receiving inspection processes (including retention) (3). Understanding of products and/or services received by the organization (3). Understanding of requirements management tools (3).

IV.E. Metrology

Process: Determine Tool & Gage Design Requirements

Process Purpose: Review regulatory, engineering, process, and production/servicing requirements (capability, maintainability, reliability, certification, and human factors) and determine requirements for design, production, and maintenance of tools and gages. This may include calibration and traceability (reference to primary standards) requirements.

Knowledge and Skill Requirements: Ability to review regulatory, engineering, process, and production/servicing requirements and determine requirements (capability, maintainability, reliability, certification, and human factors) for design, production, and maintenance of tools and gages. Ability to identify calibration and traceability (reference to primary standards) requirements. Ability to organize, document, store and integrate tooling and inspection gauging controls into the requirements hierarchy using a requirements management tool.

Task 1 (DP060-001): Review regulatory, engineering, process, and production/servicing requirements and determine capability requirements for design, production, and maintenance of tools and gages. Organize, document, store and integrate tooling and inspection gauging controls into the requirements hierarchy using a requirements management tool.

Quality Eng—Basic	Quality Eng—Senior	Q Manager	Supplier QE	CSoftQE	Program QE	Contract Review	MRO QE	Spec Process QE	Met Labs QE	Production Control	Regulatory Specialist
P	P	S	S		P	P	S	S	S	S	P

Procedural Knowledge: Ability to apply a structured process in the review of regulatory, engineering, process, and production/servicing requirements and determine capability requirements for design, production, and maintenance of tools and gages (3). Ability to follow a structured process to document and integrate the determination analysis for tools and gages into the organization's requirements management tool (3).

Conceptual Knowledge: Ability to review regulatory, engineering, process, and production/servicing requirements and determine capability requirements for design, production, and maintenance of tools and gages (3). Ability to recognize the need to obtain additional information that is not part of the program data set in order to effectively determine tool and gage capability requirements (4). Ability to organize the tool and gage capability requirements into a hierarchical structure for integration into the organization's requirements management tool (3).

Factual Knowledge: Knowledge of tool and gage standards (3). Knowledge of design processes (3). Knowledge of production processes (3). Knowledge of requirements management tools (3).

Task 2 (DP060-002): Review regulatory, engineering, process, and production/servicing requirements and determine maintainability requirements for design, production, and maintenance of tools and gages. Organize, document, store and integrate tooling and inspection gauging controls into the requirements hierarchy using a requirements management tool.

Quality Eng—Basic	Quality Eng—Senior	Q Manager	Supplier QE	CSoftQE	Program QE	Contract Review	MRO QE	Spec Process QE	Met Labs QE	Production Control	Regulatory Specialist
P	P	S	S		P	P	S	S	S	S	P

Procedural Knowledge: Ability to apply a structured process in the review of regulatory, engineering, process, and production/servicing requirements and determine maintainability requirements for design, production, and maintenance of tools and gages (3). Ability to follow a structured process to document and integrate the determination analysis for tools and gages into the organization's requirements management tool (3).

Conceptual Knowledge: Ability to review regulatory, engineering, process, and production/servicing requirements and determine maintainability requirements for design, production, and maintenance of tools and gages (3). Ability to recognize the need to obtain additional information that is not part of the program data set in order to effectively determine tool and gage maintainability requirements (4). Ability to understand the actual environment in which tools and gages will be utilized and identify any additional maintainability considerations (4). Ability to organize the tool and gage maintainability requirements into a hierarchical structure for integration into the organization's requirements management tool (3).

Factual Knowledge: Knowledge of tool and gage standards (3). Knowledge of design processes (3). Knowledge of production processes (3). Knowledge of risk measurement tools (3). Knowledge of the environment in which tools and gages are to be utilized (3). Knowledge of requirements management tools (3).

Task 3 (DP060-003): Review regulatory, engineering, process, and production/servicing requirements and determine reliability requirements for design, production, and maintenance of tools and gages. Organize, document, store and integrate tooling and inspection gauging controls into the requirements hierarchy using a requirements management tool.

Quality Eng—Basic	Quality Eng—Senior	Q Manager	Supplier QE	CSoftQE	Program QE	Contract Review	MRO QE	Spec Process QE	Met Labs QE	Production Control	Regulatory Specialist
P	P	S	S		P	P	S	S	S	S	P

Procedural Knowledge: Ability to apply a structured process in the review of regulatory, engineering, process, and production/servicing requirements and determine reliability requirements for design, production, and maintenance of tools and gages (3). Ability to follow a structured process to document and integrate the determination analysis for tools and gages into the organization's requirements management tool (3).

Conceptual Knowledge: Ability to review regulatory, engineering, process, and production/servicing requirements and determine reliability requirements for design, production, and maintenance of tools and gages (3). Ability to recognize the need to obtain additional information that is not part of the program data set in order to effectively determine tool and gage reliability requirements (4). Ability to understand the actual environment in which tools and gages will be utilized and identify any additional reliability considerations (4).

Factual Knowledge: Knowledge of tool and gage standards (3). Knowledge of design processes (3). Knowledge of production processes (3). Knowledge of risk measurement tools (3). Knowledge of the environment in which tools and gages are to be utilized (3). Knowledge of requirements management tools (3).

Task 4 (DP060-004): Review regulatory, engineering, process, and production/servicing requirements and determine certification requirements for design, production, and maintenance of tools and gages. Organize, document, store, and integrate tooling and inspection gauging controls into the requirements hierarchy using a requirements management tool.

Quality Eng—Basic	Quality Eng—Senior	Q Manager	Supplier QE	CSoftQE	Program QE	Contract Review	MRO QE	Spec Process QE	Met Labs QE	Production Control	Regulatory Specialist
P	P	S	S		P	P	S	S	S	S	P

Procedural Knowledge: Ability to apply a structured process in the review of regulatory, engineering, process, and production/servicing requirements and determine certification requirements for design, production, and maintenance of tools and gages (3). Ability to follow a structured process to document and integrate the determination analysis for tools and gages into the organization's requirements management tool (3).

Conceptual Knowledge: Ability to review regulatory, engineering, process, and production/servicing requirements and determine certification requirements for design, production, and maintenance of tools and gages (3). Ability to recognize the need to obtain additional information that is not part of the program data set in order to effectively determine tool and gage certification requirements (4).

Factual Knowledge: Knowledge of tool and gage standards (3). Knowledge of design processes (3). Knowledge of production processes (3). Knowledge of regulatory requirements (3). Knowledge of certification processes (3). Knowledge of requirements management tools (3).

Task 5 (DP060-005): Review regulatory, engineering, process, and production/servicing requirements and determine human factors requirements for design, production, and maintenance of tools and gages. Organize, document, store, and integrate tooling and inspection gauging controls into the requirements hierarchy using a requirements management tool.

Quality Eng—Basic	Quality Eng—Senior	Q Manager	Supplier QE	CSoftQE	Program QE	Contract Review	MRO QE	Spec Process QE	Met Labs QE	Production Control	Regulatory Specialist
P	P	S	S		P	P	S	S	S	S	P

Procedural Knowledge: Ability to apply a structured process in the review of regulatory, engineering, process, and production/servicing requirements and determine human factors requirements for design, production, and maintenance of tools and gages (3). Ability to follow a structured process to document and integrate the determination analysis for tools and gages into the organization's requirements management tool (3).

Conceptual Knowledge: Ability to review regulatory, engineering, process, and production/servicing requirements and determine human factors requirements for design, production, and maintenance of tools and gages (3). Ability to recognize the need to obtain additional information that is not part of the program data set in order to effectively determine requirements related to human factors in tool and gage design, production, use, and maintenance (4). Ability to organize the tool and gage human factors requirements into a hierarchical structure for integration into the organization's requirements management tool (3).

Factual Knowledge: Knowledge of tool and gage standards (3). Knowledge of design processes (3). Knowledge of production processes (3). Knowledge of human factors practices and applicable standards (3). Knowledge of requirements management tools (3). Knowledge of the environment in which tools and gages are to be utilized (3).

Task 6 (DP060-006): Identify calibration and traceability (reference to primary standards) requirements. Organize, document, store, and integrate tooling and inspection gauging controls into the requirements hierarchy using a requirements management tool.

Quality Eng—Basic	Quality Eng—Senior	Q Manager	Supplier QE	CSoftQE	Program QE	Contract Review	MRO QE	Spec Process QE	Met Labs QE	Production Control	Regulatory Specialist
P	P	S	S		P	P		P	P		P

Procedural Knowledge: Ability to apply a defined process for identifying calibration and traceability requirements for a program and/or product line (3). Ability to follow a structured process to document and integrate calibration and traceability requirements for tools and gages into the organization's requirements management tool (3).

Conceptual Knowledge: Ability to analyze program requirements, industry standards, and regulatory requirements to determine appropriate calibration and traceability requirements (4). Ability to contextualize calibration and traceability requirements and assess the global nature of the value stream and its ability to ensure compliance to stated requirements (ensure all international suppliers meet specified traceability requirements) (4). Ability to organize the tool and gage calibration and traceability requirements into a hierarchical structure for integration into the organization's requirements management tool (3).

Factual Knowledge: Knowledge of tool and gage standards (3). Knowledge of design requirements (3). Knowledge of production processes (3). Knowledge of calibration methods and standards (3). Knowledge of traceability policies (3). Knowledge of regulatory requirements related to metrology (3). Knowledge of requirements management tools (3).

Task 7 (DP060-007): Based on M&TE repeatability and reliability (R&R), identify Measuring and Test Equipment (M&TE) required for the measurement and verification of the production parts. Organize, document, store, and integrate tooling and inspection gauging controls into the requirements hierarchy using a requirements management tool.

Quality Eng—Basic	Quality Eng—Senior	Q Manager	Supplier QE	CSoftQE	Program QE	Contract Review	MRO QE	Spec Process QE	Met Labs QE	Production Control	Regulatory Specialist
P	P	S	S		P	S	S	P	P	S	S

Procedural Knowledge: Ability to apply a structured process to identify M&TE for the measurement and verification of the production parts (3). Ability to follow a structured process to document and integrate calibration and traceability requirements for tools and gages into the organization's requirements management tool (3).

Conceptual Knowledge: Ability to assess the capability of M&TE to identify appropriate application in the measurement and verification of the production parts (4). Ability to interpret R&R studies (3). Ability to organize the M&TE selections into a hierarchical structure for integration into the organization's requirements management tool (3).

Factual Knowledge: Knowledge of tool and gage standards (3). Knowledge of design requirements (3). Knowledge of production processes (3). Knowledge of the design and capability of M&TE applicable to the program/project (3). Knowledge of gage R&R methods (3). Knowledge of calibration methods and standards (3). Knowledge of traceability policies (3). Knowledge of regulatory requirements related to metrology (3). Knowledge of requirements management tools (3).

Process: Monitor Calibration Control

Process Purpose: To ensure the calibration system is in compliance with the approved system. Evaluate equipment maintenance, handling methods, calibration standards, use, and storage.

Knowledge and Skill Requirements: Understanding of metrology control processes. Understanding of the international nature of metrology control and the various national and international agreements for measurement. Ability to verify calibration systems are in compliance with the approved system requirements and applicable regulations. Ability to assess equipment maintenance, handling methods, calibration standards, use, and storage control processes and validate compliance to applicable requirements.

Task 1 (PM040-001): Verify calibration systems are in compliance with the approved system requirements and applicable regulations.

Quality Eng—Basic	Quality Eng—Senior	Q Manager	Supplier QE	CSoftQE	Program QE	Contract Review	MRO QE	Spec Process QE	Met Labs QE	Production Control	Regulatory Specialist
P	P	P	P		P	S	P	P	P	P	P

Procedural Knowledge: Ability to apply a structured and defined assessment process to verify calibration systems are in compliance with the approved system requirements and applicable regulations (3).

Conceptual Knowledge: Ability to analyze established calibration systems to verify that they are in compliance with the approved system requirements and applicable regulations (4).

Factual Knowledge: Understanding of regulatory requirements related to calibration systems (3). Understanding of contractual requirements related to calibration systems (3). Understanding of regulatory requirements related to calibration systems (3). Understanding of industry standards related to calibration systems (3). Understanding of calibration systems (3). Understanding of international agreements for metrology (3).

Task 2 (PM040-002): Assess equipment maintenance, handling methods, calibration standards, use, and storage control processes and validate compliance to applicable requirements.

Quality Eng—Basic	Quality Eng—Senior	Q Manager	Supplier QE	CSoftQE	Program QE	Contract Review	MRO QE	Spec Process QE	Met Labs QE	Production Control	Regulatory Specialist
P	P	S	P		S	S	P	P	P	P	S

Procedural Knowledge: Ability to apply a defined and structured assessment process to determine that equipment maintenance, handling methods, calibration standards, use, and storage control processes are in compliance with applicable requirements (3).

Conceptual Knowledge: Ability to analyze equipment maintenance, handling methods, calibration standards, use, and storage control processes to determine if such are in compliance with applicable requirements (4).

Factual Knowledge: Understanding of regulatory requirements related to calibration systems (3). Understanding of contractual requirements related to calibration systems (3). Understanding of industry standards related to calibration systems (3). Understanding of calibration systems (3). Understanding of applicable production/servicing processes (3).

IV.F. Measurement System Analysis (MSA)

Process: Monitor Gage R&R Control

Process Purpose: To ensure reliability of equipment used for product acceptance. To ensure the equipment used to accept product are appropriate to the level of accuracy required. When performed, ensure gage R&R studies are documented.

Knowledge and Skill Requirements: Understanding of metrology control processes. Ability to verify calibration systems are in compliance with the approved system requirements and applicable regulations. Ability to assess equipment maintenance, handling methods, calibration standards, use, and storage control processes and validate compliance to applicable requirements. Understanding of human factors in measurement. Understanding of variability in measurement processes. Ability to perform gage R & R studies. Ability to document gage R & R studies. Ability to verify that gage R & R studies have been conducted correctly throughout the value stream and appropriately documented.

Task 1 (PM045-001): Perform and document gage R & R studies.

Quality Eng—Basic	Quality Eng—Senior	Q Manager	Supplier QE	CSoftQE	Program QE	Contract Review	MRO QE	Spec Process QE	Met Labs QE	Production Control	Regulatory Specialist
P	P	S	P		S		P	P	P	P	

Procedural Knowledge: Ability to apply a defined and structured process to perform and document gage R & R studies (3).

Conceptual Knowledge: Ability to identify, understand, and analyze the various external and internal influences on gage R & R (4). Ability to perform and document gage R & R studies (3).

Factual Knowledge: Understanding of metrology control processes (3). Understanding of equipment maintenance, handling methods, calibration standards, use, and storage control processes related to measuring devices (3). Understanding of contractual requirements related to gage R & R studies (3). Understanding of human factors in measurement (3). Understanding of variability in measurement processes (3).

Task 2 (PM045-002): Verify that gage R & R studies have been conducted correctly throughout the value stream and appropriately documented.

Quality Eng—Basic	Quality Eng—Senior	Q Manager	Supplier QE	CSoftQE	Program QE	Contract Review	MRO QE	Spec Process QE	Met Labs QE	Production Control	Regulatory Specialist
P	P	S	P		P		P	S	S	P	

Procedural Knowledge: Ability to apply a defined and structured assessment process to verify that gage R & R studies have been conducted correctly throughout the value stream and appropriately documented (3).

Conceptual Knowledge: Ability to analyze established processes and applicable records to verify that gage R & R studies have been conducted correctly throughout the value stream and appropriately documented (4).

Factual Knowledge: Understanding of metrology control processes (3). Understanding of equipment maintenance, handling methods, calibration standards, use, and storage control processes related to measuring devices (3). Understanding of contractual requirements related to gage R & R studies (3). Understanding of human factors in measurement (3). Understanding of variability in measurement processes (3). Understanding of where work is performed (3). Understanding of value stream structure (3).

Process: Monitor Correlation of Metallurgical Testing

Process Purpose: To ensure the standards and equipment utilized for metallurgical testing are accurate. Ensure documentation of correlation studies is available throughout the supplier/partner base. Special emphasis on standards used on spectrometers.

Knowledge and Skill Requirements: Understanding of variance control methods in metallurgical testing. Ability to ensure the standards and equipment utilized for metallurgical testing are accurate. Ability to verify that documentation of correlation studies (e.g., "round robin") are available throughout the supplier/partner base. Understanding of special control requirements for spectrometers.

Task 1 (PM050-001): Ensure the standards and equipment utilized for metallurgical testing are accurate, throughout the value stream.

Quality Eng—Basic	Quality Eng—Senior	Q Manager	Supplier QE	CSoftQE	Program QE	Contract Review	MRO QE	Spec Process QE	Met Labs QE	Production Control	Regulatory Specialist
P	P	S	P		P		S	P	P	S	

Procedural Knowledge: Ability to apply a structured process to verify standards and equipment utilized for metallurgical testing are accurate, throughout the value stream (3).

Conceptual Knowledge: Ability to analyze control processes, quality records, and performance data to verify standards and equipment utilized for metallurgical testing are accurate, throughout the value stream (4).

Factual Knowledge: Understanding of metrology control processes (3). Understanding of equipment maintenance, handling methods, calibration standards, use, and storage control processes related to metallurgical testing equipment (3). Understanding of industry standards related to metallurgical testing and related equipment (3).

Task 2 (PM050-002): Perform correlation studies (i.e., "round robin") of metallurgical equipment and methods.

Quality Eng—Basic	Quality Eng—Senior	Q Manager	Supplier QE	CSoftQE	Program QE	Contract Review	MRO QE	Spec Process QE	Met Labs QE	Production Control	Regulatory Specialist
S	S	S	S		S		S	S	P	S	

Procedural Knowledge: Ability to apply a structured approach to perform correlation studies (i.e., "round robin") of metallurgical equipment (3).

Conceptual Knowledge: Ability to perform correlation studies (i.e., "round robin") of metallurgical equipment and methods (3). Ability to analyze results of correlation studies (i.e., "round robin") of metallurgical equipment and methods (4).

Factual Knowledge: Understanding of metrology control processes (3). Understanding of equipment maintenance, handling methods, calibration standards, use, and storage control processes related to metallurgical testing equipment (3). Understanding of industry standards related to metallurgical testing and related equipment (3). Understanding of correlation study methods (3). Understanding of product/service design requirements (3).

Task 3 (PM050-003): Verify that documentation of correlation studies (i.e., "round robin") are available throughout the supplier/partner base.

Quality Eng—Basic	Quality Eng—Senior	Q Manager	Supplier QE	CSoftQE	Program QE	Contract Review	MRO QE	Spec Process QE	Met Labs QE	Production Control	Regulatory Specialist
S	P	S	P		P		S	S	P	S	

Procedural Knowledge: Ability to apply a defined and structured assessment process to verify that documentation of correlation studies (i.e., "round robin") are available throughout the supplier/partner base (3).

Conceptual Knowledge: Ability analyze quality records and equipment to verify that documentation of correlation studies (i.e., "round robin") are available throughout the supplier/partner base (4).

Factual Knowledge: Understanding of metrology control processes (3). Understanding of equipment maintenance, handling methods, calibration standards, use, and storage control processes related to metallurgical testing equipment (3). Understanding of industry standards related to metallurgical testing and related equipment (3). Understanding of correlation study methods (3). Understanding of where work is performed (3). Understanding of value stream structure (3). Understanding of product/service design requirements (3). Understanding of how quality records are to be updated (3).

IV.G.1. Shop Assist

Process: Ensure Shop Assist Process

Process Purpose: To provide for a temporary supplier relationship to perform work that is not intended to be permanently outsourced. Ensure the scope and use of shop assist is limited to the temporary use of production capacity and is not actually an outsourcing of longer-term production and/or services. Ensure supplier/partner has systems in place to perform work in accordance with organizational procedures and process. Ensure documentation is controlled.

Knowledge and Skill Requirements: Ability to verify that shop-assist suppliers have effective procedures in place to ensure the quality of the temporary out-sourced work. Ability to determine that the scope and use of shop assist is limited to the temporary use of production capacity and is not actually an outsourcing of longer-term production and/or services. Ability to verify shop-assist quality records and production documentation are controlled.

Task 1 (PP155-001): Verify that shop-assist suppliers have effective procedures in place to ensure the quality of the temporary out-sourced work and are consistent with the organization's quality management system. Where a sub-tier supplier engages in shop-assist activities, ensure the same controls are in place.

Quality Eng—Basic	Quality Eng—Senior	Q Manager	Supplier QE	CSoftQE	Program QE	Contract Review	MRO QE	Spec Process QE	Met Labs QE	Production Control	Regulatory Specialist
P	P	S	P		S		P	S	S	P	

Procedural Knowledge: Ability to apply a structured and defined process to assess procedures and processes of shop-assist suppliers to verify adequacy and compliance to control quality of temporarily out-sourced work (3).

Conceptual Knowledge: Ability to analyze procedures and processes of shop-assist suppliers to verify adequacy and compliance to control quality of temporarily out-sourced work (4). Ability to analyze shop-assist activities to verify effectiveness of processes and procedures to ensure the quality of temporarily out-sourced work (4).

Factual Knowledge: Understanding of contractual and regulatory requirements to shop-assist (3). Understanding of shop-assist processes (3). Understanding of applicable work processes (3). Understanding of audit practices (3). Understanding of value stream structure and where work is performed (3).

Task 2 (PP155-002): Determine that the scope and use of shop assist is limited to the temporary use of production capacity and is not actually an outsourcing of longer-term production and/or services. Where a sub-tier supplier engages in shop-assist activities, ensure the same controls are in place.

Quality Eng—Basic	Quality Eng—Senior	Q Manager	Supplier QE	CSoftQE	Program QE	Contract Review	MRO QE	Spec Process QE	Met Labs QE	Production Control	Regulatory Specialist
P	P	P	P		S		P			P	

Procedural Knowledge: Ability to apply a defined and structured assessment process to verify that the scope and use of shop assist is limited to the temporary use of production capacity and is not actually an outsourcing of longer-term production and/or services (3).

Conceptual Knowledge: Ability to analyze contractual agreements, work activities, and process controls to verify that the scope and use of shop assist is limited to the temporary use of production capacity and is not actually an outsourcing of longer-term production and/or services (4).

Task 3 (PP155-003): Verify shop-assist quality records and production documentation are appropriately controlled. Where a sub-tier supplier engages in shop-assist activities, ensure the same controls are in place.

Quality Eng—Basic	Quality Eng—Senior	Q Manager	Supplier QE	CSoftQE	Program QE	Contract Review	MRO QE	Spec Process QE	Met Labs QE	Production Control	Regulatory Specialist
P	P	S	P		S		P	S	S	P	

Procedural Knowledge: Ability to apply a defined and structured process to verify shop-assist quality records and production documentation are appropriately controlled (3).

Conceptual Knowledge: Ability to review shop-assist quality records and production documentation to verify that they are appropriately controlled (3).

Factual Knowledge: Understanding of contractual and regulatory requirements related to shop-assist (3). Understanding of shop-assist processes (3). Understanding of applicable work processes (3). Understanding of audit practices (3). Understanding of value stream structure and where work is performed (3).

IV.I. Control of Tooling

Process: Ensure Process Tooling Changes—within the Organization

Process Purpose: To ensure tooling utilized in the production/servicing processes are conforming to design requirements, including any changes to such tooling. Ensure references to tooling requirements are accurately reflected in work instructions, process sheets, and other technical documents.

Knowledge and Skill Requirements: Ability to verify tooling utilized in the production/servicing processes are conforming to design requirements, including any changes to such tooling, throughout the value stream. Ability to assess and verify that references to tooling requirements are accurately reflected in work instructions, process sheets, and other technical documents.

Task 1 (PO100-001): Verify tooling utilized in the production/servicing processes are conforming to design requirements, including any changes to such tooling, throughout the organization.

Quality Eng—Basic	Quality Eng—Senior	Q Manager	Supplier QE	CSoftQE	Program QE	Contract Review	MRO QE	Spec Process QE	Met Labs QE	Production Control	Regulatory Specialist
P	P	S			S		P			P	

Procedural Knowledge: Ability to apply a defined and structured process to review tooling utilized in production/servicing to ensure conformance and compliance to design requirements (3).

Conceptual Knowledge: Ability to analyze requirements hierarchy and identify any requirements that may affect tooling (4). Ability to analyze drawings and production/servicing documents, including changes to such documents, to evaluate any potential impacts to tooling (5) Ability to analyze tooling utilized in production/servicing to ensure conformance and compliance to design requirements (4).

Factual Knowledge: Understanding of program, regulatory, design, and contractual requirements related to tooling (3). Understanding of tool design practices (3). Understanding of tooling inspection methods (3). Understanding of configuration management practices (3). Understanding of requirements management tools (3). Understanding of where work is being performed (3).

Task 2 (PO100-002): Assess and verify that references to tooling requirements are accurately reflected in work instructions, process sheets, and other technical documents throughout the organization.

Quality Eng—Basic	Quality Eng—Senior	Q Manager	Supplier QE	CSoftQE	Program QE	Contract Review	MRO QE	Spec Process QE	Met Labs QE	Production Control	Regulatory Specialist
P	P	S			S		P			P	

Procedural Knowledge: Ability to apply a structured process to assess work instructions, process sheets, and other technical documentation to verify that references to tooling requirements are accurately stated (3).

Conceptual Knowledge: Ability to analyze work instructions, process sheets, and other technical documentation to verify that references to tooling requirements are accurately stated (4).

Factual Knowledge: Understanding of program, regulatory, design, and contractual requirements related to tooling (3). Understanding of tool design practices (3). Understanding of configuration management practices (3). Understanding of work instruction control processes (3). Understanding of requirements management tools (3). Knowledge of where work is being performed (3).

Process: Ensure Process Tooling Changes—within the Value Stream

Process Purpose: To ensure tooling utilized in the production/servicing processes are conforming to design requirements, including any changes to such tooling. Ensure references to tooling requirements are accurately reflected in work instructions, process sheets, and other technical documents.

Knowledge and Skill Requirements: Ability to verify tooling utilized in the production/servicing processes are conforming to design requirements, including any changes to such tooling, throughout the value stream. Ability to assess and verify that references to tooling requirements are accurately reflected in work instructions, process sheets, and other technical documents.

Task 1 (PP100-001): Verify tooling utilized in the production/servicing processes are conforming to design requirements, including any changes to such tooling, throughout the value stream.

Quality Eng—Basic	Quality Eng—Senior	Q Manager	Supplier QE	CSoftQE	Program QE	Contract Review	MRO QE	Spec Process QE	Met Labs QE	Production Control	Regulatory Specialist
S	P	S	P		S						

Procedural Knowledge: Ability to apply a defined and structured process to review tooling utilized in production/servicing to ensure conformance and compliance to design requirements (3).

Conceptual Knowledge: Ability to analyze requirements hierarchy and identify any requirements that may affect tooling (4). Ability to analyze drawings and production/servicing documents, including changes to such documents, to evaluate any potential impacts to tooling (5) Ability to analyze tooling utilized in production/servicing to ensure conformance and compliance to design requirements (4).

Factual Knowledge: Understanding of program, regulatory, design, and contractual requirements related to tooling (3). Understanding of tool design practices (3). Understanding of tooling inspection methods (3). Understanding of configuration management practices (3). Understanding of requirements management tools (3). Understanding of the value stream structure (3). Understanding of where work is being performed (3).

Task 2 (PP100-002): Assess and verify that references to tooling requirements are accurately reflected in work instructions, process sheets, and other technical documents throughout the value stream.

Quality Eng—Basic	Quality Eng—Senior	Q Manager	Supplier QE	CSoftQE	Program QE	Contract Review	MRO QE	Spec Process QE	Met Labs QE	Production Control	Regulatory Specialist
S	P	S	P		S						

Procedural Knowledge: Ability to apply a structured process to assess work instructions, process sheets, and other technical documentation to verify that references to tooling requirements are accurately stated (3).

Conceptual Knowledge: Ability to analyze work instructions, process sheets, and other technical documentation to verify that references to tooling requirements are accurately stated (4).

Factual Knowledge: Understanding of program, regulatory, design, and contractual requirements related to tooling (3). Understanding of tool design practices (3). Understanding of configuration management practices (3). Understanding of work instruction control processes (3). Understanding of requirements management tools (3). Knowledge of the value stream structure (3). Knowledge of where work is being performed (3).

IV.K.1. Risk Based Acquisition Management

Process: Request Quotes from Suppliers

Process Purpose: Ensure a document is created that requests from suppliers/partners formal documents that define their terms, conditions, and pricing to perform work that is stated within the technical data package and defined in a statement of work.

Knowledge and Skill Requirements: Ability to ensure a document is created that requests from suppliers/partners formal documents that define their terms, conditions, and pricing to perform work that is stated within the technical data package and defined in a statement of work.

Task (SS070-001): Create a document that requests from suppliers/partners formal documents that define their terms, conditions, and pricing to perform work that is stated within the technical data package and defined in a statement of work.

Quality Eng—Basic	Quality Eng—Senior	Q Manager	Supplier QE	CSoftQE	Program QE	Contract Review	MRO QE	Spec Process QE	Met Labs QE	Production Control	Regulatory Specialist
S	P	S	P		S	S					S

Procedural Knowledge: Ability to apply a structured process to construct a document that requests from suppliers/partners formal documents that define their terms, conditions, and pricing to perform work that is stated within the technical data package and defined in a statement of work (3).

Conceptual Knowledge: Ability to construct a document that requests from suppliers/partners formal documents that define their terms, conditions, and pricing to perform work that is stated within the technical data package and defined in a statement of work (3).

Factual Knowledge: Knowledge of regulatory requirements related to supplier communication practices (3). Knowledge of quality management standard requirements related to supplier communication practices (3). Knowledge of requirements management tools (3). Knowledge of project/program requirements (3). Knowledge of value stream structure (3). Knowledge of effective communication practices (3). Knowledge of requirements related to creating a request for quote document (3). Knowledge of applicable technical requirements (3).

Process: Evaluate Quotes from Suppliers

Process Purpose: Evaluate the formal offer packages that potential suppliers/partners have provided in response for quotations. Ensure the responses accurately address all of the technical requirements stated in the scope of work. Evaluate all requested deviations/objections noted in the response from suppliers/partners, and assess the impact on quality and schedule.

Knowledge and Skill Requirements: Ability to evaluate formal offer packages that potential suppliers/partners have provided in response for quotations. Ability to assess offer packages and ensure the responses accurately address all of the technical requirements stated in the scope of work. Ability to evaluate all requested deviations/objections noted in the response from suppliers/partners, and assess the impact on quality and schedule. Ability to determine and document ratings for each supplier offer package. Ability to evaluate the proposed use of sub-tier suppliers, sources of raw materials, and logistics support providers.

Task 1 (SS075-001): Evaluate formal offer packages that potential suppliers/partners have provided in response for quotations. Assess offer packages and ensure the responses accurately address all of the technical requirements stated in the scope of work. Evaluate all requested deviations/objections noted in the response from suppliers/partners, and assess the impact on quality and schedule. Determine and document ratings for each supplier offer package.

Quality Eng—Basic	Quality Eng—Senior	Q Manager	Supplier QE	CSoftQE	Program QE	Contract Review	MRO QE	Spec Process QE	Met Labs QE	Production Control	Regulatory Specialist
P	P	S	P	S	S	S	S	S	S	S	S

Procedural Knowledge: Ability to apply a structured process to assess and apply measurement to formal offer packages of potential suppliers/partners that have provided response for quotations (3). Ability to apply a structured process to assess offer packages and ensure the responses accurately address all of the technical requirements stated in the scope of work (3). Ability to apply a structured process in evaluation of all requested deviations/objections noted in the response from suppliers/partners, and assess the impact on quality and schedule (4). Ability to apply a structured process to assign measurement ratings for each supplier offer package and document the results of the assignment (3).

Conceptual Knowledge: Ability to assess and apply measurement to formal offer packages of potential suppliers/partners that have provided response for quotations (3). Ability to evaluate offer packages and ensure the responses accurately address all of the technical requirements stated in the scope of work (5). Ability to evaluate all requested deviations/objections noted in the response from suppliers/partners, and assess the impact on quality and schedule (5). Ability to effectively communicate and document potential risk of proposed deviations and assign measurement ratings for each supplier offer package and document the results of the assignment (4).

Factual Knowledge: Understanding of request for proposal processes (3). Understanding of applicable technical processes (4). Knowledge of quotation processes (4). Knowledge of project/program requirements (3). Knowledge of requirements management tools (3). Knowledge of effective communication practices (3). Understanding of project/programming schedules (3). Knowledge of risk assessment processes (3).

Task 2 (SS075-002): Evaluate the proposed use of sub-tier suppliers, sources of raw materials, and logistics support providers. Identify any risks to capacity or capability. Document the results of analysis.

Quality Eng—Basic	Quality Eng—Senior	Q Manager	Supplier QE	CSoftQE	Program QE	Contract Review	MRO QE	Spec Process QE	Met Labs QE	Production Control	Regulatory Specialist
P	P	P	P		P			P	P		

Procedural Knowledge: Ability to apply a structured process to assess potential risks in capacity and capability of use of proposed sub-tier suppliers, sources of raw materials, and logistics support providers (3). Ability to apply a structured process in documenting the results of analysis (3).

Conceptual Knowledge: Ability to identify potential risks in capacity and capability of use of proposed sub-tier suppliers, sources of raw materials, and logistics support providers (4). Ability to quantify potential risks in capacity and capability of use of proposed sub-tier suppliers, sources of raw materials, and logistics support providers (4). Ability to effectively document the results of analysis (3).

Factual Knowledge: Knowledge of risk identification practices (3). Knowledge of risk measurement processes (3). Knowledge of the value stream structure (3). Knowledge of applicable technical processes (4). Knowledge of applicable production/servicing processes (4). Knowledge of applicable raw materials (4). Knowledge of applicable material control processes (4).

Process: Make a Supplier Selection

Process Purpose: Based on the criteria established by the organization and the ability of the potential supplier/partner to deliver to the defined scope of work, select a supplier/partner.

Knowledge and Skill Requirements: Based on the criteria established by the organization and the measured ability of the potential supplier/partner to deliver to the defined scope of work, the ability to select a supplier/partner.

Task (SS090-001): Based on the criteria established by the organization and the measured ability of the potential supplier/partner to deliver to the defined scope of work, select a supplier/partner.

Quality Eng—Basic	Quality Eng—Senior	Q Manager	Supplier QE	CSoftQE	Program QE	Contract Review	MRO QE	Spec Process QE	Met Labs QE	Production Control	Regulatory Specialist
S	S	P	P		P			S	S		

Procedural Knowledge: Ability to apply a structured process to select a supplier based on established evaluation criteria (3).

Conceptual Knowledge: Ability to gather all applicable supplier/partner evaluation information (3). As required, ability to construct a supplier evaluation matrix (3). Ability to rank order supplier capability and capacity to meet all requirements (4).

Factual Knowledge: Knowledge of supplier/partner evaluation processes (3). Knowledge of project/program requirements (3). Knowledge of capability and capacity measurement practices (3).

IV.K.3. Flow-Down of Requirements (Including Purchase Orders)

Process: Create Purchasing Documents

Process Purpose: Create formal documents to procure products or services from suppliers/partners. The documents shall contain all applicable information or references to all applicable information related to the procured scope of work.

Knowledge and Skill Requirements: Ability to create formal documents to procure products or services from supplier/partner (e.g., purchase order, contract, etc.). Ability to verify that purchasing documents contain all applicable information or references to all applicable information related to the procured scope of work (including notifications of any changes to capabilities and/or capacity).

Task 1 (PP015-001): Create formal documents to procure products or services from supplier/partner (e.g., purchase order, contract, etc.).

Quality Eng—Basic	Quality Eng—Senior	Q Manager	Supplier QE	CSoftQE	Program QE	Contract Review	MRO QE	Spec Process QE	Met Labs QE	Production Control	Regulatory Specialist
S	S		P		S			S	S		S

Procedural Knowledge: Ability to apply a defined and structured process to construct formal documents to procure products or services from supplier/partner (e.g., purchase order, contract, etc.) (3).

Conceptual Knowledge: Ability to review and identify requirements applicable to a specific procurement (4). Ability to construct formal documents to procure products or services from supplier/partner (e.g., purchase order, contract, etc.) (3).

Factual Knowledge: Knowledge of purchasing documentation practices (3). Knowledge of requirements management tools (3). Knowledge of project/program requirements (3). Knowledge of regulatory requirements related to contractual documents (3). Knowledge of standards requirements related to purchasing documents (3).

Task 2 (PP015-002): Verify that suppliers/partners have processes in place to perform reviews of purchasing documents to their sub-tier suppliers to ensure all design, regulatory, quality, and delivery requirements are flown into contracting documents, to include proper revisions.

Quality Eng—Basic	Quality Eng—Senior	Q Manager	Supplier QE	CSoftQE	Program QE	Contract Review	MRO QE	Spec Process QE	Met Labs QE	Production Control	Regulatory Specialist
			P		S			S	S		

Procedural Knowledge: Ability to apply a defined and structured process to determine the adequacy of the processes suppliers/partners have in place to perform reviews of purchasing documents to their sub-tier suppliers to ensure all applicable design, regulatory, quality, and delivery requirements are flown into contracting documents, to include proper revisions (3).

Conceptual Knowledge: Ability to review supplier/partner processes to determine adequacy of purchasing document reviews (4). Ability to assess supplier/partner processes to determine adequacy of requirements identification (4). Ability to assess supplier/partner purchasing document processes to ensure appropriate control of configuration management (4).

Factual Knowledge: Knowledge of requirements management tools (3). Knowledge of project/program requirements (3). Knowledge of applicable regulatory requirements (3). Knowledge of purchasing documents (3). Knowledge of requirements flow-down processes (4). Knowledge of supplier assessment processes (4).

Process: Provide QA/Engineering Interpretation

Process Purpose: Provide interpretation of design and quality requirements to suppliers and to other members of the organization to ensure accurate understanding of requirements. Understand cultural differences within the supplier/partner base. Ensure any formal direction is appropriately documented. Ability to understand technical requirements. Ability to understand supplier challenges and recommendations and engage with engineering, procurement, and other functional disciplines to resolve issues.

Knowledge and Skill Requirements: Ability to provide interpretation of design and quality requirements to suppliers and to other members of the organization to ensure accurate understanding of requirements. Understand cultural differences within the supplier/partner base. Ability to document formal direction, as appropriate. Ability to understand technical requirements. Ability to understand supplier challenges and recommendations and engage with engineering, procurement, and other functional disciplines to resolve issues. Ability to schedule and verify through follow up actions that interpretation provided was accurately understood.

Task 1 (PP115-001): Provide interpretation of design, quality, and regulatory requirements to suppliers and to suppliers/partners to ensure accurate understanding of requirements.

Quality Eng—Basic	Quality Eng—Senior	Q Manager	Supplier QE	CSoftQE	Program QE	Contract Review	MRO QE	Spec Process QE	Met Labs QE	Production Control	Regulatory Specialist
S	S	S	P	S	P	S	S	S	S	S	S

Procedural Knowledge: Ability to apply defined processes in providing suppliers/partners with interpretative data (3). Ability to apply defined processes for accessing organizational personnel in providing interpretative information (3).

Conceptual Knowledge: Ability to understand and evaluate unique supplier/partner inquiries (5) Ability to determine appropriate support requirements for interpretation and route to appropriate organization members for evaluation (4).

Factual Knowledge: Understanding of contractual, regulatory, design, and quality requirements as they relate to program/project requirements (3). Knowledge of technical support resources within the organization (3). Knowledge of cultural differences as they relate to technical communication (3).

Task 2 (PP115-002): Document formal direction, as appropriate. Transmit direction to supplier/partner.

Quality Eng—Basic	Quality Eng—Senior	Q Manager	Supplier QE	CSoftQE	Program QE	Contract Review	MRO QE	Spec Process QE	Met Labs QE	Production Control	Regulatory Specialist
S	S	S	P	S	P	S	S	S	S	S	S

Procedural Knowledge: Ability to apply a structured process to document formal direction to supplier/partner (3). Ability to transmit in a closed-loop fashion, the formal direction to supplier/partner (3).

Conceptual Knowledge: Ability to understand the importance of formally documenting direction to supplier/partner and the purpose of closed-loop transmittal (2).

Factual Knowledge: Knowledge of documentation processes (3). Knowledge of closed-loop data transmittal processes (3).

Task 3 (PP115-003): Schedule and verify through follow up actions that interpretation provided was accurately understood and applied, throughout the value stream.

Quality Eng—Basic	Quality Eng—Senior	Q Manager	Supplier QE	CSoftQE	Program QE	Contract Review	MRO QE	Spec Process QE	Met Labs QE	Production Control	Regulatory Specialist
			P		S			S	S		

Procedural Knowledge: Ability to apply a structured process to schedule follow-up assessments at supplier/partner to verify that formal interpretative direction was correctly applied (3). Ability to apply a structured process to assess supplier/partner application of formal interpretative direction (3).

Conceptual Knowledge: Ability to understand workflow and affectivity of formal direction so that a follow up assessment can be scheduled to verify appropriate application (4). Ability to analyze work processes at supplier/partner to determine compliance with formal direction (4).

Factual Knowledge: Understanding of workflow processes (3). Understanding of implementation affectivity processes (3). Understanding of applicable technical processes (3). Understanding of system and process audit practices (3). Understanding of where work processes are being performed (3). Understanding of the value stream structure (3).

Process: Ensure Engineering Support Is Provided—within the Value Stream

Process Purpose: To ensure suppliers/partners have accurate technical/design data to include the assurance of configuration management. To ensure understanding of requirements is accurate by assessing the implementation of technical requirements into documented work processes.

Knowledge and Skill Requirements: Understanding of configuration management practices. Understanding of design change processes. Understanding of applicable regulatory requirements related to design changes. Understanding of the relationship between design documents and production planning. Ability to verify that suppliers/partners have accurate technical/design data to include the assurance of configuration management. Ability to verify that suppliers, partners, and sub-tier suppliers possess an understanding of requirements and such is accurate by assessing the implementation of technical requirements into documented work processes.

Task 1 (PM160-001): Verify that suppliers/partners and sub-tier suppliers have accurate technical/design data to include the assurance of configuration management.

Quality Eng—Basic	Quality Eng—Senior	Q Manager	Supplier QE	CSoftQE	Program QE	Contract Review	MRO QE	Spec Process QE	Met Labs QE	Production Control	Regulatory Specialist
S	S	S	P	S	P		S	P	P	S	

Procedural Knowledge: Ability to apply a structure approach to verify that suppliers/partners and sub-tier suppliers have accurate technical/design data to include the assurance of configuration management (3).

Conceptual Knowledge: Ability to analyze established processes to verify that suppliers/partners and sub-tier suppliers have accurate technical/design data to include the assurance of configuration management (4). Ability to audit and perform analysis of work product to verify that suppliers/partners and sub-tier suppliers have accurate technical/design data to include the assurance of configuration management (4).

Factual Knowledge: Understanding of configuration management practices (3). Understanding of organizational design data processes (3). Understanding of value stream structure and where work is being performed (3).

Task 2 (PM160-002): Verify that suppliers, partners, and sub-tier suppliers possess an understanding of requirements and such is accurate by assessing the implementation of technical requirements into documented work processes.

Quality Eng—Basic	Quality Eng—Senior	Q Manager	Supplier QE	CSoftQE	Program QE	Contract Review	MRO QE	Spec Process QE	Met Labs QE	Production Control	Regulatory Specialist
S	S	S	P	S	P		S	P	P	S	

Procedural Knowledge: Ability to apply a structured approach to verify that suppliers, partners, and sub-tier suppliers possess an understanding of requirements and such is accurate by assessing the implementation of technical requirements into documented work processes (3).

Conceptual Knowledge: Ability to analyze the implementation of business processes to verify that suppliers, partners, and sub-tier suppliers possess an understanding of requirements and such is accurate by assessing the implementation of technical requirements into documented work processes (4).

Factual Knowledge: Understanding of configuration management practices (3). Understanding of organizational design data processes (3). Understanding of value stream structure and where work is being performed (3). Understanding of audit practices (3). Understanding of requirements management tools (3).

Task 3 (PM160-003): Verify that engineering support is provided to supplier/partners and sub-tier suppliers, as required.

Quality Eng—Basic	Quality Eng—Senior	Q Manager	Supplier QE	CSoftQE	Program QE	Contract Review	MRO QE	Spec Process QE	Met Labs QE	Production Control	Regulatory Specialist
S	S	P	P	S	P		S	S	S	S	

Procedural Knowledge: Ability to apply a structured approach to verify that engineering support is provided, as required (3).

Conceptual Knowledge: Ability to analyze supplier/partner and sub-tier suppliers work processes and work product to determine the appropriate level of engineering support is provided, when required (4).

Factual Knowledge: Understanding of configuration management practices (3). Understanding of organizational design data processes (3). Understanding of value stream structure and where work is being performed (3). Understanding of audit practices (3). Understanding of value stream communication practices (3).

Process: Identify Flow Down of Unique Customer Requirements

Process Purpose: To review all applicable customer requirements documents, assess and determine that any and all applicable customer and/or regulatory requirements are conveyed in the technical data and/or purchasing documents.

Knowledge and Skill Requirements: Ability to review all applicable regulatory, customer, design, production, and quality requirements, to determine that any and all applicable flow down requirements are identified. Ability to ensure applicable requirements are conveyed in the technical data and/or purchasing documents. Ability to assess partner/supplier processes to ensure applicable requirements are conveyed to their suppliers. Ability to ensure applicable requirements reside in the requirements hierarchy tool.

Task 1 (SS005-001): Review all applicable regulatory, customer, design, production, and quality requirements, to determine any and all applicable flow down requirements are identified.

Quality Eng—Basic	Quality Eng—Senior	Q Manager	Supplier QE	CSoftQE	Program QE	Contract Review	MRO QE	Spec Process QE	Met Labs QE	Production Control	Regulatory Specialist
P	P	S	P	P	P	P		P	P		P

Procedural Knowledge: Ability to follow structured process to review applicable regulations and requirements (3).

Conceptual Knowledge: Ability to determine from a regulatory or requirement source document the implications for a program (5).

Factual Knowledge: Knowledge of source material (3). Knowledge of existing program requirements and processes (3). Knowledge of requirements management tools (3).

Task 2 (SS005-002): Ensure that proper requirement hierarchies are in place to ensure complete requirement allocation. Review identified requirements list from Task 1 of this step. Determine which requirements should be used to expand existing requirements or added to the existing requirement hierarchy. Ensure that requirements are properly decomposed to ensure adequate functional and test coverage.

Quality Eng—Basic	Quality Eng—Senior	Q Manager	Supplier QE	CSoftQE	Program QE	Contract Review	MRO QE	Spec Process QE	Met Labs QE	Production Control	Regulatory Specialist
P	P	S	P	P	P	P		P	P		P

Procedural Knowledge: Ability to follow a structured process to analyze and modify the requirement database (3).

Conceptual Knowledge: Ability to analyze and decompose requirements (4). Ability to understand and reconcile requirements as they relate to the project/program (4).

Factual Knowledge: Knowledge of existing project/program requirements (3). Knowledge of requirements management tools (3).

Process: Assemble Master Technical Data and Bid Package

Process Purpose: To create a master package of technical data, terms and conditions, and other pertinent documents that accurately communicate to the supplier the expectations of the customer.

Knowledge and Skill Requirements: Ability to create a master package of technical data, terms and conditions, and other pertinent documents needed to accurately communicate to suppliers/partner the expectations of the customer.

Task (SS020-001): Create a master package of technical data, terms and conditions, and other pertinent documents needed to accurately communicate to suppliers/partner the expectations of the customer.

Quality Eng—Basic	Quality Eng—Senior	Q Manager	Supplier QE	CSoftQE	Program QE	Contract Review	MRO QE	Spec Process QE	Met Labs QE	Production Control	Regulatory Specialist
P	P	S	P	S	P	S		S	S		

Procedural Knowledge: Ability to apply a structured process to create a master package of technical data, terms and conditions, and other pertinent documents needed to accurately communicate to suppliers/partner the expectations of the customer (3).

Conceptual Knowledge: Ability to review requirements documents and identify technical data, terms and conditions, and other pertinent documents needed to accurately communicate to suppliers/partner the expectations of the customer (4).

Factual Knowledge: Knowledge of master technical data packages (3). Knowledge of project/program requirements (3). Knowledge of applicable regulatory requirements related to technical data packages (3). Knowledge of configuration management controls (3). Knowledge of requirements management tools (3).

Process: Apply Unique Requirements to Purchasing Documents

Process Purpose: To effectively communicate to the supplier/partner any special or unique requirements. This communication is to be made on applicable purchasing documents and may be included in engineering data.

Knowledge and Skill Requirements: Ability to effectively communicate to the supplier/partner through defined and documented means of communication any special or unique requirements.

Task 1 (SS040-001): Ensure applicable requirements are conveyed in the technical data and/or purchasing documents in a specific and enforceable manner.

Quality Eng—Basic	Quality Eng—Senior	Q Manager	Supplier QE	CSoftQE	Program QE	Contract Review	MRO QE	Spec Process QE	Met Labs QE	Production Control	Regulatory Specialist
P	P	S	P	S	P	S		S	S		S

Procedural Knowledge: Based on defined criteria, determine the appropriate mechanism for conveying requirements (3). Verify that applicable requirements are included in technical and/or purchasing documents (3). Ability to participate in technical and/or purchasing document review process (3).

Conceptual Knowledge: Ability to understand requirements (3). Ability to review technical and/or purchasing documents to ensure requirements are effectively documented and communicated (4). Ability to participate in technical and/or purchasing document review process (3).

Factual Knowledge: Knowledge of requirements (3). Knowledge of document review process and tools (3).

Task 2 (SS040-002): Assess partner/supplier processes to ensure applicable requirements are provided to their suppliers in a specific and enforceable manner.

Quality Eng—Basic	Quality Eng—Senior	Q Manager	Supplier QE	CSoftQE	Program QE	Contract Review	MRO QE	Spec Process QE	Met Labs QE	Production Control	Regulatory Specialist
S	S	S	P	S	S	S		S	S		S

Procedural Knowledge: Ability to obtain objective evidence of partner/supplier flow down of requirements to their suppliers (3). Ability to apply a structured process to review flow down documentation to ensure that the documentation process is sufficient and auditable (3).

Conceptual Knowledge: Ability to analyze flow down at partners/suppliers to determine adequacy of flow down process (4). Ability to analyze actual flow down requirements in contractual documents to determine that the flow down process is effective and provides objective evidence (4).

IV.K.4. Deliverable Data

Process: Issue Purchasing Document

Process Purpose: Issue the purchasing documents to the supplier/partner. Ensure the documents are formally controlled within an established system. Purchasing documents shall ensure the correct design configuration is documented.

Knowledge and Skill Requirements: Ability to issue the purchasing documents to the supplier/partner. Ability to ensure purchasing documents are formally controlled within an established system. Ability to verify that purchasing documents define the correct design configuration. Ability to verify that supplier's/partner's issue purchasing documents to sub-tier suppliers within a controlled system and ensure configuration management is maintained.

Task 1 (PP030-001): Ensure purchasing documents are formally controlled within an established system.

Quality Eng—Basic	Quality Eng—Senior	Q Manager	Supplier QE	CSoftQE	Program QE	Contract Review	MRO QE	Spec Process QE	Met Labs QE	Production Control	Regulatory Specialist
S	S		P		P						

Procedural Knowledge: Ability to apply a defined and structured process to assess the compliance of issuance of purchasing documents to established system requirements (3).

Conceptual Knowledge: Ability to assess that purchasing documents are appropriately controlled and maintained (3).

Factual Knowledge: Knowledge of compliance assessment methods (3).

Task 2 (PP030-002): Verify that purchasing documents define the correct design configuration.

Quality Eng—Basic	Quality Eng—Senior	Q Manager	Supplier QE	CSoftQE	Program QE	Contract Review	MRO QE	Spec Process QE	Met Labs QE	Production Control	Regulatory Specialist
S	S		P		P						

Procedural Knowledge: Ability to apply a structured process to determine that purchasing documents indicate the appropriate configuration revision (3).

Conceptual Knowledge: Ability to review purchasing documents and determine compliance to the appropriate configuration revision (3).

Factual Knowledge: Knowledge of purchasing documents (3). Knowledge of configuration management processes (3).

Task 3 (PP030-003): Issue the purchasing documents to the supplier/partner.

Quality Eng—Basic	Quality Eng—Senior	Q Manager	Supplier QE	CSoftQE	Program QE	Contract Review	MRO QE	Spec Process QE	Met Labs QE	Production Control	Regulatory Specialist
			S		S						

Procedural Knowledge: Ability to apply a structured and defined process to issue purchasing documents to supplier/partner (3).

Conceptual Knowledge: Ability to issue purchasing documents to supplier/partner (2).

Factual Knowledge: Knowledge of purchasing documents (3). Knowledge of configuration management processes (3). Knowledge of document management systems (3).

Task 4 (PP030-004): Verify that supplier's/partner's issue purchasing documents to sub-tier suppliers within a controlled system and ensure configuration management is maintained.

Quality Eng—Basic	Quality Eng—Senior	Q Manager	Supplier QE	CSoftQE	Program QE	Contract Review	MRO QE	Spec Process QE	Met Labs QE	Production Control	Regulatory Specialist
	S		P		P						

Procedural Knowledge: Ability to apply a defined and structured process to assess supplier/partner systems to determine compliance of purchasing document release processes (3). Ability to apply a defined and structured process to determine that suppliers/partners are flowing down the appropriate configuration revision to sub-tier suppliers (3).

Conceptual Knowledge: Ability to assess supplier/partner systems to determine compliance of purchasing document release processes to applicable requirements (4). Ability to review supplier/partner systems and documents to determine that suppliers/partners are flowing down the appropriate configuration revision to sub-tier suppliers (4). Knowledge of supplier assessment practices (4).

Factual Knowledge: Knowledge of requirements management tools (3). Knowledge of project/program requirements (3). Knowledge of regulatory requirements applicable to supplier flow-down requirements (4). Knowledge of purchasing documents (3). Knowledge of requirements flow-down processes (4). Knowledge of supplier assessment processes (3). Knowledge of configuration management practices (3).

Process: Verify Master Technical Data and Bid Package

Process Purpose: To ensure the contents of the master technical data and bid package are complete and correct.

Knowledge and Skill Requirements: Ability to ensure the contents of the technical data and bid package are complete and correct. Ability to ensure established supplier/partner processes are adequate to ensure technical data and bid packages they provide to their suppliers are complete and correct. Ability to verify that requirements are effectively flowed down to all places where value is being created. Ability to ensure the requirements flow down process is appropriately documented to provide objective evidence of compliance.

Task 1 (SS030-001): Ensure the contents of the master technical data and bid packages are complete and correct.

Quality Eng—Basic	Quality Eng—Senior	Q Manager	Supplier QE	CSoftQE	Program QE	Contract Review	MRO QE	Spec Process QE	Met Labs QE	Production Control	Regulatory Specialist
S	P	S	P	S	P			S	S		

Procedural Knowledge: Ability to apply a structured process to assess the completeness and accuracy of master technical data and bid packages (3).

Conceptual Knowledge: Ability to review master technical data and bid packages and determine the completeness and accuracy of their content and format (4).

Factual Knowledge: Knowledge of master technical data and bid packages (3). Knowledge of configuration management practices (3).

Task 2 (SS030-002): Ensure supplier/partner processes are adequate to ensure master technical data and bid packages are flowed down to their suppliers are complete and correct.

Quality Eng—Basic	Quality Eng—Senior	Q Manager	Supplier QE	CSoftQE	Program QE	Contract Review	MRO QE	Spec Process QE	Met Labs QE	Production Control	Regulatory Specialist
S	S	S	P	S	S			S	S		

Procedural Knowledge: Ability to apply a structured process to assess the processes established at suppliers and/or partners to ensure master technical data and bid packages to their suppliers are complete and correct (3).

Conceptual Knowledge: Ability to review and assess the processes established at suppliers and/or partners to ensure master technical data and bid packages to their suppliers are complete and correct (3).

Factual Knowledge: Knowledge of master technical data and bid packages (3). Knowledge of configuration management practices (3). Knowledge of value stream structure (3). Knowledge of applicable regulatory requirements related to requirements flow down (3).

IV.K.5. Pre-Award/Post-Award Evaluations

Process: Determine Supplier Capability (Assess Who Has Capability)

Process Purpose: To understand where technical capability resides in the supplier/partner base. For existing suppliers/partners, past performance may demonstrate capability. For new suppliers/partners there should be some objective evidence established to demonstrate technical capability as it relates to the design and production requirements.

Knowledge and Skill Requirements: Ability to assess the capability of work scope execution in the supplier/partner base. Ability to review various data inputs to determine supplier/partner capability. Ability to document the results of supplier/partner work scope capability assessments.

Task 1 (SS045-001): For existing suppliers/partners. review various data inputs to determine supplier/partner capability (including dependence on key personnel).

Quality Eng—Basic	Quality Eng—Senior	Q Manager	Supplier QE	CSoftQE	Program QE	Contract Review	MRO QE	Spec Process QE	Met Labs QE	Production Control	Regulatory Specialist
S	S	S	P	S	P		S	P	P	S	S

Procedural Knowledge: Ability to apply a structured process to assess the supplier's/partner's ability to perform the type of work stated within the scope of work (3).

Conceptual Knowledge: Ability to review production and/or servicing processes within the value stream and identify which existing suppliers/partners have the ability to perform the type of work defined within the work scope (4).

Factual Knowledge: Knowledge of assessment practices (3). Knowledge of project/program requirements (3). Knowledge of applicable technical processes (3).

Task 2 (SS045-002): Assess the process capability of work scope execution within the supplier/partner base (including dependence on key personnel).

Quality Eng—Basic	Quality Eng—Senior	Q Manager	Supplier QE	CSoftQE	Program QE	Contract Review	MRO QE	Spec Process QE	Met Labs QE	Production Control	Regulatory Specialist
S	S	S	P	S	P		S	P	P	S	S

Procedural Knowledge: Ability to apply a structured process to assess the process capability of work scope execution within the supplier/partner base (3).

Conceptual Knowledge: Ability to review process performance data within the value stream to determine where process capability resides to effectively execute the scope of work (4). Ability to perform process capability studies within the value stream to verify actual process capability (4). Ability to identify and quantify risks associated with process capability (4).

Factual Knowledge: Knowledge of process capability measurement practices (3). Knowledge of audit practices (3). Knowledge of project/program requirements related to work scope execution (3). Knowledge of process performance data (3). Knowledge of risk measurement practices (3).

Task 3 (SS045-003): Document the results of supplier/partner work scope capability assessments.

Quality Eng—Basic	Quality Eng—Senior	Q Manager	Supplier QE	CSoftQE	Program QE	Contract Review	MRO QE	Spec Process QE	Met Labs QE	Production Control	Regulatory Specialist
S	S	S	P	S	P		S	P	P	S	S

Procedural Knowledge: Ability to apply a structured process in the documentation of supplier/partner work scope capability assessments (3).

Conceptual Knowledge: Ability to effectively communicate the results of supplier/partner work scope capability assessments (4). Ability to effectively communicate any risks identified during capability assessments (4). Ability to archive capability results into the organization's knowledge archive (3).

Factual Knowledge: Knowledge of process capability measurement practices (3). Knowledge of audit practices (3). Knowledge of project/program requirements related to work scope execution (3). Knowledge of process performance data (3). Knowledge of risk measurement practices (3). Knowledge of organizational knowledge archives (3).

Process: Determine Supplier Capacity (Assess Who Has Capacity to Take on Additional/Changed Work)

Process Purpose: To understand where production/servicing capacity resides in the supplier/partner base. This requires recent assessment of supplier/partner workload and an accurate understanding of their total production capacity.

Knowledge and Skill Requirements: Ability to assess the production/servicing capacity in the supplier/partner base. Ability to review various data inputs to determine supplier/partner capacity. Ability to determine the effects of material shortages on production/serving capacity. Ability to determine the effects of additional work load in the supplier/partner base on overall capacity. Ability to document the results of supplier/partner work scope capacity assessments.

Task 1 (SS050-001): Assess the production/servicing capacity in the supplier/partner base. Review various data inputs to determine supplier/partner capacity.

Quality Eng—Basic	Quality Eng—Senior	Q Manager	Supplier QE	CSoftQE	Program QE	Contract Review	MRO QE	Spec Process QE	Met Labs QE	Production Control	Regulatory Specialist
S	S	S	P	S	P		S	S	S	S	

Procedural Knowledge: Ability to apply a structured process to assess the production/servicing capacity in the value stream (3).

Conceptual Knowledge: Ability to analyze various data inputs to determine the capacity of the value stream (5). Ability to review production/servicing processes within the value stream and identify concerns related to production/servicing capacity (5) Ability to quantitatively measure the production/servicing capacity in the value stream (5). Ability to recognize and identify risks associated with production/servicing capacity within the value stream (including critical path) (5).

Factual Knowledge: Knowledge of capacity planning practices (4). Knowledge of applicable technical processes (4). Knowledge of capacity measurement processes (4). Knowledge of risk measurement practices (4). Knowledge of project/program requirements related to capacity (3). Knowledge of critical path analysis (4).

Task 2 (SS050-002): Determine the effects of potential material and resource shortages on production/serving capacity.

Quality Eng—Basic	Quality Eng—Senior	Q Manager	Supplier QE	CSoftQE	Program QE	Contract Review	MRO QE	Spec Process QE	Met Labs QE	Production Control	Regulatory Specialist
S	P	P	P	S	P		S	S	S	S	

Procedural Knowledge: Ability to apply a structured process to assess the effects of potential material and resource (incl human) shortages on production/servicing capacity of the value stream (3).

Conceptual Knowledge: Ability to identify and analyze the potential effects that material and resource shortages may have on the capacity of the value stream (4). Ability to recognize and identify risks associated with potential material and resource shortages within the value stream (4).

Factual Knowledge: Knowledge of capacity planning practices (4). Knowledge of applicable technical processes (4). Knowledge of capacity measurement processes (4). Knowledge of risk measurement practices (4). Knowledge of project/program requirements related to capacity (3). Knowledge of applicable materials (4). Knowledge of the industrial sources and availability of applicable materials and resources (4). Knowledge of critical path analysis (4).

Task 3 (SS050-003): Determine the effects of additional workload in the supplier/partner base on overall capacity.

Quality Eng—Basic	Quality Eng—Senior	Q Manager	Supplier QE	CSoftQE	Program QE	Contract Review	MRO QE	Spec Process QE	Met Labs QE	Production Control	Regulatory Specialist
S	P	P	P	S	P		S	P	S	S	

Procedural Knowledge: Ability to apply a structured process to assess the effects of additional work volume in the value stream on production/servicing capacity of the value stream (3).

Conceptual Knowledge: Ability to identify and analyze the potential effects that additional work volume may have on the capacity of the value stream (5).

Factual Knowledge: Knowledge of capacity planning practices (4). Knowledge of applicable technical processes (4). Knowledge of capacity measurement processes (4). Knowledge of risk measurement practices (4). Knowledge of project/program requirements related to capacity (3). Knowledge of the industrial sources and availability of applicable materials (4). Knowledge of available sources of qualified labor (4). Knowledge of critical path analysis (4).

Task 4 (SS050-004): Document the results of supplier/partner work scope capacity assessments.

Quality Eng—Basic	Quality Eng—Senior	Q Manager	Supplier QE	CSoftQE	Program QE	Contract Review	MRO QE	Spec Process QE	Met Labs QE	Production Control	Regulatory Specialist
S	P	P	P	S	P		S	P	S	S	

Procedural Knowledge: Ability to apply a structured process to effectively document the results of supplier/partner work scope capacity assessments (3). Ability to apply a structured process to update the organizational knowledge archive with the results of the capacity assessments (3).

Conceptual Knowledge: Ability to effectively communicate the results of supplier/partner capacity assessments (4). Ability to effectively communicate any risks identified during capacity assessments (4). Ability to archive capacity results into the organization's knowledge archive (3).

Factual Knowledge: Knowledge of capacity planning practices (4). Knowledge of applicable technical processes (4). Knowledge of capacity measurement processes (4). Knowledge of risk measurement practices (4). Knowledge of project/program requirements related to capacity (3). Ability to update the organizational knowledge archive (3).

Process: Perform Pre-Award Surveys

Process Purpose: For new suppliers/partners, or existing, who are potentially assuming work for which they have no previous performance evidence, perform an evaluation at the points where value is added, which may include sub-tier suppliers. This evaluation is to determine if the supplier/partner is potentially qualified to participate in the present procurement. Scope of the evaluation may not be as extensive as full system or process approvals.

Knowledge and Skill Requirements: Ability to determine the scope of evaluation for new suppliers/partners, or existing, who are potentially assuming work for which they have no previous performance evidence. Ability to perform an evaluation at the points where value is added, which may include sub-tier suppliers. Ability to assess compliance to requirements. Ability to assess work scope execution capability. Ability to assess supplier/partner capacity and identify future risks. Ability to document the results of supplier/partner pre-selection evaluation.

Task 1 (SS055-001): Review quality requirements and determine the scope of evaluation for suppliers/partners.

Quality Eng—Basic	Quality Eng—Senior	Q Manager	Supplier QE	CSoftQE	Program QE	Contract Review	MRO QE	Spec Process QE	Met Labs QE	Production Control	Regulatory Specialist
S	P	S	P	P	S	S	S	P	P	S	S

Procedural Knowledge: Ability to apply a defined and structured process in review of project/program quality requirements to identify and document the evaluation criteria for suppliers/partners (3).

Conceptual Knowledge: Ability to elicit unstated quality requirements applicable to suppliers/partners (4). Ability to review stated project/program quality requirements to identify and document the evaluation criteria for suppliers/partners (3). Ability to create a scope of evaluation for suppliers/partners to assess their ability to meet project/program requirements (5).

Factual Knowledge: Knowledge of requirements management tools (3). Knowledge of project/program quality requirements (3). Knowledge of applicable regulatory requirements (3). Knowledge of audit practices (3). Knowledge of evaluation planning practices (3). Knowledge of applicable technical processes (4).

Task 2 (SS055-002): Perform an evaluation at the points where value is added, which may include sub-tier suppliers.

Quality Eng—Basic	Quality Eng—Senior	Q Manager	Supplier QE	CSoftQE	Program QE	Contract Review	MRO QE	Spec Process QE	Met Labs QE	Production Control	Regulatory Specialist
S	P	S	P	P	P			P	P		

Procedural Knowledge: Ability to apply a defined and structured process to evaluate (audit) supplier's/partner's ability to comply with the defined evaluation criteria (3).

Conceptual Knowledge: Ability to analyze the processes and procedures within the value stream to determine compliance to the stated evaluation criteria (4). Ability to recognize the flow down responsibilities of sub-tier suppliers within the value stream and evaluate their ability to meet the defined evaluation criteria (4). Ability to recognize where sub-tier suppliers have not yet been determined and evaluate potential risk (4). Ability to evaluate the current state of the marketplace and its ability to fill the gap of where sub-tier suppliers have not yet been identified (5).

Factual Knowledge: Knowledge of requirements management tools (3). Knowledge of project/program quality requirements (3). Knowledge of applicable regulatory requirements (3). Knowledge of audit practices (3). Knowledge of evaluation planning practices (3). Knowledge of applicable technical processes (4). Knowledge of the market-place and its ability to provide services (4).

Task 3 (SS055-003): Assess work scope execution capability at suppliers/partners (including dependence on key personnel).

Quality Eng—Basic	Quality Eng—Senior	Q Manager	Supplier QE	CSoftQE	Program QE	Contract Review	MRO QE	Spec Process QE	Met Labs QE	Production Control	Regulatory Specialist
S	P	S	P	P	P			P	P		

Procedural Knowledge: Ability to apply a structured process to assess the supplier's/partner's ability to perform the type of work stated within the scope of work (3). Ability to apply a structured process to assess the process capability of work scope execution at suppliers/partners (3).

Conceptual Knowledge: Ability to review production and/or servicing processes at potential new suppliers/partners and determine if the proposed source has the ability to perform the type of work defined within the work scope (4). Ability to review process performance data of the new proposed source to determine if adequate process capability exists to effectively execute the scope of work (4). Ability to perform process capability studies within the value stream to determine actual process capability (4). Ability to identify and quantify risks associated with process capability (4).

Factual Knowledge: Knowledge of requirements management tools (3). Knowledge of process capability measurement practices (3). Knowledge of audit practices (3). Knowledge of project/program requirements related to work scope execution (3). Knowledge of process performance data (3). Knowledge of risk measurement practices (3). Knowledge of applicable technical processes (4).

Task 4 (SS055-004): Assess supplier/partner capacity and identify risks.

Quality Eng—Basic	Quality Eng—Senior	Q Manager	Supplier QE	CSoftQE	Program QE	Contract Review	MRO QE	Spec Process QE	Met Labs QE	Production Control	Regulatory Specialist
S	P	S	P	P	S		S	P	P	S	

Procedural Knowledge: Ability to apply a structured process to assess the production/servicing capacity in the value stream (3). Ability to apply a structured process to assess the effects of potential material and resource shortages on production/servicing capacity of the value stream (3). Ability to apply a structured process to assess the effects of additional work volume in the value stream on production/servicing capacity of the value stream (3).

Conceptual Knowledge: Ability to analyze various data inputs to determine the capacity of suppliers/partners (4). Ability to review production/servicing processes within the value stream and identify concerns related to production/servicing capacity (5) Ability to quantitatively measure the production/servicing capacity in the value stream (4). Ability to recognize and identify risks associated with production/servicing capacity within the value stream (5). Ability to identify and analyze the potential effects that material and resource shortages may have on the capacity of the value stream (5). Ability to recognize and identify risks associated with potential material and resource shortages within the value stream (5). Ability to identify and evaluate the potential effects that additional work volume may have on the capacity of the value stream (5).

Factual Knowledge: Knowledge of requirements management tools (3). Knowledge of capacity planning practices (4). Knowledge of applicable technical processes (4). Knowledge of capacity measurement processes (4). Knowledge of risk measurement practices (4). Knowledge of project/program requirements related to capacity (3). Knowledge of the industrial sources and availability of applicable materials (4). Knowledge of available sources of qualified labor (4). Knowledge of applicable materials (4).

Task 5 (SS055-005): Document the results of supplier/partner pre-selection evaluation.

Quality Eng—Basic	Quality Eng—Senior	Q Manager	Supplier QE	CSoftQE	Program QE	Contract Review	MRO QE	Spec Process QE	Met Labs QE	Production Control	Regulatory Specialist
S	P	S	P	P	S		S	P	P	S	

Procedural Knowledge: Ability to apply a structured process to document the results of supplier/partner pre-selection evaluation (3). Ability to apply a structured process to update the organizational knowledge archive with the results of the capacity assessments (3).

Conceptual Knowledge: Ability to effectively document the results of pre-selection evaluation of suppliers/partners (4). Ability to archive capacity results into the organization's knowledge archive (3).

Factual Knowledge: Knowledge of effective communication practices (4). Ability to update the organizational knowledge archive (3).

Process: Perform Financial Capability Assessment on Supplier/Partner

Process Purpose: To the extent a supplier/partner will require extensive resources to perform the stated scope of work, assess the financial capability of the supplier and their sustained ability to conduct business.

Knowledge and Skills: Ability to assess the financial capability of the supplier and their sustained ability to conduct business and ensure their ability to execute the defined work scope. Ability to assess supplier/partner sub-tier suppliers to identify processes and/or products that may require significant financial capital to deliver. Ability to evaluate the financial health of key supplier/partner sub-tier suppliers. Ability to document the results of financial health analyses.

Task 1 (SS060-001): Assess the financial capability of the supplier/partner and their sustained ability to conduct business and ensure their ability to execute the defined work scope.

Quality Eng—Basic	Quality Eng—Senior	Q Manager	Supplier QE	CSoftQE	Program QE	Contract Review	MRO QE	Spec Process QE	Met Labs QE	Production Control	Regulatory Specialist
S	S	S	S		P						

Procedural Knowledge: Ability to apply a structured process to assess the financial capability of the supplier/partner and their sustained ability to conduct business and ensure their ability to execute the defined work scope (3).

Conceptual Knowledge: Ability to elicit information from various applicable sources to analyze the financial capability of suppliers/partners and their sustained ability to conduct business and ensure their ability to execute the defined work scope (4). Ability to evaluate applicable factors to determine the financial capability of the supplier/partner and their sustained ability to conduct business and ensure their ability to execute the defined work scope (5).

Factual Knowledge: Knowledge of financial capability assessments (3). Knowledge of sources of information related to financial health of organizations (3). Knowledge of industry trends related to financial sustainability of organizations (4).

Task 2 (SS060-002): Assess supplier/partner sub-tier suppliers to identify processes and/or products that may require significant financial capital to deliver.

Quality Eng—Basic	Quality Eng—Senior	Q Manager	Supplier QE	CSoftQE	Program QE	Contract Review	MRO QE	Spec Process QE	Met Labs QE	Production Control	Regulatory Specialist
S	S	S	P	S	P		S	S	S	S	

Procedural Knowledge: Ability to apply a structured process within the value stream to identify processes and/or products that may require significant financial capital to deliver (3).

Conceptual Knowledge: Ability to review production/servicing processes within the value stream to identify processes and/or products that may require significant financial capital to deliver (4). Ability to identify risks associated with capital intensive equipment and/or materials (4).

Factual Knowledge: Knowledge of applicable technical processes (4). Knowledge of evaluation practices (3).

Task 3 (SS060-003): Evaluate the financial health of key supplier/partner sub-tier suppliers.

Quality Eng—Basic	Quality Eng—Senior	Q Manager	Supplier QE	CSoftQE	Program QE	Contract Review	MRO QE	Spec Process QE	Met Labs QE	Production Control	Regulatory Specialist
S	P	S	P		P						

Procedural Knowledge: Ability to apply a structured process to assess the financial capability of key sub-tier suppliers and their sustained ability to conduct business and ensure their ability to execute the defined work scope (3).

Conceptual Knowledge: Ability to elicit information from various applicable sources to analyze the financial capability of key sub-tier suppliers and their sustained ability to conduct business and ensure their ability to execute the defined work scope (4). Ability to evaluate applicable factors to determine the financial capability of key sub-tier suppliers and their sustained ability to conduct business and ensure their ability to execute the defined work scope (5).

Factual Knowledge: Knowledge of financial capability assessments (3). Knowledge of sources of information related to financial health of organizations (3). Knowledge of industry trends related to financial sustainability of organizations (4).

Task 4 (SS060-004): Document the results of financial health analyses.

Quality Eng—Basic	Quality Eng—Senior	Q Manager	Supplier QE	CSoftQE	Program QE	Contract Review	MRO QE	Spec Process QE	Met Labs QE	Production Control	Regulatory Specialist
S	P	S	P	S	P		S	S	S	S	

Procedural Knowledge: Ability to apply a structured process to effectively document the results of financial health analysis (3). Ability to apply a structured process to update the organizational knowledge archive with the results of the financial health analyses (3).

Conceptual Knowledge: Ability to effectively document the results of value stream financial health analysis (4). Ability to archive financial health analyses results into the organization's knowledge archive (3).

Factual Knowledge: Knowledge of financial health analysis (4). Knowledge of applicable technical processes (4). Knowledge of industry trends related to financial health of organizations (4). Knowledge of risk measurement practices (4). Ability to update the organizational knowledge archive (3).

Process: Verify Suppliers' Technical Capabilities

Process Purpose: To physically evaluate and ensure the supplier/partner can perform to the requirements of the technical data and purchasing documents. This verification shall also ensure the supplier has the ability to interpret the design media provided.

Knowledge and Skill Requirements: Ability to define evaluation scope to ensure suppliers/partners have adequate technical capability to execute the scope of intended work. Ability to physically evaluate and ensure the supplier/partner can perform to the requirements of the technical data and purchasing documents. Ability to document any risks discovered during evaluation. Ability to document evaluation results.

Task 1 (SS110-001): Define evaluation scope to ensure suppliers/partners have adequate technical capability to execute the scope of intended work.

Quality Eng—Basic	Quality Eng—Senior	Q Manager	Supplier QE	CSoftQE	Program QE	Contract Review	MRO QE	Spec Process QE	Met Labs QE	Production Control	Regulatory Specialist
S	P	S	P	S	S		S	P	P	S	

Procedural Knowledge: Ability to apply a structured process in review of requirements and identify audit/evaluation scope criteria to determine if suppliers/partners have adequate technical capability to execute the scope of intended work (3). Ability to apply a structured process to document audit/evaluation scope (3).

Conceptual Knowledge: Ability to analyze requirements and identify audit/evaluation scope criteria to determine if suppliers/partners have adequate technical capability to execute the scope of intended work (4). Ability effectively plan and document an audit/evaluation. (3).

Factual Knowledge: Understanding of applicable technical processes (4). Understanding of audit scope planning processes (3). Understanding of audit practices (3). Knowledge of requirements management system (3).

Task 2 (SS110-002): Physically evaluate and ensure the supplier/partner can perform to the requirements of the technical data and purchasing documents. Document the evaluation results.

Quality Eng—Basic	Quality Eng—Senior	Q Manager	Supplier QE	CSoftQE	Program QE	Contract Review	MRO QE	Spec Process QE	Met Labs QE	Production Control	Regulatory Specialist
S	S		P	S	S		S	P	P	S	

Procedural Knowledge: Ability to apply a structured process to perform an evaluation of supplier/partner technical processes (3). Ability to apply a structured process to perform an evaluation of supplier/partner business processes (3). Ability to apply a structured process to document results of an evaluation (3).

Conceptual Knowledge: Ability to analyze supplier/partner technical processes to determine compliance and capability to perform to the requirements of the technical data and purchasing documents (4). Ability to analyze supplier/partner business processes to determine compliance and capability to perform to the requirements of the technical data and purchasing documents (4). Ability to effectively document results of an evaluation (3).

Factual Knowledge: Understanding of applicable technical processes (4). Understanding of audit scope planning processes (3). Understanding of audit practices (3). Knowledge of requirements management systems (3).

Task 3 (SS110-003): Document any risks discovered during evaluation.

Quality Eng—Basic	Quality Eng—Senior	Q Manager	Supplier QE	CSoftQE	Program QE	Contract Review	MRO QE	Spec Process QE	Met Labs QE	Production Control	Regulatory Specialist
S	S	S	P	S	P		S	P	P	S	

Procedural Knowledge: Ability to apply a structured process to identify and document risks associated with noncompliance, capability concerns, and capacity constraints found during evaluation at supplier/partner facilities (3).

Conceptual Knowledge: Ability to analyze risks associated with noncompliance, capability concerns, and capacity constraints found during evaluation at supplier/partner facilities and establish risk measurement to findings (4).

Factual Knowledge: Understanding of risk assessment and measurement processes (4). Understanding of special processes (4). Understanding of non-destructive processes (4). Understanding of audit scope planning processes (3). Understanding of audit practices (3). Knowledge of quality requirements (3).

IV.K.6. Quality and Process Approvals (ASL)

Process: Ensure Organizational Approvals Match Approved Commodity Types

Process Purpose: Review the approvals granted to the organization and ensure the approvals match the products or services the organization is capable of providing. The approvals shall note the scope of system approval, product approvals, and special process approvals. (i.e., Unique approvals may include commodities such as forgings, castings, ESD controls, NDT, special processes, etc., as well as, governmental certifications such a production and/or repair approvals.)

Knowledge and Skill Requirements: Ability to review and verify that the approvals granted to the organization match the products or services the organization is providing. The approvals shall note the scope of system approval, product approvals, and special process approvals. (i.e., Unique approvals may include commodities such as forgings, castings, ESD controls, NDT, special processes, etc.). Ability to verify that organization is validating the certifications its various departmental aspects, as well as, processes for ensuring certifications into the supplier/partner base. Knowledge of industry certification schemes and governmental certification requirements.

Task 1 (PO025-001): Review and verify that the approvals granted to the organization match the products or services the organization is providing. The approvals shall note the scope of system approval, product approvals, and special process approvals. (i.e., Unique approvals may include commodities such as forgings, castings, ESD controls, NDT, special processes, etc.).

Quality Eng—Basic	Quality Eng—Senior	Q Manager	Supplier QE	CSoftQE	Program QE	Contract Review	MRO QE	Spec Process QE	Met Labs QE	Production Control	Regulatory Specialist
S	P	P	S	S	P	P	S	S	S	S	P

Procedural Knowledge: Ability to apply a structured process to review and verify that the approvals granted to the organization match the products or services the organization is providing (3).

Conceptual Knowledge: Ability to determine special process approvals required for a specific scope of work assigned to the organization (4). Ability to review and verify that the approvals granted to the organization match the products or services the organization is providing (4).

Factual Knowledge: Knowledge of industry-managed process certification programs (4). Knowledge of regulatory certification requirements (4). Knowledge of requirements management tools (3). Knowledge of project/program requirements (3). Knowledge of organizational work processes (3). Knowledge of value stream structure (3).

Task 2 (PO025-002): Verify that organization is validating the certifications its suppliers and shop-assist facilities hold to ensure suppliers are approved to perform the scope of work assigned.

Quality Eng—Basic	Quality Eng—Senior	Q Manager	Supplier QE	CSoftQE	Program QE	Contract Review	MRO QE	Spec Process QE	Met Labs QE	Production Control	Regulatory Specialist
P	P	P	S	S	P		P	S	S	P	

Procedural Knowledge: Ability to apply a structured process to review and verify that the organization has adequate control processes to ensure the approvals granted to suppliers/partner match the products or services the organization is providing (3).

Conceptual Knowledge: Ability to assess the effectiveness and adequacy of the organization's processes to ensure suppliers/partner have adequate control processes to ensure the approvals granted to suppliers/partners match the products or services the organization (4).

Factual Knowledge: Knowledge of industry-managed process certification programs (4). Knowledge of regulatory certification requirements (4). Knowledge of requirements management tools (3). Knowledge of project/program requirements (3). Knowledge of organizational work processes (3). Knowledge of value stream structure (3).

Process: Ensure Supplier Approvals Match Approved Commodity Types

Process Purpose: Review the approvals granted to the supplier/partner and ensure the approvals match the products or services the supplier/partner is providing. The approvals shall note the scope of system approval, product approvals, and special process approvals. (i.e., Unique approvals may include commodities such as forgings, castings, ESD controls, NDT, special processes, etc.)

Knowledge and Skill Requirements: Ability to review and verify that the approvals granted to the supplier/partner match the products or services the supplier/partner is providing. The approvals shall note the scope of system approval, product approvals, and special process approvals. (i.e., Unique approvals may include commodities such as forgings, castings, ESD controls, NDT, special processes, etc.). Ability to verify that suppliers/partners are validating the certifications their sub-tier suppliers hold to ensure sub-tier suppliers are approved to perform the scope of work assigned. Knowledge of industry certification schemes.

Task 1 (PP025-001): Review and verify that the approvals granted to the supplier/partner match the products or services the supplier/partner is providing. The approvals shall note the scope of system approval, product approvals, and special process approvals. (i.e., Unique approvals may include commodities such as forgings, castings, ESD controls, NDT, special processes, etc.).

Quality Eng—Basic	Quality Eng—Senior	Q Manager	Supplier QE	CSoftQE	Program QE	Contract Review	MRO QE	Spec Process QE	Met Labs QE	Production Control	Regulatory Specialist
S	S	S	P	S	P			P	P		

Procedural Knowledge: Ability to apply a structured process to review and verify that the approvals granted to the supplier/partner match the products or services the supplier/partner is providing (3).

Conceptual Knowledge: Ability to determine special process approvals required for a specific scope of work assigned to a supplier/partner (4). Ability to review and verify that the approvals granted to the supplier/partner match the products or services the supplier/partner is providing (4).

Factual Knowledge: Knowledge of industry-managed process certification programs (4). Knowledge of requirements management tools (3). Knowledge of project/program requirements (3). Knowledge of value stream structure (3).

Task 2 (PP025-002): Verify that suppliers/partners are validating the certifications their sub-tier suppliers hold to ensure sub-tier suppliers are approved to perform the scope of work assigned.

Quality Eng—Basic	Quality Eng—Senior	Q Manager	Supplier QE	CSoftQE	Program QE	Contract Review	MRO QE	Spec Process QE	Met Labs QE	Production Control	Regulatory Specialist
S	S	S	P		P			P	P		

Procedural Knowledge: Ability to apply a structured process to review and verify that suppliers/partners have adequate control processes to ensure the approvals granted to sub-tier suppliers match the products or services the supplier/partner is providing (3).

Conceptual Knowledge: Ability to assess the effectiveness and adequacy of suppliers/partners processes to ensure sub-tier suppliers have adequate control processes to ensure the approvals granted to subsequent sub-tier suppliers match the products or services the supplier/partner is providing (4).

Factual Knowledge: Knowledge of industry-managed process certification programs (4). Knowledge of requirements management tools (3). Knowledge of project/program requirements (3). Knowledge of value stream structure (3).

Process: Add Supplier to Approved Supplier List

Process Purpose: Once a supplier/partner is formally approved, ensure they are listed on the organization's list of approved suppliers. The approved suppliers list should be made available to all points where its use is expected. This may require updating of electronic databases or the reissuance of paper documents. Approved suppliers shall be identified with their specific product and system capabilities (e.g., heat treat, castings, electronics. etc.).

Knowledge and Skill Requirements: Ability to ensure approved suppliers/partners are listed on the organization's list of approved suppliers. Ability to ensure the scope and limitations of supplier/partner approvals are well defined. Ability to assess existing processes to determine if approved suppliers lists are made available to all points where its use is expected (includes throughout the value stream).

Task 1 (SS100-001): Ensure approved suppliers/partners are listed on the organization's list of approved suppliers.

Quality Eng—Basic	Quality Eng—Senior	Q Manager	Supplier QE	CSoftQE	Program QE	Contract Review	MRO QE	Spec Process QE	Met Labs QE	Production Control	Regulatory Specialist
P	P	P	P		S			P	P		

Procedural Knowledge: Ability to apply a defined and structured process to place new and/or revised suppliers/partners on the organization's list of approved suppliers (3).

Conceptual Knowledge: Ability to verify the applicability of adding and/or revising suppliers/partners on the organization's list of approved suppliers (3).

Factual Knowledge: Knowledge of supplier approval processes (3). Knowledge of approved supplier lists (3).

Task 2 (SS100-002): Ensure the scope and limitations of supplier/partner approvals are well defined.

Quality Eng—Basic	Quality Eng—Senior	Q Manager	Supplier QE	CSoftQE	Program QE	Contract Review	MRO QE	Spec Process QE	Met Labs QE	Production Control	Regulatory Specialist
P	P	P	P		S			P	P		

Procedural Knowledge: Ability to apply a defined and structured process to identify and document the scope and applicable limitations of supplier/partner approvals (3).

Conceptual Knowledge: Ability to analyze supplier evaluation results and determine appropriate scope of approval and appropriate limitations (4). Ability to ensure scope of supplier/partner approvals are included on applicable organizational approved supplier listings (3).

Factual Knowledge: Understanding of supplier approval process (3). Knowledge of supplier evaluation processes (3). Understanding of approved supplier list control (3).

Task 3 (SS100-003): Ensure availability of access and the adequacy of established procedures to ensure approved suppliers lists are made available to all points where its use is expected (includes throughout the value stream). As required for programs with partners, edit the approved supplier listing so that a useful subset of the supplier list is provided.

Quality Eng—Basic	Quality Eng—Senior	Q Manager	Supplier QE	CSoftQE	Program QE	Contract Review	MRO QE	Spec Process QE	Met Labs QE	Production Control	Regulatory Specialist
P	P	P	P		P			P	P		P

Procedural Knowledge: Ability to apply a defined and structured process to assess the availability of access and the adequacy of established procedures to ensure approved suppliers lists are made available to all points where its use is expected (includes throughout the value stream) (3). Ability to apply a structured process to edit the approved supplier list into a subset that is relevant to a specific supplier/partner (3).

Conceptual Knowledge: Ability to audit the accessibility of approved supplier lists at all points where its usage is required (3). Ability to evaluate the adequacy of established procedures to ensure approved suppliers lists are made available to all points where its use is expected (includes throughout the value stream) (3). Ability to analyze project/program specific requirements to determine a subset of approved suppliers for a specific supplier/partner (4).

Factual Knowledge: Knowledge of regulatory requirements applicable to supplier approvals (3). Knowledge of industry standards related to supplier approvals (3). Knowledge of project/program requirements related to supplier approvals (3). Knowledge of procedural review practices (3). Knowledge of value stream structure (3). Knowledge of supplier/partner supplier management requirements (3).

IV.K.7. Sub-tier Supplier Control

Process: Ensure Supplier Has an Adequate Supplier Control System

Process Purpose: Evaluate the supplier/partner's supplier management processes and ensure the supplier/partners has effective control over the work processes sourced at sub-tier suppliers. This evaluation shall ensure that there is effective control at all levels of the supplier-base. Suppliers/Partners must have capability and capacity to ensure all work subcontracted to sub-tier suppliers. May require physical audit at sub-tier suppliers. Supplier/Partner should have formal monitoring systems in place to assess supplier performance. Supplier control processes shall also possesses controls for managing noncompliance and nonconformance, to include disposition and corrective action. (The general expectation is for the organization to substantiate through evaluation compliance to requirements throughout the entire value stream.)

Knowledge and Skill Requirements: Ability to schedule and plan the audit scope to assess a supplier's/partner's quality management system in order to determine compliance to defined requirements and to identify potential risk factors related to management system controls. Ability to evaluate the supplier/partner's supplier management processes to verify that supplier/partners have effective control over work processes sourced. Ability to evaluate the supplier/partner's supplier management processes to verify that supplier/partners have effective control over work processes sourced at sub-tier suppliers. Ability to evaluate supplier's/partner's capability and capacity to ensure conformity of work subcontracted to sub-tier suppliers. Ability to evaluate supplier/partner sub-tier supplier performance monitoring systems to verify effective oversight. Ability to evaluate supplier control processes to validate that process controls are effective for managing noncompliance and nonconformance, to include disposition and corrective action.

Task 1 (PP060-001): Schedule and plan the audit scope to assess a supplier's/partner's quality management system in order to determine compliance to defined requirements and to identify potential risk factors related to management system controls.

Quality Eng—Basic	Quality Eng—Senior	Q Manager	Supplier QE	CSoftQE	Program QE	Contract Review	MRO QE	Spec Process QE	Met Labs QE	Production Control	Regulatory Specialist
			P	S	P			S	S		

Procedural Knowledge: Ability to apply a defined and structured process to schedule, plan, and document the audit scope of assessment of a supplier's/partner's quality management system in order to determine compliance to defined requirements and to identify potential risk factors related to management system controls (3). Ability to apply a structured process to schedule an audit at supplier/partner locations, to include sub-tier supplier, if applicable (3).

Conceptual Knowledge: Ability to schedule, plan, and document the audit scope of assessment of a supplier's/partner's quality management system in order to determine compliance to defined requirements and to identify potential risk factors related to management system controls (4). Ability to interface with appropriate organizational members to schedule an audit at supplier/partner locations, to include sub-tier supplier, if applicable (3).

Factual Knowledge: Knowledge of quality management system auditing methods, to include scheduling, planning, and documenting (4). Knowledge of value stream structure (3). Knowledge of applicable technology (3). Knowledge of applicable product/service (3). Knowledge of risk assessments (4).

Task 2 (PP060-002): Conduct evaluation of the supplier/partner's supplier management processes and procedures to verify that supplier/partners have adequately defined control over work processes sourced to their sub-tier suppliers.

Quality Eng—Basic	Quality Eng—Senior	Q Manager	Supplier QE	CSoftQE	Program QE	Contract Review	MRO QE	Spec Process QE	Met Labs QE	Production Control	Regulatory Specialist
			P	S	P			S	S		

Procedural Knowledge: Ability to apply a structured process to perform an evaluation of the supplier/partner's supplier management processes and procedures to verify that supplier/partners have effective control over work processes sourced to their sub-tier suppliers (3).

Conceptual Knowledge: Ability to perform an evaluation of the supplier/partner's supplier management processes and procedures to verify that supplier/partners have effective control over work processes sourced to their sub-tier suppliers (4).

Factual Knowledge: Knowledge of quality management system auditing methods (4). Knowledge of value stream structure (3). Knowledge of applicable technology (3). Knowledge of applicable product/service (3). Knowledge of risk assessments (4). Knowledge of value stream structure (3). Understanding of applicable quality management standards (3).

Task 3 (PP060-003): Evaluate the deployed supplier/partner's supplier management processes to verify that supplier/partners have effective control over work processes sourced at sub-tier suppliers.

Quality Eng—Basic	Quality Eng—Senior	Q Manager	Supplier QE	CSoftQE	Program QE	Contract Review	MRO QE	Spec Process QE	Met Labs QE	Production Control	Regulatory Specialist
			P	S	P			S	S		

Procedural Knowledge: Ability to apply a structured process to perform an evaluation of the supplier/partner's deployed supplier management system to verify that supplier/partners have effective control over work processes sourced to their sub-tier suppliers (3).

Conceptual Knowledge: Ability to perform an evaluation of the supplier/partner's deployed supplier management system to verify that supplier/partners have effective control over work processes sourced to their sub-tier suppliers (4).

Factual Knowledge: Knowledge of quality management system auditing methods (4). Knowledge of value stream structure (3). Knowledge of applicable technology (3). Knowledge of applicable product/service (3). Knowledge of risk assessments (4). Knowledge of value stream structure (3). Knowledge of supplier quality management methods (3). Knowledge of flow down requirements for supplier management (3).

Task 4 (PP060-004): Evaluate suppliers/partners capability and capacity to ensure conformity of work subcontracted to sub-tier suppliers.

Quality Eng—Basic	Quality Eng—Senior	Q Manager	Supplier QE	CSoftQE	Program QE	Contract Review	MRO QE	Spec Process QE	Met Labs QE	Production Control	Regulatory Specialist
			P	S	S			S	S		

Procedural Knowledge: Ability to apply a structured process to evaluate the adequacy of resources at the supplier/partner to exercise effective control over their sub-tier suppliers (3). Ability to apply a structured process to evaluate the adequacy of capability of supplier/partner to exercise effective control over their sub-tier suppliers (3).

Conceptual Knowledge: Ability to evaluate the adequacy of resources at the supplier/partner to exercise effective control over their sub-tier suppliers (5). Ability to evaluate the adequacy of capability of supplier/partner to exercise effective control over their sub-tier suppliers (5).

Factual Knowledge: Knowledge of quality management system auditing methods (3). Knowledge of value stream structure (3). Knowledge of applicable technology (3). Knowledge of applicable product/service (3). Knowledge of risk assessments (3). Knowledge of value stream structure (3). Knowledge of supplier quality management methods (3). Knowledge of flow down requirements for supplier management (3).

Task 5 (PP060-005): Evaluate supplier control processes to validate that process controls are effective for managing noncompliance and nonconformance, to include disposition and corrective action.

Quality Eng—Basic	Quality Eng—Senior	Q Manager	Supplier QE	CSoftQE	Program QE	Contract Review	MRO QE	Spec Process QE	Met Labs QE	Production Control	Regulatory Specialist
			P	S	S			S	S		

Procedural Knowledge: Ability to apply a structured process to evaluate supplier control processes to validate that process controls are effective for managing noncompliance and nonconformities, to include disposition and corrective action (3).

Conceptual Knowledge: Ability to evaluate supplier control processes to determine that process controls are effective for managing noncompliance and nonconformance, to include disposition and corrective action (4).

Factual Knowledge: Knowledge of quality management system auditing methods (3). Knowledge of value stream structure (3). Knowledge of applicable technology (3). Knowledge of applicable product/service (3). Knowledge of material review processes (control of nonconforming material) (3). Knowledge of value stream structure (3). Understanding of root cause investigation tools (4). Understanding of corrective action processes (4). Knowledge of flow down requirements for supplier management (3).

Process: Ensure Supplier's Utilize Source Approval List for Sub-Tier Selection (as required)

Process Purpose: To ensure suppliers/partners only utilize approved sources when required by design or other formal document. Use of approved suppliers does not absolve suppliers/partners from the responsibility of sub-tier supplier management.

Knowledge and Skill Requirements: Ability to verify that suppliers/partners only utilize approved sources when required by design or other formal document. Ability to verify that sub-tier suppliers who source work to other sub-tier suppliers, utilize only approved sources as documented on applicable source approval lists, as required.

Task 1 (PP065-001): Verify that suppliers/partners only utilize approved sources when required by design or other formal document.

Quality Eng—Basic	Quality Eng—Senior	Q Manager	Supplier QE	CSoftQE	Program QE	Contract Review	MRO QE	Spec Process QE	Met Labs QE	Production Control	Regulatory Specialist
S	S	S	P		S			S	S		

Procedural Knowledge: Ability to apply a structured process to assess supplier/partner processes to verify that only approved sources are utilized when required by design or other formal document (3).

Conceptual Knowledge: Ability to review and assess supplier/partner processes to verify that only approved sources are utilized when required by design or other formal document (4).

Factual Knowledge: Knowledge of quality management system auditing methods (4). Knowledge of value stream structure (3). Knowledge of applicable technology (3). Knowledge of applicable product/service (3). Knowledge of process approval requirements (3). Knowledge of flow down requirements for supplier management (3).

Task 2 (PP065-002): Verify that sub-tier suppliers who source work to other sub-tier suppliers, utilize only approved sources as documented on applicable source approval lists, as required.

Quality Eng—Basic	Quality Eng—Senior	Q Manager	Supplier QE	CSoftQE	Program QE	Contract Review	MRO QE	Spec Process QE	Met Labs QE	Production Control	Regulatory Specialist
S	S	S	P		S			S	S		

Procedural Knowledge: Ability to apply a structured process to assess supplier/partner resourced work at sub-tier suppliers to verify that only approved sources are utilized when required by design or other formal document (3).

Conceptual Knowledge: Ability to review and assess supplier/partner resourced work at sub-tier suppliers to verify that only approved sources are utilized when required by design or other formal document (4).

Factual Knowledge: Knowledge of quality management system auditing methods (4). Knowledge of value stream structure (3). Knowledge of applicable technology (3). Knowledge of applicable product/service (3). Knowledge of process approval requirements (3). Knowledge of flow down requirements for supplier management (3).

Process: Monitor Supplier Sub-Tier Control

Process Purpose: To ensure suppliers/partners are effectively controlling their supply base by ensuring requirements are flowed through purchasing documents and complied with.

Knowledge and Skill Requirements: Understanding of quality management systems. Understanding of QMS standards. Understanding of applicable regulatory requirements related to supplier control and product conformity. Ability to verify that suppliers/partners are effectively controlling their supply base by ensuring requirements are flowed through purchasing documents and complied with. Ability to evaluate supplier/partner supplier control process and ensure they are adequately monitoring the quality of processes, products and services.

Task 1 (PM155-001): Verify that suppliers/partners are effectively controlling their supply base by ensuring requirements are flowed through purchasing documents and complied with.

Quality Eng—Basic	Quality Eng—Senior	Q Manager	Supplier QE	CSoftQE	Program QE	Contract Review	MRO QE	Spec Process QE	Met Labs QE	Production Control	Regulatory Specialist
S	S		P		S			S	S		

Procedural Knowledge: Ability to apply a structured approach to verify that suppliers/partners are effectively controlling their supply base by ensuring requirements are flowed through purchasing documents and complied with (3). Ability to apply a structured approach to create a hierarchy (requirements management tool) that includes all requirements flow down documents (3).

Conceptual Knowledge: Ability to establish a requirements hierarchy that defines the flow of requirements throughout the value stream (4). Ability to analyze supplier/partner purchasing and supplier quality management processes to verify that suppliers/partners are effectively controlling their supply base by ensuring requirements are flowed through purchasing documents and complied with (4). Ability to review purchasing and/or contractual documents to verify that suppliers/partners are effectively flowing requirements into the value stream (4).

Factual Knowledge: Understanding of QMS standards related to purchasing documents (3). Understanding of applicable regulatory requirements related to supplier control and product conformity (3). Understanding of value stream structure and where work is being performed (3). Understanding of requirement flow down practices (3). Understanding of requirements management tools (3).

Task 2 (PM155-002): Evaluate supplier/partner supplier control process and ensure they are adequately monitoring the quality of processes, products and services from their sub-tier suppliers (process evaluation).

Quality Eng—Basic	Quality Eng—Senior	Q Manager	Supplier QE	CSoftQE	Program QE	Contract Review	MRO QE	Spec Process QE	Met Labs QE	Production Control	Regulatory Specialist
S	S	S	P		S			S	S		

Procedural Knowledge: Ability to apply a structured assessment process to verify that suppliers/partners are adequately monitoring the quality of processes, products and services from their sub-tier suppliers (process evaluation) (3).

Conceptual Knowledge: Ability to evaluate suppliers/partners processes to verify that they are adequately monitoring the quality of processes, products and services from their sub-tier suppliers (process evaluation) (5). Ability to evaluate supplier/partner performance to verify that they are adequately monitoring the quality of processes, products and services from their sub-tier suppliers (process evaluation) (5).

Factual Knowledge: Understanding of QMS standards related to purchasing documents (3). Understanding of applicable regulatory requirements related to supplier control and product conformity (3). Understanding of value stream structure and where work is being performed (3). Understanding of requirement flow down practices (3). Understanding of requirements management tools (3). Understanding of performance measures (3). Understanding of product/services (3). Understanding of auditing practices (3).

IV.K.9. Direct/Drop Shipments

Process: Ensure Direct-Ship Approvals (Shipments Directly from Supplier to Customer)

Process Purpose: To provide for a means to control shipments of product from one supplier to another, this is not directly controlled by the receiving supplier. Ensure appropriate documented approvals are maintained to permit direct-shipments between suppliers. Ensure positive quality control is maintained throughout supplier/partner base.

Knowledge and Skill Requirements: Understanding of regulatory requirements related to direct-ship authorization. Understanding of organizational procedures related to direct-ship processes. Ability to create appropriate forms related to direct-ship processes. Ability to verify procedures are in place to control direct-ship processes. Ability to verify that appropriate documented approvals are maintained to permit direct-shipments between suppliers. Ability to verify that appropriate documented approvals are maintained to permit direct-shipments between suppliers and customers. Ability to evaluate that positive quality control is maintained throughout supplier/partner base for the management of direct ship processes.

Task 1 (PM140-001): Verify procedures are in place to control direct-ship processes in accordance with requirements.

Quality Eng—Basic	Quality Eng—Senior	Q Manager	Supplier QE	CSoftQE	Program QE	Contract Review	MRO QE	Spec Process QE	Met Labs QE	Production Control	Regulatory Specialist
P	P	P	P		P	P					P

Procedural Knowledge: Ability to apply a structured process to verify that procedures are in place to control direct-ship processes in accordance with requirements (3).

Conceptual Knowledge: Ability analyze supplier/partner procedures and process to determine compliance of defined direct-ship activities (4).

Factual Knowledge: Understanding of regulatory requirements related to direct-shipment (3). Understanding of contract and project/program requirements related to direct-shipment (3). Understanding of requirements management tools (3). Understanding of direct-shipment practices (3). Understanding of value stream structure and where work is being performed (3).

Task 2 (PM140-002): Verify that appropriate documented approvals are maintained to permit direct-shipments between suppliers.

Quality Eng—Basic	Quality Eng—Senior	Q Manager	Supplier QE	CSoftQE	Program QE	Contract Review	MRO QE	Spec Process QE	Met Labs QE	Production Control	Regulatory Specialist
S	P	P	P		P	S					P

Procedural Knowledge: Ability to apply a structured process to verify that appropriate documented approvals are maintained to permit direct-shipments between suppliers (3).

Conceptual Knowledge: Ability to assess and verify that appropriate documented approvals are maintained to permit direct-shipments between suppliers (3).

Factual Knowledge: Understanding of regulatory requirements related to direct-shipment (3). Understanding of contract and project/program requirements related to direct-shipment (3). Understanding of requirements management tools (3). Understanding of direct-shipment practices (3). Understanding of value stream structure and where work is being performed (3).

Task 3 (PM140-003): Verify that appropriate documented approvals are maintained to permit direct-shipments between suppliers and customers.

Quality Eng—Basic	Quality Eng—Senior	Q Manager	Supplier QE	CSoftQE	Program QE	Contract Review	MRO QE	Spec Process QE	Met Labs QE	Production Control	Regulatory Specialist
P	P	P	P		P	S					P

Procedural Knowledge: Ability to apply a structured process to verify that appropriate documented approvals are maintained to permit direct-shipments between suppliers and customers (3).

Conceptual Knowledge: Ability to assess and verify that appropriate documented approvals are maintained to permit direct-shipments between suppliers and customers (3).

Factual Knowledge: Understanding of regulatory requirements related to direct-shipment (3). Understanding of contract and project/program requirements related to direct-shipment (3). Understanding of requirements management tools (3). Understanding of direct-shipment practices (3). Understanding of value stream structure and where work is being performed (3).

Task 4 (PM140-004): Evaluate direct-ship processes to verify that positive quality control is maintained throughout supplier/partner base.

Quality Eng—Basic	Quality Eng—Senior	Q Manager	Supplier QE	CSoftQE	Program QE	Contract Review	MRO QE	Spec Process QE	Met Labs QE	Production Control	Regulatory Specialist
P	P	P	P		P	S					P

Procedural Knowledge: Ability to apply a structured approach to assess direct-ship processes and verify that positive quality control is maintained throughout supplier/partner base (3).

Conceptual Knowledge: Ability to analyze procedures, quality records, and work product to assess the compliance and adequacy direct-ship processes and verify that positive quality control is maintained throughout supplier/partner base (4).

Factual Knowledge: Understanding of regulatory requirements related to direct-shipment (3). Understanding of contract and project/program requirements related to direct-shipment (3). Understanding of requirements management tools (3). Understanding of direct-shipment practices (3). Understanding of value stream structure and where work is being performed (3).

Process: Ensure Direct-Ship Approvals (Shipments Directly from Organization to Customer)

Process Purpose: When the organization is not the end-item producer, to provide for a means to control direct shipments of product from the organization to the end-item customer. Ensure appropriate documented approvals are maintained to permit direct-shipments to the end-item customer or other defined destinations. Ensure positive quality control is maintained.

Knowledge and Skill Requirements: Understanding of regulatory requirements related to direct-ship authorization. Understanding of organizational procedures related to direct-ship processes. Ability to create appropriate forms related to direct-ship processes. Ability to verify procedures are in place to control direct-ship processes. Ability to verify that appropriate documented approvals are maintained to permit direct-shipments to customers and other defined destinations. Ability to verify that appropriate documented approvals are maintained to permit direct-shipments. Ability to evaluate that positive quality control is maintained for the management of direct ship processes.

Task 1 (PM141-001): Verify procedures are in place to control direct-ship processes in accordance with requirements. (Organizational)

Quality Eng—Basic	Quality Eng—Senior	Q Manager	Supplier QE	CSoftQE	Program QE	Contract Review	MRO QE	Spec Process QE	Met Labs QE	Production Control	Regulatory Specialist
P	P	P			P	P	S			S	P

Procedural Knowledge: Ability to apply a structured process to verify that procedures are in place to control direct-ship processes in accordance with requirements (3).

Conceptual Knowledge: Ability analyze organizational procedures and process to determine compliance of defined direct-ship activities (4).

Factual Knowledge: Understanding of regulatory requirements related to direct-shipment (3). Understanding of contract and project/program requirements related to direct-shipment (3). Understanding of requirements management tools (3). Understanding of direct-shipment practices (3). Understanding of value stream structure and where work is being performed (3).

Task 2 (PM141-002): Verify that appropriate documented approvals are maintained to permit direct-shipments. (Organizational)

Quality Eng—Basic	Quality Eng—Senior	Q Manager	Supplier QE	CSoftQE	Program QE	Contract Review	MRO QE	Spec Process QE	Met Labs QE	Production Control	Regulatory Specialist
P	P	P			P	S	S			S	P

Procedural Knowledge: Ability to apply a structured process to verify that appropriate documented approvals are maintained to permit direct-shipments (3).

Conceptual Knowledge: Ability to assess and verify that appropriate documented approvals are maintained to permit direct-shipments (3).

Factual Knowledge: Understanding of regulatory requirements related to direct-shipment (3). Understanding of contract and project/program requirements related to direct-shipment (3). Understanding of requirements management tools (3). Understanding of direct-shipment practices (3). Understanding of value stream structure and where work is being performed (3).

Task 3 (PM141-003): Verify that appropriate documented approvals are maintained to permit direct-shipments between suppliers and customers. (Organizational)

Quality Eng—Basic	Quality Eng—Senior	Q Manager	Supplier QE	CSoftQE	Program QE	Contract Review	MRO QE	Spec Process QE	Met Labs QE	Production Control	Regulatory Specialist
P	P	P	P		P	S					P

Procedural Knowledge: Ability to apply a structured process to verify that appropriate documented approvals are maintained to permit direct-shipments (3).

Conceptual Knowledge: Ability to assess and verify that appropriate documented approvals are maintained to permit direct-shipments (3).

Factual Knowledge: Understanding of regulatory requirements related to direct-shipment (3). Understanding of contract and project/program requirements related to direct-shipment (3). Understanding of requirements management tools (3). Understanding of direct-shipment practices (3). Understanding of value stream structure and where work is being performed (3).

Task 4 (PM141-004): Evaluate direct-ship processes to verify that positive quality control is maintained. (Organizational)

Quality Eng—Basic	Quality Eng—Senior	Q Manager	Supplier QE	CSoftQE	Program QE	Contract Review	MRO QE	Spec Process QE	Met Labs QE	Production Control	Regulatory Specialist
P	P	S	S		P						S

Procedural Knowledge: Ability to apply a structured approach to assess direct-ship processes and verify that positive quality control is maintained (3).

Conceptual Knowledge: Ability to analyze procedures, quality records, and work product to assess the compliance and adequacy direct-ship processes and verify that positive quality control is maintained (4).

Factual Knowledge: Understanding of regulatory requirements related to direct-shipment (3). Understanding of contract and project/program requirements related to direct-shipment (3). Understanding of requirements management tools (3). Understanding of direct-shipment practices (3). Understanding of value stream structure and where work is being performed (3).

IV.K.10. Supplier Support Resources

Process: Determine Support Resource Needs (IPT, SQA, Labs, etc.)

Process Purpose: To understand the actual and total resource needs to support outsourcing to suppliers/partners. Based on supplier, design, and production maturity, resource needs may vary. Need to understand differences in regional and international influences on required support resources.

Knowledge and Skill Requirements: Ability to determine actual and total resource needs to support outsourcing to suppliers/partners. Ability to quantify variation in resource requirements based on supplier/partner production maturity. Ability to quantify variation in resource requirements based on differences in regional and international influences. Ability to assess the capability of suppliers/partners to ensure they have adequate resources to oversee their suppliers. Ability to assess supplier/partner's past performance to identify potential risk factors. Ability to document results of resource analysis in terms that can be used for effective program management.

Task 1 (SS065-001): Determine actual and total technical resource needs to support outsourcing to suppliers/partners. (Assess personnel and logistics considerations, receiving inspection resources, and procurement interface resources.)

Quality Eng—Basic	Quality Eng—Senior	Q Manager	Supplier QE	CSoftQE	Program QE	Contract Review	MRO QE	Spec Process QE	Met Labs QE	Production Control	Regulatory Specialist
S	P	P	P	S	P			P	P		

Procedural Knowledge: Ability to apply a structured process to calculate technical resource requirements to support outsourcing to suppliers/partners and their associated value streams (3).

Conceptual Knowledge: Ability to analyze technical resource requirements to support outsourcing to suppliers/partners and their associated value streams (4). Ability to effectively document support resource requirements in terms that can be utilized in effective project/program management (4).

Factual Knowledge: Understanding of project/program requirements (3). Understanding of value stream structure (3). Knowledge of production schedules (3). Understanding of technical processes (4). Knowledge of existing resources (3). Understanding of effective documentation practices (3). Understanding of capacity planning practices (4).

Task 2 (SS065-002): Quantify variation in technical resource requirements based on supplier/partner production/servicing maturity (ability to consistently meet requirements). (Reviewing past performance to establish demonstrated capability).

Quality Eng—Basic	Quality Eng—Senior	Q Manager	Supplier QE	CSoftQE	Program QE	Contract Review	MRO QE	Spec Process QE	Met Labs QE	Production Control	Regulatory Specialist
S	P	P	P	S	P			P	P		

Procedural Knowledge: Ability to apply a structured process to measure variation in technical resource requirements based on supplier/partner production/servicing maturity (3).

Conceptual Knowledge: Ability to analyze the past experience of a supplier/partner and establish a measurement of production/servicing maturity (4). Ability to analyze value stream maturity and determine actual technical resource requirements (4).

Factual Knowledge: Understanding of production/servicing maturity measurement (4). Understanding of assessment processes (3). Understanding of project/program requirements (3). Understanding of requirements management systems (3). Understanding of applicable technical processes (4).

Task 3 (SS065-003): Quantify variation in technical resource requirements based on differences in regional and international influences.

Quality Eng—Basic	Quality Eng—Senior	Q Manager	Supplier QE	CSoftQE	Program QE	Contract Review	MRO QE	Spec Process QE	Met Labs QE	Production Control	Regulatory Specialist
S	P	P	P	S	P			P	P		

Procedural Knowledge: Ability to apply a structured process to measure variation in technical resource requirements based on differences in regional and international influences upon the supplier's/partner's value stream (3).

Conceptual Knowledge: Ability to assess holiday schedules and measure their impact on supplier's/partner's production/servicing capability and capacity (3). Ability to assess the impact of differences in language upon supplier's/partner's production/servicing capability and capacity (3). Ability to assess the impact on inspection and test methods based on variations in measurement standards (4). Ability to analyze broad implications of regional and international influences upon production and servicing (4).

Factual Knowledge: Knowledge of international measurement standards (4). Knowledge of technical data transfer processes (4). Knowledge of applicable technical processes (4). Knowledge of configuration management practices (3). Knowledge of value stream structure (3).

Task 4 (SS065-004): Assess the capability of suppliers/partners to ensure they have adequate resources to support their suppliers. (Is there sufficient available resources?)

Quality Eng—Basic	Quality Eng—Senior	Q Manager	Supplier QE	CSoftQE	Program QE	Contract Review	MRO QE	Spec Process QE	Met Labs QE	Production Control	Regulatory Specialist
S	P	P	P	S	P			P	P		

Procedural Knowledge: Ability to apply a structured process to evaluate the technical capability and capacity of suppliers/partners to support and control their own supplier base (3).

Conceptual Knowledge: Ability to analyze the processes and technical resources of suppliers/partners to determine if they have adequate capability and capacity to support and control their own supplier base (4).

Factual Knowledge: Knowledge of production/servicing schedules (3). Knowledge of value stream structure (3). Knowledge of assessment processes (3). Understanding of applicable technical processes (4). Knowledge of capacity planning (3).

Task 5 (SS065-005): Document results of technical resource analysis in terms that can be used for effective program management.

Quality Eng—Basic	Quality Eng—Senior	Q Manager	Supplier QE	CSoftQE	Program QE	Contract Review	MRO QE	Spec Process QE	Met Labs QE	Production Control	Regulatory Specialist
S	P	P	P	S	P			P	P		

Procedural Knowledge: Ability to apply a structured process to document and effectively communicate the results of capability and capacity analysis in terms that can be used for effective program management (3).

Conceptual Knowledge: Ability to document and effectively communicate the results of capability and capacity analysis in terms that can be used for effective program management (3).

Factual Knowledge: Knowledge of effective communication practices (3). Understanding of project/program requirements (3). Understanding of applicable technical processes (4). Understanding of risk measurement practices (3). Understanding of capacity planning processes (3).

Process: Determine Oversight Resources Needs and Plan Surveillance

Process Purpose: To understand the on-going surveillance support resource requirements to sustain the outsourcing to suppliers/partners.

Knowledge and Skill Requirements: Based on supplier's design and production maturity, and oversight requirements, ability to evaluate and identify personnel resource requirements to support outsourcing. Ability to assess the capability of suppliers/partners to ensure they have adequate resources to oversee their suppliers. Ability to document results of resource analysis in terms that can be used for effective program management. Ability to understand how regional and international (cultural) influences that may affect oversight practices.

Task 1 (SS120-001): Based on supplier's design and production maturity, and oversight requirements, evaluate and identify personnel resource requirements to support outsourcing. Assess resource requirements based on potential regional and international influences.

Quality Eng—Basic	Quality Eng—Senior	Q Manager	Supplier QE	CSoftQE	Program QE	Contract Review	MRO QE	Spec Process QE	Met Labs QE	Production Control	Regulatory Specialist
S	P	P	P	S	P			P	P		S

Procedural Knowledge: Ability to apply a structured process to calculate system oversight resources required to support outsourcing to suppliers/partners and their associated value streams (3).

Conceptual Knowledge: Ability to evaluate supplier's design and production maturity, and oversight requirements, to identify and quantify personnel resource requirements to support outsourcing (5).

Factual Knowledge: Understanding of project/program requirements (3). Understanding of value stream structure (3). Knowledge of production schedules (3). Understanding of technical processes (4). Knowledge of existing resources (3). Understanding of production/servicing maturity assessment processes (3). Understanding of regulatory requirements related to supplier oversight (3). Knowledge of requirements management tools (3). Understanding of regional and international influences on servicing/production (4).

Task 2 (SS120-002): Assess the capability of suppliers/partners to ensure they have adequate resources to oversee their suppliers. Assess potential influences that regional and international difference within the value stream may have on the capability of suppliers/partners to adequately oversee their suppliers.

Quality Eng—Basic	Quality Eng—Senior	Q Manager	Supplier QE	CSoftQE	Program QE	Contract Review	MRO QE	Spec Process QE	Met Labs QE	Production Control	Regulatory Specialist
S	P	P	P	S	P			P	P		S

Procedural Knowledge: Ability to apply a structured process to evaluate the capability and capacity of suppliers/partners to support and control their own supplier base (3).

Conceptual Knowledge: Ability to analyze the processes and resources of suppliers/partners to determine if they have adequate capability and capacity to support and control their own supplier base (4).

Factual Knowledge: Knowledge of producing/servicing schedules (3). Knowledge of value stream structure (3). Knowledge of assessment processes (3). Understanding of applicable technical processes (4). Knowledge of capacity planning (3). Understanding of regional and international influences on servicing/production (4).

Task 3 (SS120-003): Document results of resource analysis in terms that can be used for effective program management.

Quality Eng—Basic	Quality Eng—Senior	Q Manager	Supplier QE	CSoftQE	Program QE	Contract Review	MRO QE	Spec Process QE	Met Labs QE	Production Control	Regulatory Specialist
S	P	P	P	S	P			P	P		S

Procedural Knowledge: Ability to apply a structured process to document and effectively communicate the results of capability and capacity analysis in terms that can be used for effective program management (3).

Conceptual Knowledge: Ability to determine that sufficient data is available to create a report (4). Ability to document and effectively communicate the results of capability and capacity analysis in terms that can be used for effective program management (3).

Factual Knowledge: Knowledge of effective communication practices (3). Knowledge of appropriate reporting formats (3). Understanding of project/program requirements (3). Understanding of applicable technical processes (4). Understanding of risk measurement practices (3). Understanding of capacity planning processes (3). Understanding of regional and international influences on servicing/production (4).

IV.K.11. Monitoring of Suppliers

Process: Notification of Changes in Capacity and/or Capability—within the Value Stream

Process Purpose: To ensure the organization is aware of any changes (positive or negative) to the capacity and/or capability of manufacturing and/or servicing of a supplier/partner. Ensure there is a notification process in place to make all appropriate parties aware of changes.

Knowledge and Skill Requirements: Ability to assess organizational processes that ensure the organization is aware of any changes (positive or negative) to the capacity and/or capability of manufacturing and/or servicing of a supplier/partner. Ability to assess existing notification processes to ensure all appropriate parties aware of changes in supplier/partner capability/capacity.

Task (PP010-001): Assess organizational processes to ensure the organization is notified of any changes (positive or negative) to the capacity and/or capability of manufacturing and/or servicing of a supplier/partner.

Quality Eng—Basic	Quality Eng—Senior	Q Manager	Supplier QE	CSoftQE	Program QE	Contract Review	MRO QE	Spec Process QE	Met Labs QE	Production Control	Regulatory Specialist
S	P	P	P	S	P			S	S		

Procedural Knowledge: Ability to apply a structured process to review and assess processes and procedures to determine adequacy of notification of changes (3).

Conceptual Knowledge: Ability to review various streams of communication flow to determine adequacy of change notification (4). Ability to review configuration management processes and determine compliance of change notification (4).

Factual Knowledge: Knowledge of change notification processes (3). Knowledge of configuration management processes (3). Knowledge of regulatory requirements related to change notification (3).

Process: Ensure Contract Terminations—within the Value Stream

Process Purpose: To ensure when a contract is terminated with a supplier/partner that the terms and conditions of termination are complied with and quality is maintained throughout the process.

Knowledge and Skill Requirements: Understanding of contract review processes. As applicable, understanding of government requirements for contract termination. Ability to verify that when a contract is terminated with a supplier/partner that the terms and conditions of termination are complied with and quality is maintained throughout the process. Ability to verify that when a contract is terminated between a supplier/partner and sub-tier suppliers that the terms and conditions of termination are complied with and quality is maintained throughout the process. Ability to perform process, product, and QMS audits.

Task 1 (PM145-001): Verify that when a contract is terminated with a supplier/partner that the terms and conditions of termination are complied with and quality is maintained throughout the process.

Quality Eng—Basic	Quality Eng—Senior	Q Manager	Supplier QE	CSoftQE	Program QE	Contract Review	MRO QE	Spec Process QE	Met Labs QE	Production Control	Regulatory Specialist
S	P	S	P	S	P	S	S	S	S	S	S

Procedural Knowledge: Ability to apply a structured approach to verify that when a contract is terminated with a supplier/partner that the terms and conditions of termination are complied with and quality is maintained throughout the process (3).

Conceptual Knowledge: Ability to evaluate activities to verify that when a contract is terminated with a supplier/partner that the terms and conditions of termination are complied with and quality is maintained throughout the process (5).

Factual Knowledge: Understanding of contract review processes (3). Understanding of government requirements for contract termination (3). Understanding of regulatory requirements related to approved suppliers (3). Understanding of requirements management tools (3). Understanding of related work activities (3). Understanding of value stream structure and where work is being performed (3).

Task 2 (PM145-002): Verify that when a contract is terminated between a supplier/partner and sub-tier suppliers that the terms and conditions of termination are complied with and quality is maintained throughout the process.

Quality Eng—Basic	Quality Eng—Senior	Q Manager	Supplier QE	CSoftQE	Program QE	Contract Review	MRO QE	Spec Process QE	Met Labs QE	Production Control	Regulatory Specialist
S	S	S	P	S	P	S	S	S	S	S	S

Procedural Knowledge: Ability to apply a structured approach to verify that when a contract is terminated between a supplier/partner and sub-tier suppliers that the terms and conditions of termination are complied with and quality is maintained throughout the process (3).

Conceptual Knowledge: Ability to evaluate activities to verify that when a contract is terminated between a supplier/partner and sub-tier suppliers that the terms and conditions of termination are complied with and quality is maintained throughout the process (5).

Factual Knowledge: Understanding of contract review processes (3). Understanding of government requirements for contract termination (3). Understanding of regulatory requirements related to approved suppliers (3). Understanding of requirements management tools (3). Understanding of related work activities (3). Understanding of value stream structure and where work is being performed (3).

Process: Monitor Financial Condition of Key Suppliers

Process Purpose: To ensure the ability of a supplier/partner to sustain production/servicing. Maintain awareness of the financial condition of suppliers/partners by review public information and discussions within industry. Establish awareness through observation in the supplier base of conditions such as material changes, changes in suppliers, equipment condition, employee turn-overs, etc.

Knowledge and Skill Requirements: Understanding of financial health indicators. Ability to evaluate the ability of a supplier/partner to sustain production/servicing. Ability to maintain awareness of the financial condition of suppliers/partners by review of public information and discussions within industry. Ability to maintain an awareness through observation in the supplier base of conditions such as material changes, changes in suppliers, equipment condition, employee turn-overs, etc., which may be indicators of financial health stress at suppliers/partners and key sub-tier suppliers. Ability to effectively communicate financial health measurements to management.

Task 1 (PM200-001): Maintain awareness of the financial condition of key suppliers/partners by review of public information and discussions within industry. Through analysis of collected data and physical review of operations evaluate the ability of a supplier/partner to sustain production/servicing.

Quality Eng—Basic	Quality Eng—Senior	Q Manager	Supplier QE	CSoftQE	Program QE	Contract Review	MRO QE	Spec Process QE	Met Labs QE	Production Control	Regulatory Specialist
		S	P		P						

Procedural Knowledge: Ability to apply a structured approach to ensure an awareness is maintained of the financial condition of suppliers/partners by review of public information and discussions within industry (3). Ability to apply a structured approach to perform a physical assessment of a supplier/partner to verify capability to financially sustain production/servicing (3).

Conceptual Knowledge: Ability to monitor and analyze public information and engage in discussions within industry to ensure an awareness is maintained of the financial condition of key suppliers/partners (4). Ability to plan and perform a physical assessment of key suppliers/partners to verify capability to financially sustain production/servicing (4). Ability to evaluate data and assessment results to determine the financial health of key suppliers/partners and determine their capability to financially sustain production/servicing (5).

Factual Knowledge: Understanding of financial health indicators (3). Understanding of where to obtain financial health data (3). Understanding of industry material flows (3). Understanding of capital requirements to provide products and services (3). Understanding of value stream structure and where work is being performed (3). Understanding of supplier/partner performance (3). Understanding of audit practices (3).

Task 2 (PM200-002): Maintain an awareness through observation in the supplier base of conditions such as material changes, changes in suppliers, equipment condition, timely payment to sub-contractors, employee turn-overs, etc., which may be indicators of financial health stress at suppliers/partners and key sub-tier suppliers.

Quality Eng—Basic	Quality Eng—Senior	Q Manager	Supplier QE	CSoftQE	Program QE	Contract Review	MRO QE	Spec Process QE	Met Labs QE	Production Control	Regulatory Specialist
		S	P		P						

Procedural Knowledge: Ability to apply a structured approach to maintain an awareness through observation in the supplier base of conditions such as material changes, changes in suppliers, timely payment to sub-contractors, equipment condition, employee turn-overs, etc., which may be indicators of financial health stress at suppliers/partners and key sub-tier suppliers (3).

Conceptual Knowledge: Ability to maintain an awareness through observation in the supplier base of conditions such as material changes, changes in suppliers, equipment condition, timely payment to sub-contractors, employee turn-overs, etc., which may be indicators of financial health stress at suppliers/partners and key sub-tier suppliers (4). Ability to understand the capital requirements to successfully execute a project/program (4). Ability to evaluate potential financial health stress indicators to identify potential impact to project/program performance (5).

Factual Knowledge: Understanding of financial health indicators (3). Understanding of where to obtain financial health data (3). Understanding of industry material flows (3). Understanding of capital requirements to provide products and services (3). Understanding of value stream structure and where work is being performed (3). Understanding of supplier/partner performance (3) Understanding of audit practices (3).

Task 3 (PM200-003): Effectively communicate financial health measurements of the value stream to management.

Quality Eng—Basic	Quality Eng—Senior	Q Manager	Supplier QE	CSoftQE	Program QE	Contract Review	MRO QE	Spec Process QE	Met Labs QE	Production Control	Regulatory Specialist
	S	P	P		P						

Procedural Knowledge: Ability to apply a structured approach to effectively communicate financial health measurements of the value stream to management (3).

Conceptual Knowledge: Ability to prepare written and verbal communications to effectively communicate financial health measurements of the value stream to management (4).

Factual Knowledge: Understanding of financial health indicators (3). Understanding of industry material flows (3). Understanding of capital requirements to provide products and services (3). Understanding of value stream structure and where work is being performed (3). Understanding of supplier/partner performance (3). Understanding of effective communication practices (3).

Process: Monitor Risk of Ultra-Small Key Suppliers

Process Purpose: In the event that the organization engages an ultra-small supplier for services or products that are considered key, ensure risk management plans are in place. (Ultra-small suppliers are organizations that will be adversely affected by the absence of one to two persons.)

Knowledge and Skill Requirements: Ability to identify ultra-small suppliers who provide key products and services throughout the value stream. Ability to develop contingency plans in the event that ultra-small suppliers are not able to support project/program needs. Understanding of relevant program requirements. Understanding of knowledge management practices. Knowledge of the relevant value stream structure.

Task (PM205-002): Establish and document a contingency plan in the event that ultra-small key suppliers are not able to support project/program needs.

Quality Eng—Basic	Quality Eng—Senior	Q Manager	Supplier QE	CSoftQE	Program QE	Contract Review	MRO QE	Spec Process QE	Met Labs QE	Production Control	Regulatory Specialist
S	S	S	S	S	P						

Procedural Knowledge: Ability to apply a structured approach to establish a contingency plan, in the event that ultra-small suppliers are not able to support project/program needs (3).

Conceptual Knowledge: Ability to evaluate potential scenarios related to the absence of 1 or 2 key persons at ultra-small key suppliers and establish a contingency plan, in the event that ultra-small suppliers are not able to support project/program needs (5) Ability to integrate effective risk mitigation into the contingency plan (4). Ability to effectively communicate risk measurement and management to the organization so that appropriate actions may be taken (4).

Factual Knowledge: Understanding of audit practices (3). Understanding of the value stream structure (3). Understanding of knowledge management practices (3). Understanding of risk measurement and management practices (3). Understanding of project/program requirements (3). Understanding of requirements management tools (3). Understanding of contingency planning practices (3). Ability to document and communicate contingency plans to management (3).

Process: Monitor Supplier-QMS Corrective and Preventative Action Process

Process Purpose: As QMS noncompliance is identified at suppliers/partners, ensure procedures are in place and the process of effective closed-loop corrective and preventative action is deployed. May include managing the implementation timing of corrective action and the management of deviations.

Knowledge and Skill Requirements: Ability to verify and validate that suppliers/partners have effective procedures in place to ensure effective closed-loop corrective and preventative action is deployed as a result of findings and observations documented during a QMS evaluation. Ability to manage the implementation timing of corrective action and the management of deviations. Ability to identify and measure risks associated with the delay of corrective action implementation. Ability to assess and determine that closed-loop corrective and preventative action is applied, as required, throughout the supplier/partner's supplier base.

Task 1 (SS080-001): Verify and validate that suppliers/partners have effective procedures in place to ensure effective root cause, closed-loop corrective and preventative action is deployed as a result of findings and observations documented during a QMS evaluation.

Quality Eng—Basic	Quality Eng—Senior	Q Manager	Supplier QE	CSoftQE	Program QE	Contract Review	MRO QE	Spec Process QE	Met Labs QE	Production Control	Regulatory Specialist
S	S	S	P		S						

Procedural Knowledge: Ability to apply a structured process to verify and validate that suppliers/partners have effective procedures in place to ensure effective closed-loop corrective and preventative action is deployed as a result of findings and observations documented during a QMS evaluation (3).

Conceptual Knowledge: Ability to analyze established root cause, closed-loop corrective and preventative action processes to determine adequacy to established regulations, standards and requirements (4). Ability to analyze objective evidence to verify and validate that suppliers/partners have effective processes in place to ensure effective root cause, closed-loop corrective and preventative action deployment (4).

Factual Knowledge: Knowledge of root cause tools (4). Knowledge of closed-loop corrective action processes (3). Knowledge of applicable regulations, standards, and requirements related to root cause, closed-loop corrective and preventative action (3). Knowledge of problem definition practices (3).

Task 2 (SS080-002): Identify and measure risks associated with the delay of corrective action implementation. Manage the implementation timing of corrective action and the management of deviations.

Quality Eng—Basic	Quality Eng—Senior	Q Manager	Supplier QE	CSoftQE	Program QE	Contract Review	MRO QE	Spec Process QE	Met Labs QE	Production Control	Regulatory Specialist
S	P	S	P	S	S	S	S	S	S	S	S

Procedural Knowledge: Ability to apply a structured process to identify and apply risk measurement to delays in corrective action implementation (3). Ability to apply a structured process to manage implementation of corrective action (3).

Conceptual Knowledge: Ability to identify potential risks associated with delays in corrective action implementation (4). Ability to evaluate the potential risk associated with delays in corrective action implementation (4). Ability to maintain a schedule of corrective action implementation (3).

Factual Knowledge: Knowledge of corrective action processes (3). Understanding of product use and application (3). Understanding of risk identification practices (3). Knowledge of risk measurement processes (3). Understanding of schedule management to include critical path analysis (3).

Task 3 (SS080-003): Assess and determine that closed-loop corrective and preventative action is applied, as required, throughout the supplier/partner's supplier base.

Quality Eng—Basic	Quality Eng—Senior	Q Manager	Supplier QE	CSoftQE	Program QE	Contract Review	MRO QE	Spec Process QE	Met Labs QE	Production Control	Regulatory Specialist
S	S	S	P		P						

Procedural Knowledge: Ability to apply a structured process in the evaluation of appropriate implementation of closed-loop corrective action and preventative action throughout the supplier's/partner's value stream (3).

Conceptual Knowledge: Ability to assess the implementation of closed-loop corrective and preventative action throughout the supplier's/partner's value stream (3). Ability to determine the adequacy of closed-loop corrective action and preventative action implementation in supplier's/partner's value stream (4). Ability to assess the effectiveness of closed-loop corrective and preventative action in the supplier's/partner's value stream (4).

Factual Knowledge: Understanding of value stream structure (3). Understanding of closed-loop corrective and preventative action (3). Understanding of objective evidence (3). Understanding of auditing practices (3). Understanding of corrective action effectiveness (3).

IV.K.13. Supplier Quality Plans

Process: Develop Alternate Supplier Approval Process (as required)

Process Purpose: To provide a means of outsourcing work to a potential supplier this has not been approved under the required processes. Create a quality plan that defines the specific controls to be in place to effectively manage work to a supplier that is not approved under normal processes. If the quality plan is subject to customer approval prior to implementation, ensure such approval is obtained.

Knowledge and Skill Requirements: Ability to interpret design, production, and quality requirements. Ability to create a quality plan that defines the specific controls to be in place to effectively manage work to a supplier that is not approved under normal processes. Ability to validate the effective implementation of quality plans. Ability to verify that changes to quality plans are subject to customer approval prior to implementation, as required, and ensure such approval is obtained.

Task 1 (PP160-001): In review of design, program, contract, and quality requirements, create a quality plan that defines the specific controls to be in place to effectively manage work to a supplier that is not approved under normal processes.

Quality Eng—Basic	Quality Eng—Senior	Q Manager	Supplier QE	CSoftQE	Program QE	Contract Review	MRO QE	Spec Process QE	Met Labs QE	Production Control	Regulatory Specialist
S	S	S	P	S	S	S	S	S	S	S	S

Procedural Knowledge: Ability to apply a defined process to create a quality plan that defines the specific controls to be in place to effectively manage work to a supplier that is not approved under normal processes (3).

Conceptual Knowledge: Ability to analyze design, program, contractual, and quality requirements to identify any special considerations for a supplier quality plan (4). Ability to create a quality plan that defines the specific controls to be in place to effectively manage work to a supplier that is not approved under normal processes (5).

Factual Knowledge: Understanding of regulatory requirements related to control of suppliers (3). Understanding of quality plan development (3). Understanding of applicable work activities (3).

Task 2 (PP160-002): When quality plans are required, validate the effective implementation of quality plans at suppliers/partners.

Quality Eng—Basic	Quality Eng—Senior	Q Manager	Supplier QE	CSoftQE	Program QE	Contract Review	MRO QE	Spec Process QE	Met Labs QE	Production Control	Regulatory Specialist
S	S	S	P	S	S			S	S		

Procedural Knowledge: Ability apply a defined and structured assessment process to verify the effective implementation of quality plans at suppliers/partners (3).

Conceptual Knowledge: Ability to analyze supplier work processes to verify the effective implementation of the quality plan (4).

Factual Knowledge: Understanding of regulatory requirements related to control of suppliers (3). Understanding of quality plan development (3). Understanding of applicable work activities (3).

Task 3 (PP160-003): Verify that changes to quality plans are subject to customer/regulatory approval prior to implementation, as required, and ensure such approval is obtained.

Quality Eng—Basic	Quality Eng—Senior	Q Manager	Supplier QE	CSoftQE	Program QE	Contract Review	MRO QE	Spec Process QE	Met Labs QE	Production Control	Regulatory Specialist
S	S	S	P	S	P	P	S	S	S	S	P

Procedural Knowledge: Ability to apply a defined process to verify that changes to quality plans are subject to customer/regulatory approval prior to implementation, as required, and ensure such approval is obtained (3).

Conceptual Knowledge: Ability to analyze proposed changes to quality plans to ensure changes are in compliance with applicable requirements (4). Ability to verify that changes to quality plans are subject to customer/regulatory approval prior to implementation, as required, and ensure such approval is obtained (2).

Factual Knowledge: Understanding of regulatory and contractual requirements related to control of suppliers (3). Understanding of quality plan development (3). Understanding of applicable work activities (3).

IV.M. Special Process Control

Process: Ensure Chemical/ Metallurgical Processes Are Conforming—within the Organization

Process Purpose: Ensure chemical and metallurgical processes are conforming to design and process requirements. May require destructive and/or nondestructive evaluation methods. Personnel must be qualified to perform special processes.

Knowledge and Skill Requirements: Understanding of relevant standards related to chemical and metallurgical processes. Ability to evaluate processes and verify that chemical and metallurgical processes are conforming to design and process requirements. Understanding of destructive and/or nondestructive evaluation methods. Ability to perform over-sight inspection to verify process conformance. Ability to assess personnel capabilities to verify assigned persons are qualified, and certified as required, to perform special processes.

Task 1 (PO110-002): Verify that chemical and metallurgical processes are verified.

Quality Eng—Basic	Quality Eng—Senior	Q Manager	Supplier QE	CSoftQE	Program QE	Contract Review	MRO QE	Spec Process QE	Met Labs QE	Production Control	Regulatory Specialist
S	S	S			S		S	P	P	S	

Procedural Knowledge: Ability to apply a structured assessment processes to establish compliance of the organization's chemical and metallurgical processes to design and process requirements (4).

Conceptual Knowledge: Ability to evaluate the organization's chemical and metallurgical processes to establish conformance and compliance to design and process requirements (5).

Factual Knowledge: Understanding of chemical and metallurgical requirements as they relate to project/program design (4). Understanding of industry standards as they relate to chemical and metallurgical verification processes (4). Understanding of assessment practices (3). Understanding of where chemical and metallurgical processes are being performed (3).

Task 2 (PO110-003): Perform over-sight inspection of work activities to verify conformance of chemical and metallurgical processes to design and process requirements.

Quality Eng—Basic	Quality Eng—Senior	Q Manager	Supplier QE	CSoftQE	Program QE	Contract Review	MRO QE	Spec Process QE	Met Labs QE	Production Control	Regulatory Specialist
S	S	S			S		S	P	P	S	

Procedural Knowledge: Ability to apply a structured process in the over-sight inspection of work activities to verify conformance of chemical and metallurgical processes to design and process requirements (3).

Conceptual Knowledge: Ability to evaluate process performance and plan over-sight inspection of work activities to verify conformance of chemical and metallurgical processes to design and process requirements (5). Ability to determine compliance and conformance of work activities as they relate to chemical and metallurgical processes (5). Ability to document the results of over-sight inspection activities, to include findings of nonconformance and noncompliance (3).

Factual Knowledge: Understanding of chemical and metallurgical requirements as they relate to project/program design (4). Understanding of industry standards as they relate to chemical and metallurgical verification processes (4). Understanding of assessment practices (3). Understanding of where chemical and metallurgical processes are being performed (3).

Task 3 (PO110-004): Assess personnel capabilities and verify assigned persons are qualified, and certified as required, to perform special processes.

Quality Eng—Basic	Quality Eng—Senior	Q Manager	Supplier QE	CSoftQE	Program QE	Contract Review	MRO QE	Spec Process QE	Met Labs QE	Production Control	Regulatory Specialist
S	S	S			S		S	P	P	S	

Procedural Knowledge: Ability to apply a structured process to identify personnel qualifications for the performance of special processes (3). Ability to apply a structured process to assess special process personnel qualifications (3).

Conceptual Knowledge: Ability to analyze special process work activities and determine qualification requirements for personnel (4). Ability to assess special process personnel qualifications records to determine compliance to qualification requirements (includes certification) (4). Ability to evaluate the capabilities of special process personnel to determine adequacy (5).

Factual Knowledge: Understanding of special process methods (3). Understanding of project/program special process requirements (3). Understanding of technical processes (3). Understanding of personnel capability assessments (3). Understanding of applicable special process technologies (3). Knowledge of applicable industry standards related to special processes (3). Understanding of where chemical and metallurgical processes are being performed (3).

Process: Ensure Chemical/ Metallurgical Processes Are Conforming—within the Value Stream

Process Purpose: Ensure chemical and metallurgical processes are conforming to design and process requirements. May require destructive and/or nondestructive evaluation methods. Ensure chemical and metallurgical processes are verified by the supplier and may require over-sight inspection to verify. Personnel must be qualified to perform special processes.

Knowledge and Skill Requirements: Understanding of various standards related to chemical and metallurgical processes. Ability to evaluate processes and verify that chemical and metallurgical processes are conforming to design and process requirements. Understanding of destructive and/or nondestructive evaluation methods. Ability to verify that chemical and metallurgical processes are verified by supplier. Ability to perform over-sight inspection to verify process conformance. Ability to assess personnel capabilities to verify assigned persons are qualified, and certified as required, to perform special processes.

Task 1 (PP110-001): Evaluate supplier/partner processes and verify that chemical and metallurgical processes are conforming to design and process requirements, throughout the value stream.

Quality Eng—Basic	Quality Eng—Senior	Q Manager	Supplier QE	CSoftQE	Program QE	Contract Review	MRO QE	Spec Process QE	Met Labs QE	Production Control	Regulatory Specialist
S	S		S		S			P	P		

Procedural Knowledge: Ability to apply defined and structured assessment methods to determine conformance of chemical and metallurgical control and verification processes to design and process requirements (4).

Conceptual Knowledge: Ability to evaluate chemical and metallurgical processes to determine conformance to design and process requirements (5).

Factual Knowledge: Understanding of chemical and metallurgical requirements as they relate to project/program design (4). Understanding of industry standards as they relate to chemical and metallurgical verification processes (4). Understanding of assessment practices (3). Understanding of where chemical and metallurgical processes are being performed (3). Understanding of the value stream structure (3).

Task 2 (PP110-002): Verify that chemical and metallurgical processes are verified by suppliers/partners.

Quality Eng—Basic	Quality Eng—Senior	Q Manager	Supplier QE	CSoftQE	Program QE	Contract Review	MRO QE	Spec Process QE	Met Labs QE	Production Control	Regulatory Specialist
S	S	S	P		S			P	P		

Procedural Knowledge: Ability to apply a structured assessment processes to establish compliance of supplier's/partner's chemical and metallurgical processes to design and process requirements (4).

Conceptual Knowledge: Ability to evaluate supplier's/partner's chemical and metallurgical processes to establish conformance and compliance to design and process requirements (5).

Factual Knowledge: Understanding of chemical and metallurgical requirements as they relate to project/program design (4). Understanding of industry standards as they relate to chemical and metallurgical verification processes (4). Understanding of assessment practices (3). Understanding of where chemical and metallurgical processes are being performed (3). Understanding of the value stream structure (3).

Task 3 (PP110-003): Perform over-sight inspection of supplier/partner work activities to verify conformance of chemical and metallurgical processes to design and process requirements, throughout the value stream.

Quality Eng—Basic	Quality Eng—Senior	Q Manager	Supplier QE	CSoftQE	Program QE	Contract Review	MRO QE	Spec Process QE	Met Labs QE	Production Control	Regulatory Specialist
S	S		P		S			P	P		

Procedural Knowledge: Ability to apply a structured process in the over-sight inspection of supplier/partner work activities to verify conformance of chemical and metallurgical processes to design and process requirements (3).

Conceptual Knowledge: Ability to evaluate supplier performance and plan over-sight inspection of supplier/partner work activities to verify conformance of chemical and metallurgical processes to design and process requirements (5). Ability to determine compliance and conformance of supplier/partner work activities as they relate to chemical and metallurgical processes (5). Ability to document the results of over-sight inspection activities, to include findings of nonconformance and noncompliance (3).

Factual Knowledge: Understanding of chemical and metallurgical requirements as they relate to project/program design (4). Understanding of industry standards as they relate to chemical and metallurgical verification processes (4). Understanding of assessment practices (3). Understanding of where chemical and metallurgical processes are being performed (3). Understanding of the value stream structure (3).

Task 4 (PP110-004): Assess personnel capabilities to verify assigned persons are qualified, and certified as required, to perform special processes, throughout the value stream.

Quality Eng—Basic	Quality Eng—Senior	Q Manager	Supplier QE	CSoftQE	Program QE	Contract Review	MRO QE	Spec Process QE	Met Labs QE	Production Control	Regulatory Specialist
S	S		S		S			P	P		

Procedural Knowledge: Ability to apply a structured process to identify personnel qualifications for the performance of special processes (3). Ability to apply a structured process to assess special process personnel qualifications (3).

Conceptual Knowledge: Ability to analyze special process work activities and determine qualification requirements for personnel (4). Ability to assess special process personnel qualifications records to determine compliance to qualification requirements (includes certification) (4). Ability to evaluate the capabilities of special process personnel to determine adequacy (5).

Factual Knowledge: Understanding of special process methods (3). Understanding of project/program special process requirements (3). Understanding of technical processes (3). Understanding of personnel capability assessments (3). Understanding of applicable special process technologies (3). Knowledge of applicable industry standards related to special processes (3). Understanding of where chemical and metallurgical processes are being performed (3). Understanding of the value stream structure (3).

Process: Perform Periodic Special Processes (Weld, Peen, Plating, etc.) Evaluations

Process Purpose: To assure control, capability, compliance and conformity of special processes performed in production/servicing processes. Plan, schedule, perform, and document periodic evaluations of special processes. Utilize risk based methods as appropriate.

Knowledge and Skill Requirements: Understanding of various NDT methods. Understanding of special process controls. Ability to plan, schedule, perform, and document periodic evaluations of special processes at supplier/partner locations. Ability to apply risk based product assurance methods as appropriate.

Task 1 (PF050-001): Evaluate personnel performing special process controls to ensure they have the appropriate skills and experience to understand and assess product/service conformity.

Quality Eng—Basic	Quality Eng—Senior	Q Manager	Supplier QE	CSoftQE	Program QE	Contract Review	MRO QE	Spec Process QE	Met Labs QE	Production Control	Regulatory Specialist
S	S	S	S		S		S	P	P	S	

Procedural Knowledge: Ability to apply a structured assessment process to determine if personnel performing special processes have the appropriate skills and experience to effectively execute assigned processes (3).

Conceptual Knowledge: Ability to analyze special processes and determine skill requirements for inspection personnel (4). Ability to analyze special process personnel training records to determine if assigned personnel have adequate training (4). Ability to analyze quality records of previous work product to determine the effectiveness of special process personnel (4).

Factual Knowledge: Understanding of various NDT methods (3). Understanding of special process controls (3). Understanding of skill requirements related to personnel performing special processes

(3). Understanding of physical requirements for personnel performing special processes (3). Understanding of personnel training records processes (3). Knowledge of training materials and courses for personnel performing special processes (3). Understanding of where work processes are being performed (3). Understanding of the value stream structure (3).

Task 2 (PF050-002): Plan and schedule periodic evaluations of special processes within the organization and at supplier/partner and sub-tier supplier locations.

Quality Eng—Basic	Quality Eng—Senior	Q Manager	Supplier QE	CSoftQE	Program QE	Contract Review	MRO QE	Spec Process QE	Met Labs QE	Production Control	Regulatory Specialist
P	P	S	P		P			P	P		

Procedural Knowledge: Ability to apply a structured process to plan and schedule special process evaluations within the value stream (3).

Conceptual Knowledge: Ability to analyze the performance of special process suppliers to determine the need to perform oversight evaluations (4) Ability to assess governmental, contractual, and/or project/program requirements to determine oversight evaluation requirements at special process suppliers (4).

Factual Knowledge: Understanding of various NDT methods (3). Understanding of special process controls (3). Understanding of regulatory, contractual and/or project/program requirements related to evaluations of special process suppliers (3). Understanding of where work processes are being performed (3). Understanding of the value stream structure (3). Understanding of audit practices (3). Understanding of requirements management tools (3).

Task 3 (PF050-003): Perform periodic special process evaluations.

Quality Eng—Basic	Quality Eng—Senior	Q Manager	Supplier QE	CSoftQE	Program QE	Contract Review	MRO QE	Spec Process QE	Met Labs QE	Production Control	Regulatory Specialist
S	S	S	S		S		S	P	P	S	

Procedural Knowledge: Ability to apply a defined and structured assessment process of special process within the value stream (3).

Conceptual Knowledge: Ability to analyze the processes, equipment, procedures, work product, personnel capabilities, and facilities of the value stream to ensure compliance of special process requirements (4).

Factual Knowledge: Understanding of various NDT methods (3). Understanding of special process controls (3). Understanding of regulatory, contractual and/or project/program requirements related to evaluations of special process suppliers (3). Understanding of process evaluation practices (3). Understanding of where work processes are being performed (3). Understanding of the value stream structure (3). Understanding of audit practices (3). Understanding of requirements management tools (3).

Task 4 (PF050-004): Apply risk based oversight methods to special process oversight, as appropriate.

Quality Eng—Basic	Quality Eng—Senior	Q Manager	Supplier QE	CSoftQE	Program QE	Contract Review	MRO QE	Spec Process QE	Met Labs QE	Production Control	Regulatory Specialist
S	S	P	P		P		S	P	P	S	

Procedural Knowledge: Ability to apply a structured approach to analyzing special process supplier performance and product characteristics in application of risk based oversight methods. (3).

Conceptual Knowledge: Ability to analyze design and production requirements to determine the appropriateness of risk based oversight methods. (4) Apply to analyze special process supplier performance to determine the appropriateness of risk based oversight methods. (4).

Factual Knowledge: Understanding of product design requirements related to special processes (3). Understanding of quality performance measurements (3). Understanding of various NDT methods (3). Understanding of special process controls (3). Understanding of regulatory, contractual and/or project/program requirements related to evaluations of special process suppliers (3). Understanding of risk measurement processes (3). Understanding of where work processes are being performed (3). Understanding of the value stream structure (3). Understanding of audit practices (3). Understanding of requirements management tools (3).

IV.M.1. Monitoring of Second and Third Party Certifications (e.g, Nadcap)

Process: Monitor Second and Third Party Surveillance Process

Process Purpose: To monitor suppliers/partners who have processes approved by a second or third-party (e.g., Nadcap) to ensure they are maintaining their approval as current. For suppliers who allow their certification to expire, engage an alternative means of special process surveillance.

Knowledge and Skill Requirements: Understanding of regulatory requirements related to second and other party oversight of suppliers. Understanding of unique certification schemes (e.g., RMC/ICOP, Nadcap, etc.). Understanding of controls for special processes. Understanding of product conformity implications related to out-of-control processes. Ability to monitor suppliers/partners who have processes approved by second or third party and verify their approval as current. Ability to engage alternative means of special process over-sight for suppliers who allow their certification to expire. For sub-tier suppliers, ability to verify that first tier suppliers have processes in place to notify the organization in the event of changes in the sub-tier supplier certification status.

Task 1 (PM245-001): Monitor suppliers/partners who have special processes approved by a second or third party and verify their approval as current.

Quality Eng—Basic	Quality Eng—Senior	Q Manager	Supplier QE	CSoftQE	Program QE	Contract Review	MRO QE	Spec Process QE	Met Labs QE	Production Control	Regulatory Specialist
S	P	S	P		P		S	P	P	S	

Procedural Knowledge: Ability to apply a structured approach to monitor suppliers/partners who have special processes approved by a second or third party and verify their approval as current (3). Ability to apply a structured process to take appropriate action when a supplier is in default of their second or third party certification (3).

Conceptual Knowledge: Ability to review the second or third party's database to monitor suppliers/partners who have special processes approved and verify their approval as current (3). Ability to evaluate the conditions of a supplier's/partner's default in second or third party certification to determine impact to product (5). Ability to perform analysis of the product impact evaluation and determine if notifications are required (4).

Factual Knowledge: Understanding of regulatory requirements related to second and other party oversight of suppliers (3). Understanding of requirements management tools (3). Understanding of unique certification schemes (e.g., RMC/ICOP, Nadcap, etc.). (3). Understanding of controls for special processes and NDT (3). Understanding of product conformity implications related to special processes (3). Understanding of audit practices (3). Understanding of the value stream structure and where work is performed (3).

Task 2 (PM245-002): As required, engage second/thirty party scheme (e.g., Nadcap) and/or special process experts as an alternative means of special process over-sight for suppliers. This may occur if suppliers allow their certification to expire, have their certification revoked, or as an additional sampling method.

Quality Eng—Basic	Quality Eng—Senior	Q Manager	Supplier QE	CSoftQE	Program QE	Contract Review	MRO QE	Spec Process QE	Met Labs QE	Production Control	Regulatory Specialist
S	P	P	P		P		S	P	P	S	

Procedural Knowledge: Ability to apply a defined and structured process to assess the compliance of special process suppliers who do not have Nadcap and/or special process certification (3).

Conceptual Knowledge: Ability to evaluate the unique variations at a special process supplier and identify the conditions that prevent certification to second or third party process certification scheme (5), Ability to analyze the results of the noncompliance evaluation and develop oversight processes to ensure process conformance and compliance (4).

Factual Knowledge: Understanding of regulatory requirements related to second and other party oversight of suppliers (3). Understanding of requirements management tools (3). Understanding of unique certification schemes (e.g., RMC/ICOP, Nadcap, etc.) (3). Understanding of controls for special processes and NDT (3). Understanding of product conformity implications related to special processes (3). Understanding of audit practices (3). Understanding of the value stream structure and where work is performed (3).

Task 3 (PM245-003): For sub-tier suppliers, verify that first tier suppliers have processes in place to notify the organization in the event of changes in the sub-tier supplier second or third party process certification status.

Quality Eng—Basic	Quality Eng—Senior	Q Manager	Supplier QE	CSoftQE	Program QE	Contract Review	MRO QE	Spec Process QE	Met Labs QE	Production Control	Regulatory Specialist
S	S	S	P		S		S	P	P	S	

Procedural Knowledge: Ability to apply a structured assessment process to ensure suppliers/partners have adequate processes in place to monitor and manage changes in their sub-tier supplier base related to second and third party process certification (3).

Conceptual Knowledge: Ability to analyze the capability and effectiveness of established supplier/partner processes that monitor and manage changes in their sub-tier supplier base related to second and third party process certifications (4).

Factual Knowledge: Understanding of regulatory requirements related to second and other party oversight of suppliers (3). Understanding of requirements management tools (3). Understanding of unique certification schemes (e.g., RMC/ICOP, Nadcap, etc.) (3). Understanding of controls for special processes and NDT (3). Understanding of product conformity implications related to special processes (3). Understanding of audit practices (3). Understanding of the value stream structure and where work is performed (3).

IV.N.1. Deliverable Software

Process: Ensure Software Conformance—Deliverable

Process Purpose: To ensure deliverable software (software that is embedded in a product and is delivered to a customer) conforms to design requirements. Ensure verification processes are performed (e.g., code comparison, run tests, peer review) to verify deliverable software is conforming. Ensure quality records reflect software configuration, conformity, and acceptance. As required, ensure testing tools are provided with the deliverable code or software program.

Knowledge and Skill Requirements: Understanding of Software Quality principals and methods. Ability to verify that deliverable software (software that is embedded in a product and is delivered to a customer) conforms to design requirements. Ability to verify the performance of verification processes (e.g., run tests, peer review) to verify deliverable software is conforming. Ability to verify that quality records reflect software configuration, conformity, and acceptance. Ability to verify that code includes the appropriate level of comments and that user and reference documentation are provided.

Task 1 (PF040-001): Verify that deliverable software (software that is embedded in a product and is delivered to a customer) conforms to design requirements.

Quality Eng—Basic	Quality Eng—Senior	Q Manager	Supplier QE	CSoftQE	Program QE	Contract Review	MRO QE	Spec Process QE	Met Labs QE	Production Control	Regulatory Specialist
P	P	P	P	P	S		S			S	

Procedural Knowledge: Ability to apply a structured approach to verify that deliverable software (software that is embedded in a product and is delivered to a customer) conforms to design requirements (3).

Conceptual Knowledge: Ability to evaluate requirements, code, quality records, and product test results to verify that deliverable software (software that is embedded in a product and is delivered to a customer) conforms to design requirements (5).

Factual Knowledge: Understanding of Software Quality principals and methods (3). Understanding of design requirements related to deliverable software (software that is embedded in a product and is delivered to a customer)(3). Understanding of personnel qualification requirements related to software verification (3). Understanding of requirements management tools (3).

Task 2 (PF040-002): Verify the performance of verification processes (e.g., run tests, peer review) to verify deliverable software is conforming to applicable requirements. Verify that code includes the appropriate level of comments and that user and reference documentation are provided. Verify that appropriate program documentation is provided in SYSML and/or DODAF, or other required documentation technique.

Quality Eng—Basic	Quality Eng—Senior	Q Manager	Supplier QE	CSoftQE	Program QE	Contract Review	MRO QE	Spec Process QE	Met Labs QE	Production Control	Regulatory Specialist
S	S	S	S	P	S		S			S	

Procedural Knowledge: Ability to apply a structured approach to verify that verification activities on deliverable software (software that is embedded in a product and is delivered to a customer) are being performed to ensure software conforms to design requirements (3).

Conceptual Knowledge: Ability to evaluate requirements, code, quality records, and product test results to verify the performance of verification activities of deliverable software (software that is embedded in a product and is delivered to a customer) to ensure conformance to design requirements (5).

Factual Knowledge: Understanding of Software Quality principals and methods (3). Understanding of design requirements related to deliverable software (software that is embedded in a product and is delivered to a customer)(3). Understanding of personnel qualification requirements related to software verification (3). Understanding of where work processes are being performed (3). Understanding of the value stream structure (3). Understanding of SYSML and/or DODAF, or other applicable documentation techniques (3) Knowledge of requirements management tools (3).

Task 3 (PF040-003): Verify that quality records reflect software configuration, conformity, and acceptance.

Quality Eng—Basic	Quality Eng—Senior	Q Manager	Supplier QE	CSoftQE	Program QE	Contract Review	MRO QE	Spec Process QE	Met Labs QE	Production Control	Regulatory Specialist
S	P	S	P	P	P		S			S	

Procedural Knowledge: Ability to apply a structured assessment process to verify that quality records reflect software configuration, conformity, and acceptance (3).

Conceptual Knowledge: Ability to analyze quality records to verify that software configuration, conformity, and acceptance is appropriately documented (4).

Factual Knowledge: Understanding of Software Quality principals and methods (3). Understanding of design requirements related to deliverable software (software that is embedded in a product and is delivered to a customer)(3). Understanding of personnel qualification requirements related to software verification (3). Understanding of where work processes are being performed (3). Understanding of the value stream structure (3). Knowledge of requirements management tools (3).

Process: Monitor Software Quality Plan

Process Purpose: To ensure software quality plans are being applied as required.

Knowledge and Skill Requirements: Understanding of software quality practices. Ability to review technical program requirements and identify specific software control requirements. Ability to create a software quality plan that will verify the technical and functional capabilities of embedded software. Ability to verify that software quality plans are being applied as required. Ability to verify that results of software quality control activities are appropriately documented. Ability to identify software quality responsibilities as they apply to subcontractors and ensure that information is distributed and understood throughout the value stream.

Task 1 (PM150-001): Identify requirements and create software quality plans that will verify the technical and functional capabilities of embedded and deliverable software.

Quality Eng—Basic	Quality Eng—Senior	Q Manager	Supplier QE	CSoftQE	Program QE	Contract Review	MRO QE	Spec Process QE	Met Labs QE	Production Control	Regulatory Specialist
S	S	S	S	P	S						S

Procedural Knowledge: Ability to apply a structured approach to identify requirements and create software quality plans that will verify the technical and functional capabilities of embedded and deliverable software (3).

Conceptual Knowledge: Ability to evaluate regulatory, contractual, and project/program requirements to identify control requirements for software quality plans (5). Ability to develop software quality plans that will verify the technical and functional capabilities of embedded and deliverable software (4).

Factual Knowledge: Knowledge of industry standards related to software verification (e.g., DO-105) (3). Knowledge of inspection planning processes (3). Knowledge of work processes related to program/project requirements (3). Understanding of requirements management tools (3). Knowledge of software quality assurance methods (3). Knowledge of software quality planning practices (3). Understanding of where work processes are being performed (3). Understanding of the value stream structure (3).

Task 2 (PM150-002): Identify software quality requirements as they relate to individual suppliers throughout the value stream. Ensure supplier/partners are informed and understand their software quality responsibilities. Verify that software quality plans are being applied as required, throughout the value stream.

Quality Eng—Basic	Quality Eng—Senior	Q Manager	Supplier QE	CSoftQE	Program QE	Contract Review	MRO QE	Spec Process QE	Met Labs QE	Production Control	Regulatory Specialist
S	S	P	P	P	P						S

Procedural Knowledge: Ability to apply a structured approach to identify software quality requirements as they relate to individual suppliers throughout the value stream (3). Ability to apply a structured approach to distribute that information throughout the value stream (3). Ability to apply structured assessment practices to verify that software quality plans are being applied as required (3).

Conceptual Knowledge: Ability to analyze project/program requirements and identify software quality requirements as they relate to individual suppliers throughout the value stream (4). Ability to distribute software quality control information throughout the value stream (3). Ability analyze software related processes to verify that software quality plans are being applied as required (4).

Factual Knowledge: Knowledge of industry standards related to software verification (e.g., DO-105) (3). Knowledge of inspection planning processes (3). Knowledge of work processes related to program/project (3). Understanding of requirements management tools (3). Knowledge of software quality assurance methods (3). Knowledge of software quality planning practices (3). Understanding of where work processes are being performed (3). Understanding of the value stream structure (3).

Task 3 (PM150-003): Verify deliverable/embedded software meets design requirements and that production systems will assure conformity to applicable standards and program requirements, throughout the value stream.

Quality Eng—Basic	Quality Eng—Senior	Q Manager	Supplier QE	CSoftQE	Program QE	Contract Review	MRO QE	Spec Process QE	Met Labs QE	Production Control	Regulatory Specialist
S	S	S	S	P	S						S

Procedural Knowledge: Ability to apply a structured assessment process to verify that supplier/partner has documented the verification requirements for deliverable/embedded software (3).

Conceptual Knowledge: Ability to analyze supplier/partner inspection planning to verify planning addresses verification requirements for deliverable/embedded software (4).

Factual Knowledge: Knowledge of industry standards related to software verification (e.g., DO-105) (3). Knowledge of inspection planning processes (3). Knowledge of work processes related to program/project (3). Understanding of requirements management tools (3). Knowledge of software quality assurance methods (3). Understanding of where work processes are being performed (3). Understanding of the value stream structure (3).

IV.O. Packaging, Handling, & Shipping

Process: Ensure Organizational Packaging Controls Are in Place

Process Purpose: Ensure the organization has processes to protect products and services in packaging to prevent damage during transport. May require familiarity with various packaging standards. Ability to understand packaging specifications and drawing requirements. Ensure shipping documents, to include labeling are legible, applied appropriately, and contain all applicable data required.

Knowledge and Skill Requirements: Ability to verify that organizational processes protect products and services in packaging to prevent damage during transport. Understanding of various packaging standards, including international and government standards, as applicable. Ability to verify that shipping documents meet applicable requirements and that labeling is legible, applied appropriately, and contain all applicable data required. Ability to ensure the organization has adequate processes in place to ensure appropriate packaging requirements are applied. Understanding of direct-ship processes and related regulatory requirements, as applicable.

Task 1 (PO070-001): Verify that organization has defined processes to protect products in packaging to prevent damage during transport and in compliance with applicable requirements.

Quality Eng—Basic	Quality Eng—Senior	Q Manager	Supplier QE	CSoftQE	Program QE	Contract Review	MRO QE	Spec Process QE	Met Labs QE	Production Control	Regulatory Specialist
P	P	P			P		S			S	P

Procedural Knowledge: Ability to apply a structured process to assess organizational processes to verify that defined processes protect products in packaging to prevent damage during transport (3). Ability to apply a structured process to assess organizational processes to verify that defined processes are in compliance with applicable customer and regulatory requirements (3).

Conceptual Knowledge: Ability to assess organizational defined processes to verify that defined processes protect products in packaging to prevent damage during transport (4).

Factual Knowledge: Knowledge of quality management system auditing methods (4). Knowledge of value stream structure (3). Knowledge of applicable technology (3). Knowledge of applicable product (3). Knowledge of applicable packaging requirements (international, customer, and governmental) (4).

Task 2 (PO070-002): Verify the organization's shipping documents meet applicable requirements (international, customer, and governmental) and that labeling is legible, applied appropriately, and contains all applicable data required.

Quality Eng—Basic	Quality Eng—Senior	Q Manager	Supplier QE	CSoftQE	Program QE	Contract Review	MRO QE	Spec Process QE	Met Labs QE	Production Control	Regulatory Specialist
P	P	S			P						P

Procedural Knowledge: Ability to apply a structured process to assess and verify that shipping documents meet applicable requirements and that labeling is legible, applied appropriately, and contains all applicable data required (3).

Conceptual Knowledge: Ability to interpret project/program requirements to determine applicable packaging, marking, and documentation requirements (4). Ability to assess and verify that shipping documents meet applicable requirements and that labeling is legible, applied appropriately, and contains all applicable data required (4).

Factual Knowledge: Knowledge of quality management system auditing methods (3). Knowledge of requirements management tool (3). Knowledge of project/program requirements related to packaging, marking, and documentation (4). Knowledge of applicable packaging requirements (4).

Task 3 (PP070-003): Ensure supplier/partner has adequate processes in place to ensure sub-tier suppliers apply packaging requirements (international, customer, and governmental).

Quality Eng—Basic	Quality Eng—Senior	Q Manager	Supplier QE	CSoftQE	Program QE	Contract Review	MRO QE	Spec Process QE	Met Labs QE	Production Control	Regulatory Specialist
S	S	S	P		S	S					S

Procedural Knowledge: Ability to apply a structured process to assess and determine that supplier/partner has adequate processes in place to ensure sub-tier suppliers apply packaging requirements (3).

Conceptual Knowledge: Ability to assess and determine that supplier/partner has adequate processes in place to ensure sub-tier suppliers apply packaging requirements (4).

Factual Knowledge: Knowledge of quality management system auditing methods (3). Knowledge of requirements management tool (3). Knowledge of project/program requirements related to packaging, marking, and documentation (3). Knowledge of applicable packaging requirements (3). Knowledge of value stream structure (3).

Process: Ensure Packaging & Preservation

Process Purpose: To prevent deterioration of product during shipping and storage. Evaluate compliance to packaging, handling and preservation requirements.

Knowledge and Skill Requirements: Understanding of various government packaging standards. Understanding of product protection methods. Ability to evaluate supplier/partner packaging and handling procedures to determine compliance with applicable requirements. Ability to evaluate packaging to ensure adequacy to prevent deterioration of product during shipping and storage. Ability to evaluate compliance to packaging, handling and preservation requirements. Ability to verify that sub-tier suppliers are in compliance with applicable packaging and handling requirements.

Task 1 (PM195-001): Evaluate organizational/supplier/partner packaging and handling procedures to determine compliance with applicable requirements.

Quality Eng—Basic	Quality Eng—Senior	Q Manager	Supplier QE	CSoftQE	Program QE	Contract Review	MRO QE	Spec Process QE	Met Labs QE	Production Control	Regulatory Specialist
S	P	S	P		P	P	S			S	

Procedural Knowledge: Ability to apply defined and structured assessment process to verify that supplier/partner packaging and handling procedures are in compliance with applicable requirements (3).

Conceptual Knowledge: Ability to analyze supplier/partner packaging and handling procedures to verify that such are in compliance with applicable requirements (4).

Factual Knowledge: Understanding of government packaging standard requirements (3). Understanding of contractual and project/program packaging standard requirements (3). Understanding of requirements management tools (3). Understanding of product protection methods (3). Understanding of delivery documentation requirements (3).

Task 2 (PM195-002): Evaluate packaging to ensure adequacy to prevent deterioration of product during shipping and storage. Evaluate compliance to packaging, handling, and preservation requirements (may include testing of packaging such as vibration and water resistance).

Quality Eng—Basic	Quality Eng—Senior	Q Manager	Supplier QE	CSoftQE	Program QE	Contract Review	MRO QE	Spec Process QE	Met Labs QE	Production Control	Regulatory Specialist
S	P	S	P		P	S	S			S	

Procedural Knowledge: Ability to apply a structured approach to verify that packaging is adequate to prevent deterioration of product during shipping and storage (3).

Conceptual Knowledge: Ability to analyze product, materials, and processes to verify that packaging is adequate to prevent deterioration of product during shipping and storage (4).

Factual Knowledge: Understanding of government packaging standard requirements (3). Understanding of contractual and project/program packaging standard requirements (3). Understanding of requirements management tools (3). Understanding of product protection methods (3). Understanding of delivery documentation requirements (3). Understanding of product/service delivery conditions (3). Understanding of factors that may affect product/service during shipping, handling, and storage (3).

Task 3 (PM195-003): Verify that sub-tier suppliers are in compliance with applicable packaging and handling requirements.

Quality Eng—Basic	Quality Eng—Senior	Q Manager	Supplier QE	CSoftQE	Program QE	Contract Review	MRO QE	Spec Process QE	Met Labs QE	Production Control	Regulatory Specialist
S	S	S	P		S						

Procedural Knowledge: Ability to apply a structured assessment process to verify that sub-tier suppliers are in compliance with applicable packaging and handling requirements (3).

Conceptual Knowledge: Ability to analyze established processes at sub-tier suppliers to verify that they are in compliance with applicable packaging and handling requirements (4).

Factual Knowledge: Understanding of government packaging standard requirements (3). Understanding of contractual and project/program packaging standard requirements (3). Understanding of requirements management tools (3). Understanding of product protection methods (3). Understanding of delivery documentation requirements (3). Understanding of the value stream structure and where work is being performed (3).

IV.O.2. Foreign Object Debris

Process: Monitor Handling Practices/Foreign Object Damage Controls

Process Purpose: To ensure products are not damaged during handling and processing. Evaluate material handling process to ensure product is effectively protected during processing. Evaluate work practices and ensure there are positive practices to prevent FOD.

Knowledge and Skill Requirements: Understanding of foreign object debris prevention and detection methods. Understanding of potential product conformity impact related to FOD. Understanding of handling and packaging methods. Ability to evaluate supplier/partner material handling processes to verify product is effectively protected during processing. Ability to evaluate supplier/partner work practices and ensure there are positive practices to prevent and detect FOD. Ability to verify that FOD prevention practices are engaged at all levels of the supplier base. Ability to ensure that personnel interacting with product and services are familiar with FOD controls. Understanding of potential impact of FOD on product and services.

Task 1 (PM250-001): Evaluate material handling processes (established planning and procedures) to verify product is effectively protected during processing.

Quality Eng—Basic	Quality Eng—Senior	Q Manager	Supplier QE	CSoftQE	Program QE	Contract Review	MRO QE	Spec Process QE	Met Labs QE	Production Control	Regulatory Specialist
P	P	S	P		S		P	S	S	P	

Procedural Knowledge: Ability to apply a structured process to assess supplier/partner material handling processes to verify products/services are effectively protected during processing (3).

Conceptual Knowledge: Ability to analyze established handling processes to verify there are adequate controls planned to effectively protect products and services during processing and delivery (4).

Factual Knowledge: Understanding of foreign object debris prevention and detection methods. Understanding of potential product conformity impact related to FOD. Understanding of handling and packaging methods. Understanding of contractual requirements related to packaging (3). Understanding of procedural development (3). Understanding of the value stream structure and where work is performed (3). Understanding of relevant manufacturing processes (3).

Task 2 (PM250-002): Evaluate work practices and ensure there are positive practices (including training and awareness) to prevent and detect FOD.

Quality Eng—Basic	Quality Eng—Senior	Q Manager	Supplier QE	CSoftQE	Program QE	Contract Review	MRO QE	Spec Process QE	Met Labs QE	Production Control	Regulatory Specialist
P	P	P	P		S		P	S	S	P	

Procedural Knowledge: Ability to apply defined and structured assessment processes to verify that work practices effectively protect product and services during processing, handling, and delivery, as well as, ensure there are positive practices to prevent and detect FOD (3).

Conceptual Knowledge: Ability to analyze established work practices to verify that established material handling practices effectively protect product and services during processing, handling, and delivery, as well as, ensure there are positive practices to prevent and detect FOD (4).

Factual Knowledge: Understanding of foreign object debris prevention and detection methods (3). Understanding of potential product conformity impact related to FOD (3). Understanding of handling and packaging methods (3). Understanding of contractual requirements related to packaging (3). Understanding of audit practices (3). Understanding of the value stream structure and where work is performed (3). Understanding of relevant manufacturing processes (3). Ability to assess and audit FOD training practices (3).

Task 3 (PM250-003): Verify that FOD prevention practices are engaged in the organization and at all levels of the value stream.

Quality Eng—Basic	Quality Eng—Senior	Q Manager	Supplier QE	CSoftQE	Program QE	Contract Review	MRO QE	Spec Process QE	Met Labs QE	Production Control	Regulatory Specialist
P	P	P	P		P		P	S	S	P	

Procedural Knowledge: Ability to apply a defined and structured assessment process to verify that FOD prevention practices are engaged in the organization and at all levels of the value stream (3).

Conceptual Knowledge: Ability to analyze work processes throughout the value stream to verify that FOD prevention practices are engaged in the organization and at all levels of the supplier base (4). Ability to analyze quality records to determine the effectiveness of established work processes to verify FOD prevention practices (4).

Factual Knowledge: Understanding of foreign object debris prevention and detection methods (3). Understanding of potential product conformity impact related to FOD (3). Understanding of handling and packaging methods (3). Understanding of contractual requirements related to packaging (3). Understanding of audit practices (3). Understanding of the value stream structure and where work is performed (3). Understanding of requirements management tools (3). Understanding of relevant manufacturing processes (3).

IV.P. Product Pedigree Assurance

Process: Ensure Buy Back Processes Are Controlled

Process Purpose: To provide a means to control product to re-enter a production system this has previously been certified as airworthy. Assurances such as state of airworthiness, regulatory authority approvals, customer approvals, and other product and system requirements must be addressed.

Knowledge and Skill Requirements: Understanding of applicable regulatory requirements related to product buy-back. Understanding of contract review processes. Ability to assess the applicability for buy-back and positive confirmation of airworthiness. Ability to verify that appropriate documentation is obtained for buy-back materials/product. Ability to determine if appropriate approvals have been obtained (e.g., regulatory authority approvals, customer approvals, etc.).

Task 1 (PM190-001): Assess the applicability for buy-back of product into production/servicing activities and verify positive confirmation of conformity to design and production requirements.

Quality Eng—Basic	Quality Eng—Senior	Q Manager	Supplier QE	CSoftQE	Program QE	Contract Review	MRO QE	Spec Process QE	Met Labs QE	Production Control	Regulatory Specialist
P	P	P	S		P						

Procedural Knowledge: Ability to apply a structure approach to assess the applicability of product for buy-back into production/servicing activities and verify positive confirmation of positive confirmation of airworthiness (3).

Conceptual Knowledge: Ability to physically evaluate product to determine applicability of product for buy-back into production/servicing activities and verify positive confirmation of positive confirmation of airworthiness (5). Ability to evaluate product records to determine applicability of product for buy-back into production/servicing activities and verify positive confirmation of positive confirmation of airworthiness (5).

Factual Knowledge: Understanding of regulatory requirements related to reintegration of products into production/servicing (3). Understanding of contractual and project/program requirements related to reintegration of products into production/servicing (3). Understanding of regulatory requirements related to airworthiness certification (3). Understanding of production processes (3). Understanding of positive recall processes (3).

Task 2 (PM190-002): Verify that appropriate documentation is obtained for buy-back materials/product.

Quality Eng—Basic	Quality Eng—Senior	Q Manager	Supplier QE	CSoftQE	Program QE	Contract Review	MRO QE	Spec Process QE	Met Labs QE	Production Control	Regulatory Specialist
P	P	S	S		S		S			S	S

Procedural Knowledge: Ability to apply structured processes to verify that appropriate documentation is obtained for buy-back materials/product (3).

Conceptual Knowledge: Ability to evaluate product documentation to ensure appropriate documentation is obtained for buy-back materials/product (5).

Factual Knowledge: Understanding of regulatory requirements related to reintegration of products into production/servicing. Understanding of contractual and project/program requirements related to reintegration of products into production/servicing (3). Understanding of regulatory requirements related to airworthiness certification (3). Understanding of production processes (3). Understanding of positive recall processes (3). Understanding of requirements management tools (3).

Task 3 (PM190-003): Determine if appropriate approvals have been obtained for buy-back product and materials (e.g., regulatory authority approvals, customer approvals, etc.).

Quality Eng—Basic	Quality Eng—Senior	Q Manager	Supplier QE	CSoftQE	Program QE	Contract Review	MRO QE	Spec Process QE	Met Labs QE	Production Control	Regulatory Specialist
P	P	P	S		P	P	S			S	P

Procedural Knowledge: Ability to apply a structured approach to verify that appropriate approvals have been obtained (e.g., regulatory authority approvals, customer approvals, etc.) for the acquisition of buy-back products (3).

Conceptual Knowledge: Ability to analyze contractual documents and applicable regulatory requirements to determine if approvals are required for buy-back products (4). Ability to assess if appropriate approvals have been obtained (e.g., regulatory authority approvals, customer approvals, etc.) for the acquisition of buy-back products (3).

Factual Knowledge: Understanding of regulatory requirements related to reintegration of products into production/servicing (3). Understanding of contractual and project/program requirements related to reintegration of products into production/servicing (3). Understanding of regulatory requirements related to airworthiness certification (3). Understanding of production processes (3). Understanding of positive recall processes (3). Understanding of requirements management tools (3).

Task 4 (PM190-004): Ensure buy-back materials/product are effectively controlled during re-integration into the organization's QMS.

Quality Eng—Basic	Quality Eng—Senior	Q Manager	Supplier QE	CSoftQE	Program QE	Contract Review	MRO QE	Spec Process QE	Met Labs QE	Production Control	Regulatory Specialist
P	P	S	S		P		S			S	S

Procedural Knowledge: Ability to apply a structured process to verify that buy-back materials/product are effectively controlled during re-integration into the organization's QMS (3).

Conceptual Knowledge: Ability to analyze organizational process to verify that buy-back materials/product are effectively controlled during re-integration into the organization's QMS (4). Ability to analyze buy-back work packages to verify that buy-back materials/product are effectively controlled during re-integration into the organization's QMS (4).

Factual Knowledge: Understanding of regulatory requirements related to reintegration of products into production/servicing (3). Understanding of contractual and project/program requirements related to reintegration of products into production/servicing (3). Understanding of regulatory requirements related to airworthiness certification (3). Understanding of production processes (3). Understanding of positive recall processes (3). Understanding of receiving inspection processes (3). Understanding of requirements management tools (3).

IV.R. Certification of Flight Readiness/Airworthiness Certification

Process: Complete Governmental Forms Related to Production and/or Design Conformity—within the Organization

Process Purpose: To create the appropriate documentation for product design and production approval by governmental agencies. Ensure authorized persons complete the official forms and that supporting documentation is provided.

Knowledge and Skill Requirements: Understanding of regulatory direction (e.g., FAA/EASA Orders and Advisory Circulars) and applicable regulations. Ability to verify that organizational business processes adequately control the issuance of governmental forms. Ability to create the appropriate documentation for product design and production approval by governmental agencies. Ability to ensure authorized persons complete the official forms and that supporting documentation is provided. Understanding of resource and schedule requirements to process governmental forms.

Task 1 (PO165-001): Identify which governmental forms are required for the various parts, components, and systems of the project/program. Verify that organizational business processes adequately control and plan for the issuance of governmental forms related to quality.

Quality Eng—Basic	Quality Eng—Senior	Q Manager	Supplier QE	CSoftQE	Program QE	Contract Review	MRO QE	Spec Process QE	Met Labs QE	Production Control	Regulatory Specialist
P	P	S			P						S

Procedural Knowledge: Ability to review the project/program design and requirements to determine which governmental forms are required for the various parts, components, and systems of the project/program. Ability to apply a structured process to verify that organizational business processes adequately control the issuance of governmental forms related to quality (3).

Conceptual Knowledge: Ability to understand which governmental forms are required for quality related activities (3). Ability to analyze organizational business processes to ensure they adequately plan for and control the issuance of governmental forms related to quality (4).

Factual Knowledge: Understanding of regulatory requirements related to government quality documentation (3). Understanding of resource and schedule requirements to process governmental forms (3). Understanding of where work processes are being performed (3). Understanding of the value stream structure (3).

Task 2 (PO165-002): Create the appropriate documentation for product design and production approval by governmental agencies.

Quality Eng—Basic	Quality Eng—Senior	Q Manager	Supplier QE	CSoftQE	Program QE	Contract Review	MRO QE	Spec Process QE	Met Labs QE	Production Control	Regulatory Specialist
P	P	S		S	S		S			S	P

Procedural Knowledge: Ability to apply a structured process to create the appropriate documentation for product design and production approval by governmental agencies (3).

Conceptual Knowledge: Ability to create the appropriate documentation for product design and production approval by governmental agencies, in accordance with applicable requirements (3).

Factual Knowledge: Understanding of regulatory requirements related to government quality documentation (3). Knowledge of formal guidelines related to the completion of governmental forms (3).

Task 3 (PO165-003): Verify that authorized persons complete the official forms and that supporting documentation is provided.

Quality Eng—Basic	Quality Eng—Senior	Q Manager	Supplier QE	CSoftQE	Program QE	Contract Review	MRO QE	Spec Process QE	Met Labs QE	Production Control	Regulatory Specialist
P	P	P		S	S		S			S	P

Procedural Knowledge: Ability to apply a defined process to verify that authorized persons complete the official forms and that supporting documentation is provided (3).

Conceptual Knowledge: Ability to analyze work packages and associated processes to verify that authorized persons complete the official forms and that supporting documentation is provided (4).

Factual Knowledge: Understanding of regulatory requirements related to government quality documentation (3). Understanding of audit practices (3). Knowledge of formal guidelines related to the completion of governmental forms (3).

Task 4 (PO165-004): Ensure appropriate governmental forms flow with the product throughout applicable processes and delivery.

Quality Eng—Basic	Quality Eng—Senior	Q Manager	Supplier QE	CSoftQE	Program QE	Contract Review	MRO QE	Spec Process QE	Met Labs QE	Production Control	Regulatory Specialist
P	P	P		S	S		S			S	P

Procedural Knowledge: Ability to apply a structured and defined process to verify that appropriate governmental forms flow with the product throughout applicable processes and delivery (3).

Conceptual Knowledge: Ability to assess work product to verify that appropriate governmental forms flow with the product throughout applicable processes and delivery (3).

Factual Knowledge: Understanding of regulatory requirements related to government quality documentation (3). Understanding of audit practices (3). Knowledge of formal guidelines related to the completion of governmental forms (3).

Process: Complete Governmental Forms Related to Production and/or Design Conformity—within the Value Stream

Process Purpose: To create the appropriate documentation for product design and production approval by governmental agencies. Ensure authorized persons complete the official forms and that supporting documentation is provided.

Knowledge and Skill Requirements: Understanding of regulatory direction (e.g., FAA/EASA Orders and Advisory Circulars) and applicable regulations. Ability to verify that organizational and supplier/partner business processes adequately control the issuance of governmental forms. Ability to create the appropriate documentation for product design and production approval by governmental agencies. Ability to ensure authorized persons complete the official forms and that supporting documentation is provided. Understanding of resource and schedule requirements to process governmental forms.

Task 1 (PP165-001): Identify which governmental forms are required for the various parts, components, and systems of the project/program. Verify that organizational and supplier/partner business processes adequately control and plan for the issuance of governmental forms related to quality, throughout the value stream.

Quality Eng—Basic	Quality Eng—Senior	Q Manager	Supplier QE	CSoftQE	Program QE	Contract Review	MRO QE	Spec Process QE	Met Labs QE	Production Control	Regulatory Specialist
S	P	S	P		P	S	S	S	S	S	P

Procedural Knowledge: Ability to review the project/program design and requirements to determine which governmental forms are required for the various parts, components, and systems of the project/program. Ability to apply a structured process to verify that organizational and supplier/partner business processes adequately control the issuance of governmental forms related to quality (3).

Conceptual Knowledge: Ability to understand which governmental forms are required for quality related activities (3). Ability to analyze organizational and supplier/partner business processes to ensure they adequately plan for and control the issuance of governmental forms related to quality (4).

Factual Knowledge: Understanding of regulatory requirements related to government quality documentation (3). Understanding of resource and schedule requirements to process governmental forms (3). Understanding of where work processes are being performed (3). Understanding of the value stream structure (3).

Task 2 (PP165-002): At suppliers, ensure the appropriate documentation is published for product design and production approval by governmental agencies.

Quality Eng—Basic	Quality Eng—Senior	Q Manager	Supplier QE	CSoftQE	Program QE	Contract Review	MRO QE	Spec Process QE	Met Labs QE	Production Control	Regulatory Specialist
S	S	S	P		P						P

Procedural Knowledge: Ability to apply a structured process to create the appropriate documentation for product design and production approval by governmental agencies (3).

Conceptual Knowledge: Ability to create the appropriate documentation for product design and production approval by governmental agencies, in accordance with applicable requirements (3).

Factual Knowledge: Understanding of regulatory requirements related to government quality documentation (3). Knowledge of formal guidelines related to the completion of governmental forms (3).

Task 3 (PP165-003): At suppliers/partners, verify that authorized persons complete the official forms and that supporting documentation is provided.

Quality Eng—Basic	Quality Eng—Senior	Q Manager	Supplier QE	CSoftQE	Program QE	Contract Review	MRO QE	Spec Process QE	Met Labs QE	Production Control	Regulatory Specialist
S	S	S	P		S						P

Procedural Knowledge: Ability to apply a defined process to verify that authorized persons complete the official forms and that supporting documentation is provided (3).

Conceptual Knowledge: Ability to analyze work packages and associated processes to verify that authorized persons complete the official forms and that supporting documentation is provided (4).

Factual Knowledge: Understanding of regulatory requirements related to government quality documentation (3). Understanding of audit practices (3). Knowledge of formal guidelines related to the completion of governmental forms (3).

Task 4 (PP165-004): At suppliers/partners, ensure appropriate governmental forms flow with the product throughout applicable processes and delivery.

Quality Eng—Basic	Quality Eng—Senior	Q Manager	Supplier QE	CSoftQE	Program QE	Contract Review	MRO QE	Spec Process QE	Met Labs QE	Production Control	Regulatory Specialist
S	P	S	P		S						P

Procedural Knowledge: Ability to apply a structured and defined process to verify that appropriate governmental forms flow with the product throughout applicable processes and delivery (3).

Conceptual Knowledge: Ability to assess work product to verify that appropriate governmental forms flow with the product throughout applicable processes and delivery (3).

Factual Knowledge: Understanding of regulatory requirements related to government quality documentation (3). Understanding of audit practices (3). Knowledge of formal guidelines related to the completion of governmental forms (3).

IV.R.1. Civil Aviation

Process: Issue Governmental Forms for Documentation of Major Repairs and Alteration (e.g., FAA Form 337)

Process Purpose: Where appropriate, produce governmental documentation to record and approve major repairs and alternations (e.g., FAA Form 337).

Knowledge and Skill Requirements: Understanding of applicable regulations, advisory materials, and Orders that relate to the documentation of major repairs and alterations. Ability to verify that suppliers, partners, and sub-tier suppliers have defined processes in place to effectively control the issuance of governmental documentation to record and approve major repairs and alterations. Ability to verify, that when required, governmental documentation to record and approve major repairs and alterations is created. Ability to substantiate that approved personnel have made entries on governmental documentation to record and approve major repairs and alterations.

Task 1 (PF120-001): Verify defined processes are in place to effectively control the issuance of governmental documentation to record and approve major repairs and alterations.

Quality Eng—Basic	Quality Eng—Senior	Q Manager	Supplier QE	CSoftQE	Program QE	Contract Review	MRO QE	Spec Process QE	Met Labs QE	Production Control	Regulatory Specialist
S	P	S	P		S	P	P				P

Procedural Knowledge: Ability to apply a defined and structured assessment process to verify that defined processes are in place to effectively control the issuance of governmental documentation to record and approve major repairs and alterations (3).

Conceptual Knowledge: Ability to assess the processes, procedures, and work product to verify that defined processes are in place to effectively control the issuance of governmental documentation to record and approve major repairs and alterations (3).

Factual Knowledge: Understanding of applicable regulations, advisory materials, and Orders that relate to the documentation of major repairs and alterations (3).

Task 2 (PF120-002): Verify, that when required, governmental documentation to record and approve major repairs and alterations are created. Substantiate that approved personnel have made entries on governmental documentation to record and approve major repairs and alterations.

Quality Eng—Basic	Quality Eng—Senior	Q Manager	Supplier QE	CSoftQE	Program QE	Contract Review	MRO QE	Spec Process QE	Met Labs QE	Production Control	Regulatory Specialist
S	P	S	P		S		P				P

Procedural Knowledge: Ability to apply a structured approach to verify that when required, governmental documentation to record and approve major repairs and alterations are created and entries are made by approved personnel (3).

Conceptual Knowledge: Ability to assess processes, procedures, and work product to verify that when required, governmental documentation to record and approve major repairs and alterations are created and entries are made by approved personnel (4).

Factual Knowledge: Understanding of applicable regulations, advisory materials, and Orders that relate to the documentation of major repairs and alterations (3). Understanding of audit practices (3). Understanding of personnel qualification records (3).

Process: Issue Airworthiness Certificate (e.g., FAA Form 8130-series in USA)

Process Purpose: When applicable, produce the appropriate civil aviation airworthiness certificate.

Knowledge and Skill Requirements: Understanding of applicable civil regulations, advisory materials, and orders that relate to the certification of product. Ability to verify that suppliers/partners have adequate process controls to establish airworthiness and release of applicable regulatory forms. Ability to verify that authorized personnel issue relevant civil aviation regulatory forms.

Task 1 (PF130-001): Verify that defined processes are in place to effectively control the issuance of governmental documentation to record and approve the conformity (e.g., airworthiness) of products and materials.

Quality Eng—Basic	Quality Eng—Senior	Q Manager	Supplier QE	CSoftQE	Program QE	Contract Review	MRO QE	Spec Process QE	Met Labs QE	Production Control	Regulatory Specialist
S	P	S	P		S	P	S			S	P

Procedural Knowledge: Ability to apply a defined and structured assessment process to verify that defined processes are in place to effectively control the issuance of governmental documentation to record and approve the airworthiness of products and materials. (3).

Conceptual Knowledge: Ability to assess the processes, procedures, and work product to verify they have defined processes in place to effectively control the issuance of governmental documentation to record and approve the airworthiness of products and materials. (3).

Factual Knowledge: Understanding of applicable regulations, advisory materials, and orders that relate to the documentation of the airworthiness of products and materials (3). (Examples of forms include FAA Form 8130-3, 8130-9, EASA Form 1, etc.) (FAA Advisory Circular 00-56 provides guidance on the FAA's voluntary distributor accreditation program.)

Task 2 (PF130-002): Verify adequacy of process controls to establish airworthiness and release of applicable civil aviation regulatory airworthiness forms. Verify that authorized personnel issue governmental forms.

Quality Eng—Basic	Quality Eng—Senior	Q Manager	Supplier QE	CSoftQE	Program QE	Contract Review	MRO QE	Spec Process QE	Met Labs QE	Production Control	Regulatory Specialist
S	P	S	P		S	P	S			S	P

Procedural Knowledge: Ability to apply a structured approach to verify the adequacy of process controls to establish airworthiness and release of applicable civil aviation regulatory airworthiness forms. Ability to apply a structured approach to verify that authorized personnel issue governmental forms (3).

Conceptual Knowledge: Ability to analyze the existing processes, procedures, and work product to verify that adequate process controls are in place to ensure product airworthiness and create the release of applicable civil aviation regulatory airworthiness forms (4). Ability to assess the supplier/partner process, procedures, and work product to verify that authorized personnel issue applicable civil aviation regulatory airworthiness forms (3).

Factual Knowledge: Understanding of applicable civil aviation regulations, advisory materials, and orders that relate to the certification of product (3). Understanding of audit practices (3). Understanding of personnel qualification records (3).

IV.R.2. Military Products—Special Requirements

Process: Issue Governmental Product Acceptance Documentation (e.g., DD Form 250 (USA))

Process Purpose: When applicable, produce the U.S. Government documentation to accept products/services and transfer ownership (e.g., DD 250).

Knowledge and Skill Requirements: Understanding of applicable U.S. Department of Defense requirements for the documentation to accept products/services and transfer ownership. As required by contract, ability to verify suppliers/partners provide governmental product acceptance documents (e.g., DD-250) with receiving documentation packages. An awareness and understanding of governmental policies, procedures, and regulations related to product and system certification (e.g., U.S. Air Force certification requirements are provided in AFPD 62-6 and AFI 62-601.)

Task 1 (PF125-001): Verify that defined processes are in place to effectively control the issuance of governmental documentation to record and approve the transfer of ownership of products and materials.

Quality Eng—Basic	Quality Eng—Senior	Q Manager	Supplier QE	CSoftQE	Program QE	Contract Review	MRO QE	Spec Process QE	Met Labs QE	Production Control	Regulatory Specialist
P	P	P	P		P	P	S			S	

Procedural Knowledge: Ability to apply a defined and structured assessment process to verify that defined processes are in place to effectively control the issuance of governmental documentation to record and approve the transfer of ownership of products and materials. (3).

Conceptual Knowledge: Ability to assess the processes, procedures, and work product to verify that defined processes are in place to effectively control the issuance of governmental documentation to record and approve the transfer of ownership of products and materials. (3).

Factual Knowledge: Understanding of applicable regulations, advisory materials, and orders that relate to the documentation and transfer of ownership of products and materials (3).

Task 2 (PF125-002): Verify that applicable governmental forms (e.g., DD-250, relevant FAA forms, etc.) with receiving documentation packages, as required.

Quality Eng—Basic	Quality Eng—Senior	Q Manager	Supplier QE	CSoftQE	Program QE	Contract Review	MRO QE	Spec Process QE	Met Labs QE	Production Control	Regulatory Specialist
S	P	S	P		S		S			S	

Procedural Knowledge: Ability to apply a structured approach to verify that governmental product acceptance documents (e.g., DD-250) are provided with receiving documentation packages, as required.

Conceptual Knowledge: Ability to assess the processes, procedures, and work product to verify that an appropriately completed governmental product acceptance documents (e.g., DD-250) are provided with receiving documentation packages, as required (4).

Factual Knowledge: Understanding of applicable U.S. Government requirements for the documentation to accept products/services and transfer ownership (3).

V.A. Quality Control Tools

Process: Apply Continuous Improvement Support

Process Purpose: To ensure process waste is removed and quality is improved, utilize data and various quality tools to understand process performance and improvement opportunity. Ensure improvement action does not compromise approved production controls.

Knowledge and Skill Requirements: Understanding of various continuous improvement (CI) methods. Ability to apply CI methods to remove waste and improve quality. Ability to utilize data and various quality tools to understand process performance and improvement opportunity. Ability to ensure improvement actions do not compromise approved production controls. Ability to integrate best practices and lessons learned into organizational memory. Ability to develop and present a business case for CI activities/projects.

Task 1 (PM010-001): Apply CI methods to remove waste and improve quality. Ability to utilize data and various quality tools to understand process performance and improvement opportunity. Ability to develop a business case in support of CI methods/project and present to management.

Quality Eng—Basic	Quality Eng—Senior	Q Manager	Supplier QE	CSoftQE	Program QE	Contract Review	MRO QE	Spec Process QE	Met Labs QE	Production Control	Regulatory Specialist
P	P	P	P	S	P	S	P	S	S	P	S

Procedural Knowledge: Ability to apply structured processes in the application of CI methods to remove waste and improve quality (3). Ability to apply structured processes to utilize data and various quality tools to understand process performance and improvement opportunity (3). Ability to apply a structured process to develop a business case in support of CI methods/project and present to management (3).

Conceptual Knowledge: Ability to apply CI methods to remove waste and improve quality (3). Ability to utilize data and various quality tools to understand process performance and improvement opportunity (3). Ability to analyze data to understand organizational, process, and product performance and identify improvement opportunities (4). Ability to conduct analysis of data and business conditions to develop a business case in support of CI methods/project and present to management (4).

Factual Knowledge: Understanding of various CI tools (3). Understanding of data analysis (3). Understanding of production/servicing work processes (3). Understanding of business case development processes (3).

Task 2 (PM010-002): Ensure best practices are identified, documented, and integrated into organizational memory.

Quality Eng—Basic	Quality Eng—Senior	Q Manager	Supplier QE	CSoftQE	Program QE	Contract Review	MRO QE	Spec Process QE	Met Labs QE	Production Control	Regulatory Specialist
P	P	P	P	P	P	P	P	P	P	P	P

Procedural Knowledge: Ability to apply a structured process to ensure best practices are identified, documented, and integrated into organizational memory (3).

Conceptual Knowledge: Ability to document best practices so they can be understood and reapplied (4). Ability to ensure best practices are integrated into organization memory archives (3).

Factual Knowledge: Understanding of production/servicing work processes (3). Understanding of best practice documentation processes (3).

V.B. Quality Management and Planning Tools

Process: Establish Appropriate Performance Metrics & Data Analysis

Process Purpose: To provide visibility of product, process, and organizational performance establish appropriate performance measurements to assess process performance and quality.

Knowledge and Skill Requirements: Ability to establish appropriate performance measurements to assess process performance and quality. Ability to integrate performance measurement processes into business, production, and service processes.

Task 1 (PM015-001): Ensure the personnel within the value stream who are establishing performance measurements have the appropriate knowledge to successfully execute the process. Review value stream performance to ensure effectiveness of performance measurement selection and effectiveness of improvement/corrective actions.

Quality Eng—Basic	Quality Eng—Senior	Q Manager	Supplier QE	CSoftQE	Program QE	Contract Review	MRO QE	Spec Process QE	Met Labs QE	Production Control	Regulatory Specialist
S	S	P	P		P			S	S		

Procedural Knowledge: Ability to apply a structured assessment process to ensure the personnel within the value stream who are establishing performance measurements have the appropriate knowledge to successfully execute the process (3). Ability to apply a structured review process to ensure effectiveness of performance measurement selection and effectiveness of improvement/corrective actions within the value stream (3).

Conceptual Knowledge: Ability to evaluate work processes and work product to ensure that persons within the value stream who are establishing performance measurements have the appropriate knowledge to successfully execute the process (5). Ability to analyze work processes and work product to ensure effectiveness of performance measurement selection and effectiveness of improvement/corrective actions within the value stream (4).

Factual Knowledge: Understanding of performance measurements (3). Understanding of effective communication (3). Understanding of applicable measured processes (3). Understanding of program/project requirements (3). Understanding of requirements management tools (3). Understanding of audit practices (3). Understanding of value stream structure and where work is being performed (3). Understanding of personnel training records (3).

Task 2 (PM015-002): Establish appropriate performance measurements to assess process performance and quality.

Quality Eng—Basic	Quality Eng—Senior	Q Manager	Supplier QE	CSoftQE	Program QE	Contract Review	MRO QE	Spec Process QE	Met Labs QE	Production Control	Regulatory Specialist
S	P	P	P	S	P		S	S	S	S	

Procedural Knowledge: Ability to apply a structured process to establish appropriate performance measurements in assessing process performance and quality (3).

Conceptual Knowledge: Ability to analyze project/program requirements to identify appropriate performance measurements in assessing process performance and quality (4). Ability to analyze product and process performance to identify appropriate performance measurements in assessing process performance and quality (4).

Factual Knowledge: Understanding of various CI tools (3). Understanding of data analysis (3). Understanding of production/servicing work processes (3). Understanding of measurement processes (3).

Task 3 (PM015-003): Integrate performance measurement processes into business, production, and service processes.

Quality Eng—Basic	Quality Eng—Senior	Q Manager	Supplier QE	CSoftQE	Program QE	Contract Review	MRO QE	Spec Process QE	Met Labs QE	Production Control	Regulatory Specialist
S	P	P	P	S	P		S	S	S	S	

Procedural Knowledge: Ability to apply a structured process to integrate performance measurement processes into business, production, and service processes (3).

Conceptual Knowledge: Ability to analyze business, production, and service processes to determine how and where to integrate performance measurement processes (4). Ability to coordinate with process owners to integrate performance measurement processes into business, production, and service processes (4).

Factual Knowledge: Understanding of various CI tools (3). Understanding of data analysis (3). Understanding of production/servicing work processes (3). Understanding of measurement processes (3). Understanding of effective communication methods (3).

Task 4 (PM015-004): Effectively communicate performance measurements to applicable parties (e.g., management, customer, and value stream). (Note: This task is for all types of performance measurement communication—internal and external.)

Quality Eng—Basic	Quality Eng—Senior	Q Manager	Supplier QE	CSoftQE	Program QE	Contract Review	MRO QE	Spec Process QE	Met Labs QE	Production Control	Regulatory Specialist
P	P	P	P	P	P	P	P	P	P	P	P

Procedural Knowledge: Ability to apply a structured process to effectively communicate performance measurements (3).

Conceptual Knowledge: Ability to effectively communicate performance measurements so that they may understood and action taken (4).

Factual Knowledge: Understanding of performance measurements (3). Understanding of effective communication (3). Understanding of applicable measured processes (3). Understanding of program/project requirements (3).

V.D. Corrective Action

Process: Monitor/Approve Root Cause Analysis—Product/Process

Process Purpose: Ensure appropriate root cause analysis tools are utilized during analysis of documented findings. Ensure authorized (e.g., MRB) personnel have approved the identification of root cause. Ensure product/services previously produced/provided are reviewed for applicability.

Knowledge and Skill Requirements: Understanding of regulatory requirements related to the documentation of product and/or process nonconformities. Understanding of industry standard and/or customer requirements related to the documentation of product and/or process nonconformities. Understanding of material review board processes. Understanding of contract requirements. Understanding of root cause analysis. Ability to understand the scope of root cause applicability. Understanding of closed-loop corrective action. Understanding of the product and/or services. Ability to verify that appropriate root cause analysis tools are utilized during analysis of nonconformance. Ability to verify that authorized (e.g., MRB) personnel have approved the identification of root cause. Ability to substantiate that product/services previously produced/provided are reviewed for applicability. Ability to verify that root cause documentation is maintained as quality records.

Task 1 (PF080-001): Verify that appropriate root cause analysis tools are utilized during analysis of nonconformance. Ability to identify the required scope of root cause analysis.

Quality Eng—Basic	Quality Eng—Senior	Q Manager	Supplier QE	CSoftQE	Program QE	Contract Review	MRO QE	Spec Process QE	Met Labs QE	Production Control	Regulatory Specialist
P	P	P	P	P	P	S	P	P	P	P	P

Procedural Knowledge: Ability to apply a structured approach to verify that appropriate root cause analysis tools are utilized during analysis of nonconformance (3).

Conceptual Knowledge: Ability to analyze the root cause analysis applied to specific nonconformities and noncompliance to determine appropriateness and effectiveness (4).

Factual Knowledge: Understanding of regulatory requirements related to the documentation of product and/or process nonconformities (3). Understanding of industry standard and/or customer requirements related to the documentation of product and/or process nonconformities (3). Understanding of contract requirements related to root cause analysis (3). Understanding of root cause analysis tools (3). Understanding of scope identification processes related to root cause analysis (3). Understanding of requirements management tools (3).

Task 2 (PF080-002): Verify that authorized (e.g., MRB) personnel have approved the identification of root cause.

Quality Eng—Basic	Quality Eng—Senior	Q Manager	Supplier QE	CSoftQE	Program QE	Contract Review	MRO QE	Spec Process QE	Met Labs QE	Production Control	Regulatory Specialist
P	P	P	P	P	P	S	P	S	S	P	P

Procedural Knowledge: Ability to apply a structured approach to verify that authorized (e.g., MRB) personnel have approved the identification of root cause (3)

Conceptual Knowledge: Ability to assess quality records to verify that authorized (e.g., MRB) personnel have approved the identification of root cause (3).

Factual Knowledge: Understanding of regulatory requirements related to the documentation of product and/or process nonconformities (3). Understanding of industry standards and/or customer requirements related to the documentation of product and/or process nonconformities (3). Understanding of contract requirements related to root cause analysis (3). Understanding of root cause analysis tools (3). Understanding of requirements management tools (3).

Task 3 (PF080-003): Substantiate that product/services previously produced/provided have been reviewed for applicability of the same or similar root cause.

Quality Eng—Basic	Quality Eng—Senior	Q Manager	Supplier QE	CSoftQE	Program QE	Contract Review	MRO QE	Spec Process QE	Met Labs QE	Production Control	Regulatory Specialist
P	P	P	P	P	P	S	P	P	P	P	P

Procedural Knowledge: Ability to apply a structured approach to substantiate that product/services previously produced/provided are reviewed for applicability of the same or similar root cause (3).

Conceptual Knowledge: Ability to analyze product-families, processes, and value stream to substantiate that product/services previously produced/provided are reviewed for applicability of the same or similar root cause (4).

Factual Knowledge: Understanding of regulatory requirements related to the documentation of product and/or process nonconformities (3). Understanding of industry standards and/or customer requirements related to the documentation of product and/or process nonconformities (3). Understanding of contract requirements related to root cause analysis (3). Understanding of root cause analysis tools (3). Understanding of product recall processes (3). Understanding of requirements management tools (3).

Task 4 (PF080-004): Verify that root cause documentation is maintained as quality records.

Quality Eng—Basic	Quality Eng—Senior	Q Manager	Supplier QE	CSoftQE	Program QE	Contract Review	MRO QE	Spec Process QE	Met Labs QE	Production Control	Regulatory Specialist
P	P	P	P	S	P	S	S	S	S	S	P

Procedural Knowledge: Ability to apply a defined and structured assessment process to verify that root cause documentation is maintained as quality records (3).

Conceptual Knowledge: Ability to assess records throughout the value stream to verify that root cause documentation is maintained as quality records (3).

Factual Knowledge: Understanding of regulatory requirements related to the documentation of product and/or process nonconformities (3). Understanding of industry standards and/or customer requirements related to the documentation of product and/or process nonconformities (3). Understanding of contract requirements related to root cause analysis (3). Understanding of root cause analysis tools (3). Understanding of requirements management tools (3).

Process: Monitor/Approve Corrective & Preventive Action Processes—Product/Process

Process Purpose: To ensure effective corrective and preventative action processes are in place and applied to the root cause of noncompliance/nonconformance in a timely fashion. Verify effectiveness after implementation and prior to closure of the corrective action records.

Knowledge and Skill Requirements: Understanding of regulatory requirements related to the documentation of product and/or process nonconformities. Understanding of industry standard and/or customer requirements related to the documentation of product and/or process nonconformities. Understanding of material review board processes. Understanding of contract requirements. Understanding of root cause analysis. Understanding of closed-loop corrective action. Understanding of the product and/or services. Ability to verify that effective corrective and preventative action processes are in place and applied to the root cause of noncompliance/nonconformance in a timely fashion. Ability to verify that corrective and preventative action plans have been approved, as required. Ability to verify effectiveness after implementation and prior to closure of the corrective action records (follow-up).

Task 1 (PF085-001): Verify that effective corrective and preventative action processes are in place and applied to the root cause of noncompliance/nonconformance in a timely fashion.

Quality Eng—Basic	Quality Eng—Senior	Q Manager	Supplier QE	CSoftQE	Program QE	Contract Review	MRO QE	Spec Process QE	Met Labs QE	Production Control	Regulatory Specialist
P	P	P	P	P	P	P	P	P	P	P	P

Procedural Knowledge: Ability to apply a structured approach to verify that effective corrective and preventative action processes are in place and applied to the root cause of noncompliance/nonconformance in a timely fashion (3).

Conceptual Knowledge: Ability to analyze corrective and preventative action events to determine effectiveness and timeliness (4).

Factual Knowledge: Understanding of regulatory requirements related to the documentation of product and/or process nonconformities (3). Understanding of industry standards and/or customer requirements related to the documentation of product and/or process nonconformities (3). Understanding of contract requirements related to corrective action of nonconformities and noncompliance (3). Understanding of corrective action tools (3). Understanding of corrective action effectiveness (3). Understanding of corrective action urgency (3). Understanding of requirements management tools (3).

Task 2 (PF085-002): Verify that corrective and preventative action plans have been approved, as required.

Quality Eng—Basic	Quality Eng—Senior	Q Manager	Supplier QE	CSoftQE	Program QE	Contract Review	MRO QE	Spec Process QE	Met Labs QE	Production Control	Regulatory Specialist
P	P	P	P	S	P	S	P	S	S	P	S

Procedural Knowledge: Ability to apply a structured process to determine if corrective and preventative action require approval(s) (3). Ability to apply a structured approach to verify that corrective and preventative action plans have been approved, as required (3).

Conceptual Knowledge: Ability to review and assess various applicable requirements to determine if approvals are required for corrective and preventative actions (3). Ability to assess quality records to verify that corrective and preventative action plans have been approved, as required (3).

Factual Knowledge: Understanding of regulatory requirements related to the documentation of product and/or process nonconformities (3). Understanding of industry standards and/or customer requirements related to the documentation of product and/or process nonconformities (3). Understanding of contract requirements related to corrective action of nonconformities and noncompliance (3). Understanding of requirements management tools (3).

Task 3 (PF085-003): Verify effectiveness after implementation and prior to closure of the corrective action records (follow up).

Quality Eng—Basic	Quality Eng—Senior	Q Manager	Supplier QE	CSoftQE	Program QE	Contract Review	MRO QE	Spec Process QE	Met Labs QE	Production Control	Regulatory Specialist
P	P	P	P	P	P	S	P	P	P	P	S

Procedural Knowledge: Ability to apply a structured approach to verify effectiveness after implementation and prior to closure of the corrective action records (follow up).

Conceptual Knowledge: Ability to analyze process and product performance (to include data) to verify effectiveness after implementation and prior to closure of the corrective action records (follow up) (4).

Factual Knowledge: Understanding of regulatory requirements related to the documentation of product and/or process nonconformities (3). Understanding of industry standards and/or customer requirements related to the documentation of product and/or process nonconformities (3). Understanding of contract requirements related to corrective action of nonconformities and noncompliance (3). Understanding of corrective action tools (3). Understanding of corrective action effectiveness (3). Understanding of corrective action urgency (3). Understanding of requirements management tools (3).

Process: Take Action Based on Data Analysis

Process Purpose: Based on analysis of data, take improvement action to achieve objectives.

Knowledge and Skill Requirements: Understanding of performance measurement processes. Ability to evaluate course of action based on analysis of data to take desired improvement action. Ability to coordinate proposed action with key constituents. Ability to ensure action is taken. Ability to document executed action and apply to lessons learned and standard work, as applicable.

Task 1 (PM025-001): Ensure the personnel within the value stream who are conducting analysis of performance measurements have the appropriate knowledge to successfully execute the process. Review value stream performance to ensure effectiveness of performance measurement selection and effectiveness of improvement/corrective actions.

Quality Eng—Basic	Quality Eng—Senior	Q Manager	Supplier QE	CSoftQE	Program QE	Contract Review	MRO QE	Spec Process QE	Met Labs QE	Production Control	Regulatory Specialist
S	P	P	P	S	P		P	S	S	P	

Procedural Knowledge: Ability to apply a structured assessment process to ensure the personnel within the value stream who are establishing performance measurements have the appropriate knowledge to successfully execute the process (3). Ability to apply a structured review process to ensure effectiveness of performance measurement selection and effectiveness of improvement/corrective actions within the value stream (3).

Conceptual Knowledge: Ability to evaluate work processes and work product to ensure that persons within the value stream who are establishing performance measurements have the appropriate knowledge to successfully execute the process (5). Ability to analyze work processes and work product to ensure effectiveness of performance measurement selection and effectiveness of improvement/corrective actions within the value stream (4).

Factual Knowledge: Understanding of performance measurements (3). Understanding of effective communication (3). Understanding of applicable measured processes (3). Understanding of program/project requirements (3). Understanding of requirements management tools (3). Understanding of audit practices (3). Understanding of value stream structure and where work is being performed (3). Understanding of personnel training records (3).

Task 2 (PM025-002): Evaluate analysis of data to establish a proposed course of improvement action. Ability to develop a business case in support of product/process improvement and present to management.

Quality Eng—Basic	Quality Eng—Senior	Q Manager	Supplier QE	CSoftQE	Program QE	Contract Review	MRO QE	Spec Process QE	Met Labs QE	Production Control	Regulatory Specialist
P	P	P	P	S	P		S	S	S	S	

Procedural Knowledge: Ability to apply a structure approach in the evaluation of data analysis to establish a proposed course of improvement action (3). Ability to apply a structured process to create a business case to support the proposed improvement action (3)

Conceptual Knowledge: Ability to evaluate data analysis to establish a proposed course of improvement action (5). Ability to conduct evaluation on performance measurements and proposed improvement actions to develop and present a business case (5). Ability to document proposed action into a plan that others can understand and apply (4).

Factual Knowledge: Understanding of data collection methods (3). Understanding of data analysis (3). Understanding of production/servicing work processes (3). Understanding of measurement processes (3). Understanding of project/program performance requirements (3). Understanding of business case development practices for improvement projects (3).

Task 3 (PM025-003): Effectively coordinate proposed action with key constituents.

Quality Eng—Basic	Quality Eng—Senior	Q Manager	Supplier QE	CSoftQE	Program QE	Contract Review	MRO QE	Spec Process QE	Met Labs QE	Production Control	Regulatory Specialist
P	P	P	P	P	P		P	P	P	P	

Procedural Knowledge: Ability to apply a structured approach to coordinating proposed action with key constituents (3).

Conceptual Knowledge: Ability to identify key constituents of a process (3). Ability to effectively coordinate proposed action with key constituents (4).

Factual Knowledge: Understanding of production/servicing work processes (3). Understanding of where work is performed (3). Understanding of the value stream structure (3). Understanding of effective communication processes (3). Understanding of measurement processes (3). Understanding of project/program performance requirements (3). Understanding of who key constituents are (3).

Task 4 (PM025-004): Ensure performance improvement action is undertaken and effective.

Quality Eng—Basic	Quality Eng—Senior	Q Manager	Supplier QE	CSoftQE	Program QE	Contract Review	MRO QE	Spec Process QE	Met Labs QE	Production Control	Regulatory Specialist
P	P	P	P	P	P		P	P	P	P	

Procedural Knowledge: Ability to apply a structured process to verify that performance improvement action is undertaken and effective (3). Ability to apply a structured and defined process to document in-effective performance actions (3).

Conceptual Knowledge: Ability to monitor performance improvement plans and verify that action is undertaken (3). Ability to analyze performance improvement action to verify that action is effective (4). Ability to ensure in-effective performance actions are appropriately documented and submitted to the organizations corrective action processes (4).

Factual Knowledge: Understanding of performance measurements (3). Understanding of process effectiveness measurements (3). Understanding of applicable measured processes (3). Understanding of program/project requirements (3). Understanding of corrective action processes (3).

Task 5 (PM025-005): Document executed performance improvement action and apply to lessons learned and standard work, as applicable.

Quality Eng—Basic	Quality Eng—Senior	Q Manager	Supplier QE	CSoftQE	Program QE	Contract Review	MRO QE	Spec Process QE	Met Labs QE	Production Control	Regulatory Specialist
P	P	P	P	P	P		P	P	P	P	

Procedural Knowledge: Ability to apply a structured approach to document executed action and apply to lessons learned and standard work, as applicable (3).

Conceptual Knowledge: Ability to evaluate executed action and ensure it is adequately descriptive for inclusion into lessons learned and standard work, as applicable (5).

Factual Knowledge: Knowledge of lessons learned processes (3). Knowledge of applicable technologies (4). Knowledge of interrelatedness of business and technical processes (4).

Process: Monitor QMS Findings—Value Stream

Process Purpose: To ensure noncompliance is documented when found during QMS assessments and are managed for root cause and corrective action, in the value stream.

Knowledge and Skill Requirements: Understanding of quality management systems. Understanding of QMS standards. Understanding of applicable regulatory requirements related to internal audits. Ability to perform QMS evaluations. Understanding of root cause analysis. Understanding of closed-loop corrective action. Ability to verify when required that noncompliance is documented when found during self-assessments and are managed for root cause and corrective action. Ability to verify that appropriate follow up actions occur as scheduled. Ability to verify at all levels of the organization and supplier/partner base that value stream-based QMS findings are appropriately managed and closed. As required, ability to verify that review and acceptance by the procuring authority is obtained and documented.

Task 1 (PM092-001): Verify that noncompliance is documented when found during self-assessments and are managed for root cause and corrective action, throughout the organization and value stream.

Quality Eng—Basic	Quality Eng—Senior	Q Manager	Supplier QE	CSoftQE	Program QE	Contract Review	MRO QE	Spec Process QE	Met Labs QE	Production Control	Regulatory Specialist
S	P	P	P	S	P		S	S	S	S	

Procedural Knowledge: Ability to apply a defined and structured assessment process to verify that noncompliance is documented when found during self-assessments and are managed for root cause and corrective action (3).

Conceptual Knowledge: Ability to review and analyze organizational and supplier/partner quality records to verify that noncompliance is documented when found during self-assessments (4). Ability to review and analyze supplier/partner quality records to verify that noncompliance is managed for root cause and corrective action (4).

Factual Knowledge: Understanding of requirements management systems (3). Understanding of regulatory requirements related to internal audits (3). Understanding of contractual and project/program requirements related to internal audit (3). Understanding of industry standards related to internal audit (3). Understanding of organizational/supplier audit practices (3). Understanding of the value stream structure and where work is being executed (3). Understanding of noncompliance documentation practices (3).

Task 2 (PM092-002): Verify that appropriate corrective action follow up actions occur as scheduled.

Quality Eng—Basic	Quality Eng—Senior	Q Manager	Supplier QE	CSoftQE	Program QE	Contract Review	MRO QE	Spec Process QE	Met Labs QE	Production Control	Regulatory Specialist
S	P	P	P	S	P		S	S	S	S	

Procedural Knowledge: Ability to apply a defined and structured assessment process to verify that appropriate corrective action follow up actions occur as scheduled (3).

Conceptual Knowledge: Ability to audit quality records and actual deployed actions to verify that appropriate corrective action follow up actions occur as scheduled (3).

Factual Knowledge: Understanding of industry standards related to internal audit (3). Understanding of supplier audit practices (3). Understanding of the value stream structure and where work is being executed (3). Understanding of noncompliance documentation practices (3). Understanding of RCCA tools (4).

Task 3 (PM092-003): Verify that at all levels of the organization and supplier/partner base that QMS findings are appropriately managed and closed.

Quality Eng—Basic	Quality Eng—Senior	Q Manager	Supplier QE	CSoftQE	Program QE	Contract Review	MRO QE	Spec Process QE	Met Labs QE	Production Control	Regulatory Specialist
S	P	P	P	S	P		S	S	S	S	

Procedural Knowledge: Ability to apply a structured approach to verify that at all levels of the organization and supplier/partner base that QMS findings are appropriately managed and closed (3).

Conceptual Knowledge: Ability to review quality records to verify that at all levels of the organization and supplier/partner base that QMS findings are appropriately managed and closed (3).

Factual Knowledge: Understanding of industry standards related to internal audit (3). Understanding of supplier audit practices (3). Understanding of the value stream structure and where work is being executed (3). Understanding of noncompliance documentation practices (3). Understanding of RCCA tools (4).

Process: Monitor QMS Findings—Organizational

Process Purpose: To ensure noncompliance is documented when found during organizational QMS assessments and are managed for root cause and corrective action.

Knowledge and Skill Requirements: Understanding of quality management systems. Understanding of QMS standards. Understanding of applicable regulatory requirements related to internal audits. Ability to perform QMS evaluations. Understanding of root cause analysis. Understanding of closed-loop corrective action. Ability to verify when required that noncompliance is documented when found during self-assessments and are managed for root cause and corrective action. Ability to verify that appropriate follow up actions occur as scheduled. Ability to verify at all levels of the organization and supplier/partner base that QMS findings are appropriately managed and closed. As required, ability to verify that review and acceptance by the procuring authority is obtained and documented.

Task 1 (PM100-001): Verify that noncompliance is documented when found during self-assessments and are managed for root cause and corrective action, throughout the organization and value stream. (Organizational focus)

Quality Eng—Basic	Quality Eng—Senior	Q Manager	Supplier QE	CSoftQE	Program QE	Contract Review	MRO QE	Spec Process QE	Met Labs QE	Production Control	Regulatory Specialist
P	P	P	S	P	P		P	P	P	P	S

Procedural Knowledge: Ability to apply a defined and structured assessment process to verify that noncompliance is documented when found during self-assessments and managed for root cause and corrective action (3).

Conceptual Knowledge: Ability to review and analyze organizational and supplier/partner quality records to verify that noncompliance is documented when found during self-assessments (4). Ability to review and analyze supplier/partner quality records to verify that noncompliance is managed for root cause and corrective action (4).

Factual Knowledge: Understanding of requirements management systems (3). Understanding of regulatory requirements related to internal audits (3). Understanding of contractual and project/program requirements related to internal audit (3). Understanding of industry standards related to internal audit (3). Understanding of organizational/supplier audit practices (3). Understanding of the value stream structure and where work is being executed (3). Understanding of noncompliance documentation practices (3).

Task 2 (PM100-002): Verify that appropriate corrective action follow up actions occur as scheduled. (Organizational focus)

Quality Eng—Basic	Quality Eng—Senior	Q Manager	Supplier QE	CSoftQE	Program QE	Contract Review	MRO QE	Spec Process QE	Met Labs QE	Production Control	Regulatory Specialist
P	P	P	S	P	P		P	P	P	P	S

Procedural Knowledge: Ability to apply a defined and structured assessment process to verify that appropriate corrective action follow up actions occur as scheduled (3).

Conceptual Knowledge: Ability to audit quality records and actual deployed actions to verify that appropriate corrective action follow up actions occur as scheduled (3).

Factual Knowledge: Understanding of industry standards related to internal audit (3). Understanding of supplier audit practices (3). Understanding of the value stream structure and where work is being executed (3). Understanding of noncompliance documentation practices (3). Understanding of RCCA tools (4).

Task 3 (PM100-003): Verify that at all levels of the organization and supplier/partner base that QMS findings are appropriately managed and closed. (Organizational focus)

Quality Eng—Basic	Quality Eng—Senior	Q Manager	Supplier QE	CSoftQE	Program QE	Contract Review	MRO QE	Spec Process QE	Met Labs QE	Production Control	Regulatory Specialist
P	P	P	P	S	P		S	S	S	S	S

Procedural Knowledge: Ability to apply a structured approach to verify that at all levels of the organization and supplier/partner base that QMS findings are appropriately managed and closed (3).

Conceptual Knowledge: Ability to review quality records to verify that at all levels of the organization and supplier/partner base that QMS findings are appropriately managed and closed (3).

Factual Knowledge: Understanding of industry standards related to internal audit (3). Understanding of supplier audit practices (3). Understanding of the value stream structure and where work is being executed (3). Understanding of noncompliance documentation practices (3). Understanding of RCCA tools (4).

Process: Monitor Supplier Root Cause Analysis—QMS (Root Cause Process)

Process Purpose: To ensure suppliers/partners utilized appropriate root cause analysis tools and such are applied during analysis of documented findings.

Knowledge and Skill Requirements: Understanding of quality management systems. Understanding of QMS standards. Understanding of applicable regulatory requirements related to internal audits. Ability to perform QMS evaluations. Understanding of root cause analysis. Understanding of closed-loop corrective action. Understanding of the business processes and/or technical processes under audit. Ability to verify that suppliers/partners are utilizing appropriate root cause analysis tools and such are applied during analysis of documented findings. Ability to evaluate the effectiveness of root cause analysis tools utilized in the sub-tier supplier base.

Task 1 (PM115-001): Verify that suppliers/partners are utilizing appropriate root cause analysis tools and such are applied during analysis of documented QMS findings.

Quality Eng—Basic	Quality Eng—Senior	Q Manager	Supplier QE	CSoftQE	Program QE	Contract Review	MRO QE	Spec Process QE	Met Labs QE	Production Control	Regulatory Specialist
S	P	S	P	S	P			S			

Procedural Knowledge: Ability to apply a structured approach to assess and verify that suppliers/partners are utilizing appropriate root cause analysis tools and such are applied during analysis of documented QMS findings (3).

Conceptual Knowledge: Ability to analyze supplier/partner root cause analysis tools to verify such are appropriate, effective, and applied during analysis of documented QMS findings (4).

Factual Knowledge: Understanding of root cause analysis tools (3). Understanding of closed-loop corrective action (3). Understanding of QMS (3). Understanding of the business processes and/or technical processes under audit (3). Understanding of value stream structure and where work is being performed (3).

Task 2 (PM115-002): Analyze the effectiveness of root cause analysis tools utilized in the sub-tier supplier base.

Quality Eng—Basic	Quality Eng—Senior	Q Manager	Supplier QE	CSoftQE	Program QE	Contract Review	MRO QE	Spec Process QE	Met Labs QE	Production Control	Regulatory Specialist
S	P	S	P	S	P			S			

Procedural Knowledge: Ability to apply a structured approach to assess the effectiveness of root cause analysis tools utilized in the sub-tier supplier base (3).

Conceptual Knowledge: Ability to analyze quality records and supplier/partner performance to assess the effectiveness of root cause analysis tools utilized in the sub-tier supplier base (4).

Factual Knowledge: Understanding of root cause analysis tools (3). Understanding of closed-loop corrective action (3). Understanding of QMS (3). Understanding of the business processes and/or technical processes under audit (3). Understanding of value stream structure and where work is being performed (3).

Process: Monitor Organizational Root Cause Analysis—QMS (Root Cause Process)

Process Purpose: To ensure appropriate root cause analysis tools and such are applied during analysis of documented findings.

Knowledge and Skill Requirements: Understanding of quality management systems. Understanding of QMS standards. Understanding of applicable regulatory requirements related to internal audits. Ability to perform QMS evaluations. Understanding of root cause analysis. Understanding of closed-loop corrective action. Understanding of the business processes and/or technical processes under audit. Ability to verify appropriate root cause analysis tools and such are applied during analysis of documented findings. Ability to evaluate the effectiveness of root cause analysis tools.

Task 1 (PM116-001): Verify appropriate root cause analysis tools and such are applied during analysis of documented QMS findings.

Quality Eng—Basic	Quality Eng—Senior	Q Manager	Supplier QE	CSoftQE	Program QE	Contract Review	MRO QE	Spec Process QE	Met Labs QE	Production Control	Regulatory Specialist
P	P	P	S	P	P	S	P	P	P	P	S

Procedural Knowledge: Ability to apply a structured approach to assess and verify that appropriate root cause analysis tools and such are applied during analysis of documented QMS findings (3).

Conceptual Knowledge: Ability to analyze organizational root cause analysis tools to verify such are appropriate, effective, and applied during analysis of documented QMS findings (4).

Factual Knowledge: Understanding of root cause analysis tools (3). Understanding of closed-loop corrective action (3). Understanding of QMS (3). Understanding of the business processes and/or technical processes under audit (3). Understanding of where work is being performed (3).

Task 2 (PM116-002): Analyze the effectiveness of root cause analysis tools utilized in the organization.

Quality Eng—Basic	Quality Eng—Senior	Q Manager	Supplier QE	CSoftQE	Program QE	Contract Review	MRO QE	Spec Process QE	Met Labs QE	Production Control	Regulatory Specialist
P	P	P	S	S	P	S	S	S	S	S	S

Procedural Knowledge: Ability to apply a structured approach to assess the effectiveness of root cause analysis tools (3).

Conceptual Knowledge: Ability to analyze quality records and organizational performance to assess the effectiveness of root cause analysis tools (4).

Factual Knowledge: Understanding of root cause analysis tools (3). Understanding of closed-loop corrective action (3). Understanding of QMS (3). Understanding of the business processes and/or technical processes under audit (3). Understanding of value stream structure and where work is being performed (3).

Process: Approve Supplier Root Cause Analysis—QMS (Individual Root Cause Analysis)

Process Purpose: To ensure the supplier/partner has identified the true root cause of QMS noncompliance.

Knowledge and Skill Requirements: Understanding of quality management systems. Understanding of QMS standards. Understanding of applicable regulatory requirements related to internal audits. Understanding of root cause analysis. Understanding of closed-loop corrective action. Understanding of the business processes and/or technical processes under audit. Ability to validate that the supplier/partner has identified the true root cause of QMS noncompliance. Ability to evaluate root cause analysis and ensure the root cause is addressed in the sub-tier supplier base.

Task 1 (PM120-001): Evaluate that the supplier/partner has identified the true root cause of QMS noncompliance.

Quality Eng—Basic	Quality Eng—Senior	Q Manager	Supplier QE	CSoftQE	Program QE	Contract Review	MRO QE	Spec Process QE	Met Labs QE	Production Control	Regulatory Specialist
S	S	S	P	S	P		S	S	S	S	

Procedural Knowledge: Ability to apply a structured approach to validate that the supplier/partner has identified the true root cause of QMS noncompliance (3).

Conceptual Knowledge: Ability to apply evaluate supplier/partner quality records and performance to validate that the true root cause of QMS noncompliance has been appropriately identified (5).

Factual Knowledge: Understanding of root cause analysis tools (3). Understanding of closed-loop corrective action (3). Understanding of QMS (3). Understanding of the business processes and/or technical processes under audit (3). Understanding of value stream structure and where work is being performed (3).

Task 2 (PM120-002): Evaluate root cause analysis and ensure the true root cause is addressed in the sub-tier supplier base.

Quality Eng—Basic	Quality Eng—Senior	Q Manager	Supplier QE	CSoftQE	Program QE	Contract Review	MRO QE	Spec Process QE	Met Labs QE	Production Control	Regulatory Specialist
S	S	S	P	S	P		S	S	S	S	

Procedural Knowledge: Ability to apply a structured approach to verify that the true root cause is addressed in the sub-tier supplier base (3).

Conceptual Knowledge: Ability to evaluate quality records and value stream performance to verify that true root cause is addressed in the sub-tier supplier base (4).

Factual Knowledge: Understanding of root cause analysis tools (3). Understanding of closed-loop corrective action (3). Understanding of QMS (3). Understanding of the business processes and/or technical processes under audit (3). Understanding of value stream structure and where work is being performed (3).

Process: Approve Organizational Root Cause Analysis—QMS (Individual Root Cause Analysis)

Process Purpose: To ensure the organization has identified the true root cause of QMS noncompliance.

Knowledge and Skill Requirements: Understanding of quality management systems. Understanding of QMS standards. Understanding of applicable regulatory requirements related to internal audits. Understanding of root cause analysis. Understanding of closed-loop corrective action. Understanding of the business processes and/or technical processes under audit. Ability to validate that the organization has identified the true root cause of QMS noncompliance. Ability to evaluate root cause analysis and ensure the root cause is addressed.

Task 1 (PM121-001): Evaluate that the organization has identified the true root cause of QMS noncompliance.

Quality Eng—Basic	Quality Eng—Senior	Q Manager	Supplier QE	CSoftQE	Program QE	Contract Review	MRO QE	Spec Process QE	Met Labs QE	Production Control	Regulatory Specialist
P	P	P	S	S	P	S	P	P	P	P	S

Procedural Knowledge: Ability to apply a structured approach to validate that the organization has identified the true root cause of QMS noncompliance (3).

Conceptual Knowledge: Ability to apply evaluate organizational quality records and performance to validate that the true root cause of QMS noncompliance has been appropriately identified (5).

Factual Knowledge: Understanding of root cause analysis tools (3). Understanding of closed-loop corrective action (3). Understanding of QMS (3). Understanding of the business processes and/or technical processes under audit (3). Understanding of value stream structure and where work is being performed (3).

Task 2 (PM121-002): Evaluate root cause analysis and ensure the true root cause is addressed in the organization.

Quality Eng—Basic	Quality Eng—Senior	Q Manager	Supplier QE	CSoftQE	Program QE	Contract Review	MRO QE	Spec Process QE	Met Labs QE	Production Control	Regulatory Specialist
P	P	P	S	S	P	S	P	P	P	P	S

Procedural Knowledge: Ability to apply a structured approach to verify that the true root cause is addressed (3).

Conceptual Knowledge: Ability to evaluate quality records and value stream performance to verify that true root cause is addressed (4).

Factual Knowledge: Understanding of root cause analysis tools (3). Understanding of closed-loop corrective action (3). Understanding of QMS (3). Understanding of the business processes and/or technical processes under audit (3). Understanding of value stream structure and where work is being performed (3).

Process: Monitor Supplier Corrective & Preventive Action Processes—QMS

Process Purpose: To ensure suppliers/partners have effective corrective and preventative action processes in place and such are applied to the root cause of noncompliance. Ensure process identifies and contains any potential impact on product.

Knowledge and Skill Requirements: Understanding of quality management systems. Understanding of QMS standards. Understanding of applicable regulatory requirements related to internal audits. Understanding of root cause analysis. Understanding of closed-loop corrective action. Understanding of the business processes and/or technical processes under audit. Ability to verify that suppliers/partners have effective corrective and preventative action processes in place and such are applied to the root cause of noncompliance. Ability to verify that corrective action process identifies and contains any potential impact on product. Ability to verify that effective corrective action processes are in place within the sub-tier supplier base.

Task 1 (PM125-001): Verify that suppliers/partners have effective corrective and preventative action processes in place and such are applied to the root cause of noncompliance.

Quality Eng—Basic	Quality Eng—Senior	Q Manager	Supplier QE	CSoftQE	Program QE	Contract Review	MRO QE	Spec Process QE	Met Labs QE	Production Control	Regulatory Specialist
S	S	S	P		S						

Procedural Knowledge: Ability to apply a structured process to verify that suppliers/partners have effective corrective and preventative action processes in place and such are applied to the root cause of noncompliance (3).

Conceptual Knowledge: Ability to analyze suppliers/partners corrective and preventative action processes to verify that such processes are effective and are systemically applied to the root cause of noncompliance (4).

Factual Knowledge: Understanding of regulatory requirements related to corrective action (3). Understanding of contractual and project/program requirements related to corrective action (3). Understanding of requirements management tools (3). Understanding of root cause analysis tools (3). Understanding of closed-loop corrective action (3). Understanding of the business processes and/or technical processes under audit (3). Understanding of value stream structure and where work is being performed (3).

Task 2 (PM125-002): Verify that corrective action processes identify and mitigates any potential impact on product. (Value stream)

Quality Eng—Basic	Quality Eng—Senior	Q Manager	Supplier QE	CSoftQE	Program QE	Contract Review	MRO QE	Spec Process QE	Met Labs QE	Production Control	Regulatory Specialist
S	P	S	P	S	P		S	S	S	S	S

Procedural Knowledge: Ability to apply a structured approach to verify that corrective action processes identify and mitigate any potential impact on product (3).

Conceptual Knowledge: Ability to analyze QMS corrective action processes to ensure they evaluate potential impact on product and engage in appropriate mitigation activities (4).

Factual Knowledge: Understanding of regulatory requirements related to corrective action (3). Understanding of contractual and project/program requirements related to corrective action (3). Understanding of requirements management tools (3). Understanding of root cause analysis tools (3). Understanding of closed-loop corrective action (3). Understanding of the business processes and/or technical processes under audit (3). Understanding of value stream structure and where work is being performed (3). Understanding of QMS and product inter-relationships (3).

Task 3 (PM125-003): Verify that effective corrective action processes are in place within the sub-tier supplier base.

Quality Eng—Basic	Quality Eng—Senior	Q Manager	Supplier QE	CSoftQE	Program QE	Contract Review	MRO QE	Spec Process QE	Met Labs QE	Production Control	Regulatory Specialist
S	S	S	P		S						

Procedural Knowledge: Ability to apply a structured approach to verify that effective corrective action processes are in place within the sub-tier supplier base (3).

Conceptual Knowledge: Ability to analyze corrective action activities within the value stream to verify that effective corrective action processes are in place (4).

Factual Knowledge: Understanding of regulatory requirements related to corrective action (3). Understanding of contractual and project/program requirements related to corrective action (3). Understanding of requirements management tools (3). Understanding of root cause analysis tools (3). Understanding of closed-loop corrective action (3). Understanding of the business processes and/or technical processes under audit (3). Understanding of value stream structure and where work is being performed (3).

Process: Monitor Organizational Corrective & Preventive Action Processes—QMS

Process Purpose: To ensure the organization has effective corrective and preventative action processes in place and such are applied to the root cause of noncompliance. Ensure process identifies and contains any potential impact on product.

Knowledge and Skill Requirements: Understanding of quality management systems. Understanding of QMS standards. Understanding of applicable regulatory requirements related to internal audits. Understanding of root cause analysis. Understanding of closed-loop corrective action. Understanding of the business processes and/or technical processes under audit. Ability to verify effective corrective and preventative action processes in place and such are applied to the root cause of noncompliance. Ability to verify that corrective action process identifies and contains any potential impact on product. Ability to verify that effective corrective action processes are in place within the organization, as appropriate.

Task 1 (PM126-001): Verify the organization has effective corrective and preventative action processes in place and such are applied to the root cause of noncompliance.

Quality Eng—Basic	Quality Eng—Senior	Q Manager	Supplier QE	CSoftQE	Program QE	Contract Review	MRO QE	Spec Process QE	Met Labs QE	Production Control	Regulatory Specialist
P	P	P	S		P	S					S

Procedural Knowledge: Ability to apply a structured process to verify that effective corrective and preventative action processes in place and such are applied to the root cause of noncompliance (3).

Conceptual Knowledge: Ability to analyze corrective and preventative action processes to verify that such processes are effective and are systemically applied to the root cause of noncompliance (4).

Factual Knowledge: Understanding of regulatory requirements related to corrective action (3). Understanding of contractual and project/program requirements related to corrective action (3). Understanding of requirements management tools (3). Understanding of root cause analysis tools (3). Understanding of closed-loop corrective action (3). Understanding of the business processes and/or technical processes under audit (3). Understanding of value stream structure and where work is being performed (3).

Task 2 (PM126-002): Verify that corrective action processes identify and mitigates any potential impact on product. (Organizational)

Quality Eng—Basic	Quality Eng—Senior	Q Manager	Supplier QE	CSoftQE	Program QE	Contract Review	MRO QE	Spec Process QE	Met Labs QE	Production Control	Regulatory Specialist
P	P	P	S	S	P	S	S	S	S	S	S

Procedural Knowledge: Ability to apply a structured approach to verify that corrective action processes identify and mitigate any potential impact on product (3).

Conceptual Knowledge: Ability to analyze QMS corrective action processes to ensure they evaluate potential impact on product and engage in appropriate mitigation activities (4).

Factual Knowledge: Understanding of regulatory requirements related to corrective action (3). Understanding of contractual and project/program requirements related to corrective action (3). Understanding of requirements management tools (3). Understanding of root cause analysis tools (3). Understanding of closed-loop corrective action (3). Understanding of the business processes and/or technical processes under audit (3). Understanding of value stream structure and where work is being performed (3). Understanding of QMS and product inter-relationships (3).

Task 3 (PM127-001): Verify that effective corrective action processes are in place within the organization.

Quality Eng—Basic	Quality Eng—Senior	Q Manager	Supplier QE	CSoftQE	Program QE	Contract Review	MRO QE	Spec Process QE	Met Labs QE	Production Control	Regulatory Specialist
P	P	P	S	S	P	S	S	S	S	S	S

Procedural Knowledge: Ability to apply a structured approach to verify that effective corrective action processes are in place within the organization (3).

Conceptual Knowledge: Ability to analyze corrective action activities to verify that effective corrective action processes are in place (4).

Factual Knowledge: Understanding of regulatory requirements related to corrective action (3). Understanding of contractual and project/program requirements related to corrective action (3). Understanding of requirements management tools (3). Understanding of root cause analysis tools (3). Understanding of closed-loop corrective action (3). Understanding of the business processes and/or technical processes under audit (3). Understanding of value stream structure and where work is being performed (3).

Process: Approve Supplier Corrective & Preventive Action—QMS

Process Purpose: To ensure the documented corrective and preventative action determinations effectively address the root cause in a timely fashion.

Knowledge and Skill Requirements: Understanding of quality management systems. Understanding of QMS standards. Understanding of applicable regulatory requirements related to internal audits. Understanding of root cause analysis. Understanding of closed-loop corrective action. Understanding of the business processes and/or technical processes under audit. Ability to verify that the documented corrective and preventative action determinations effectively address the root cause in a timely fashion. Ability to document acceptance, in the appropriate manner, to indicate acceptance of corrective action.

Task (PM130-001): Evaluate the documented corrective and preventative action determinations, and verify that they effectively address the root cause in a timely fashion, throughout the value stream.

Quality Eng—Basic	Quality Eng—Senior	Q Manager	Supplier QE	CSoftQE	Program QE	Contract Review	MRO QE	Spec Process QE	Met Labs QE	Production Control	Regulatory Specialist
S	P	P	P	S	P		S	S	S	S	

Procedural Knowledge: Ability to apply a structured approach to verify that the documented corrective and preventative action determinations effectively address the root cause in a timely fashion (3).

Conceptual Knowledge: Ability to evaluate individual noncompliance and verify that corrective and preventative actions effectively address the root cause in a timely fashion (5).

Factual Knowledge: Understanding of regulatory requirements related to corrective action (3). Understanding of contractual and project/program requirements related to corrective action (3). Understanding of requirements management tools (3). Understanding of root cause analysis tools (3). Understanding of closed-loop corrective action (3). Understanding of the business processes and/or technical processes under audit (3). Understanding of value stream structure and where work is being performed (3).

Process: Approve Organizational Corrective & Preventive Action—QMS

Process Purpose: To ensure the documented corrective and preventative action determinations effectively address the root cause in a timely fashion.

Knowledge and Skill Requirements: Understanding of quality management systems. Understanding of QMS standards. Understanding of applicable regulatory requirements related to internal audits. Understanding of root cause analysis. Understanding of closed-loop corrective action. Understanding of the business processes and/or technical processes under audit. Ability to verify that the documented corrective and preventative action determinations effectively address the root cause in a timely fashion. Ability to document acceptance, in the appropriate manner, to indicate acceptance of corrective action.

Task (PM131-001): Evaluate the documented corrective and preventative action determinations, and verify that they effectively address the root cause in a timely fashion, in the organization.

Quality Eng—Basic	Quality Eng—Senior	Q Manager	Supplier QE	CSoftQE	Program QE	Contract Review	MRO QE	Spec Process QE	Met Labs QE	Production Control	Regulatory Specialist
P	P	P	S	S	P	S	P	S	S	P	S

Procedural Knowledge: Ability to apply a structured approach to verify that the documented corrective and preventative action determinations effectively address the root cause in a timely fashion (3).

Conceptual Knowledge: Ability to evaluate individual noncompliance and verify that corrective and preventative actions effectively address the root cause in a timely fashion (5).

Factual Knowledge: Understanding of regulatory requirements related to corrective action (3). Understanding of contractual and project/program requirements related to corrective action (3). Understanding of requirements management tools (3). Understanding of root cause analysis tools (3). Understanding of closed-loop corrective action (3). Understanding of the business processes and/or technical processes under audit (3). Understanding of value stream structure and where work is being performed (3).

Process: Follow Up of Corrective & Preventative Action

Process Purpose: To verify the effectiveness of corrective and preventative actions by assessing implemented solutions after actions had been taken. Establish appropriate interval for effectiveness assessment; audit the implemented solutions to ensure they reflect the agreed-to arrangements; and that the solution prevents recurrence of the root cause.

Knowledge and Skill Requirements: Understanding of quality management systems. Understanding of QMS standards. Understanding of applicable regulatory requirements related to internal audits. Understanding of root cause analysis. Understanding of closed-loop corrective action. Understanding of the business processes and/or technical processes under audit. Ability to verify the effectiveness of corrective and preventative actions by assessing implemented solutions after actions had been taken. Ability to establish appropriate interval for follow up effectiveness assessment. Ability to perform a follow up audit of the implemented solutions to ensure they reflect the agreed-to arrangements; and that the solution prevents recurrence of the root cause. Ability to verify that corrective action follow up audits are effectively planned, performed, and documented within the sub-tier supplier base.

Task (PM135-001): Verify the effectiveness of corrective and preventative actions by assessing implemented solutions after actions had been taken, throughout the organization and value stream.

Quality Eng—Basic	Quality Eng—Senior	Q Manager	Supplier QE	CSoftQE	Program QE	Contract Review	MRO QE	Spec Process QE	Met Labs QE	Production Control	Regulatory Specialist
P	P	P	P	P	P	S	P	P	P	P	S

Procedural Knowledge: Ability to apply a structured assessment process to verify the effectiveness of corrective and preventative actions by assessing implemented solutions after actions had been taken (3).

Conceptual Knowledge: Ability to plan follow up verification activities of implemented corrective action to substantiate effectiveness (3). Ability to analyze implemented corrective action activities to substantiate effectiveness (4).

Factual Knowledge: Understanding of regulatory requirements related to corrective action (3). Understanding of contractual and project/program requirements related to corrective action (3). Understanding of requirements management tools (3). Understanding of root cause analysis tools (3). Understanding of closed-loop corrective action (3). Understanding of the business processes and/or technical processes under audit (3). Understanding of value stream structure and where work is being performed (3).

V.G. Lessons Learned

Process: Apply Lessons Learned—Conceptual Phase

Process Purpose: Based on the lessons learned in previous programs/projects, apply lessons learned to ensure previous issues are not repeated within the Conceptual Phase.

Knowledge and Skill Requirements: Understanding of the business processes of the Conceptual Phase. Ability to access lessons learned libraries and apply the experience gained from previous programs/projects to the new program/projects.

Task 1 (CP010-001): Review lessons learned libraries and determine applicability based on the program/project requirements.

Quality Eng—Basic	Quality Eng—Senior	Q Manager	Supplier QE	CSoftQE	Program QE	Contract Review	MRO QE	Spec Process QE	Met Labs QE	Production Control	Regulatory Specialist
S	S	S	S	S	P	S	S	S	S	S	P

Procedural Knowledge: Ability to follow structured process to access existing lessons learned libraries and identify applicable lessons learned for the program/project (3). (Libraries may reside in and outside of the organization.)

Conceptual Knowledge: Ability to evaluate existing lessons learned and identify which may be applicable to the new program/project (5). Evaluate existing lessons learned libraries and determine if additional sources of information may be required (5).

Factual Knowledge: Knowledge of lessons learned methods (3). Knowledge of existing lessons learned libraries (3). Knowledge of existing program/project requirements (3).

Task 2 (CP010-002): Evaluate current program/project requirements against existing lessons learned libraries and determine applicability of application. Determine which lessons learned should be applied to the new program/project. Ensure identified lessons learned are documented so that they can be integrated into the program/project plan.

Quality Eng—Basic	Quality Eng—Senior	Q Manager	Supplier QE	CSoftQE	Program QE	Contract Review	MRO QE	Spec Process QE	Met Labs QE	Production Control	Regulatory Specialist
S	S	S	S	S	P	S	S	S	S	S	P

Procedural Knowledge: Ability to follow a structured process in performing an applicability assessment of lessons learned to the new program/project (3). Ability to follow a structured process for measuring and documenting potential risk associated with not implementing identified lessons learned (3).

Conceptual Knowledge: Ability to evaluate existing lessons learned to determine which should be applied to the new program/project (5). Ability to understand and recognize commonality of business and technical processes (4). Ability to analyze and measure risks associated with not implementing lessons learned (5). Ability to communicate risk factors into program/project planning (4).

Factual Knowledge: Understanding of lessons learned processes (3). Knowledge of applicable technologies (4). Understanding of business and technical process relationships (4). Understanding of process interrelationships (4). Knowledge of risk measurement and management processes (4).

Task 3 (CP010-003): Implement lessons learned into program/project planning and/or document initial risk management practices for risk associated with not implementing lessons learned.

Quality Eng—Basic	Quality Eng—Senior	Q Manager	Supplier QE	CSoftQE	Program QE	Contract Review	MRO QE	Spec Process QE	Met Labs QE	Production Control	Regulatory Specialist
S	S	S	S	S	P	S	S	S	S	S	P

Procedural Knowledge: Ability to follow a structured process in the implementation of applicable lessons learned into a new program/project (3).

Conceptual Knowledge: Ability to understand where lessons learned are most appropriately applied (3). Ability to define and quantify risks associated with not applying lessons learned (5).

Factual Knowledge: Knowledge of lessons learned processes (3). Knowledge of applicable technologies (3). Knowledge of interrelatedness of business and technical processes (4). Knowledge of risk assessment methods (4).

Process: Capture, Document, and Preserve Lessons Learned—Conceptual Phase

Process Purpose: To recognize, understand, and document lessons learned during the application of the previous processes in the Conceptual phase so that such may be used in future programs/projects.

Knowledge and Skill Requirements: Ability to identify and document lessons learned and integrate such into organizational lessons learned libraries so that the experience gained from previous processes in the Conceptual Phase can be applied to future program/projects.

Task 1 (CP035-001): Review the processes applied during the Conceptual Phase and identify lessons learned for future use.

Quality Eng—Basic	Quality Eng—Senior	Q Manager	Supplier QE	CSoftQE	Program QE	Contract Review	MRO QE	Spec Process QE	Met Labs QE	Production Control	Regulatory Specialist
S	S	S	S	S	P	S	S	S	S	S	P

Procedural Knowledge: Ability to follow a structured process to ensure a post-Conceptual Phase review is performed and potential lessons learned are identified (3).

Conceptual Knowledge: Ability to apply evaluation to the previous processes applied during the Conceptual Phase and to recognize events and issues that may qualify as lessons learned (5). Ability to analyze events and issues and determine root cause (4). Ability to create the lessons learned for future use (6).

Factual Knowledge: Knowledge of the lessons learned identification process (3). Knowledge of Conceptual Phase processes (3). Understand how lessons learned for future use are documented (3).

Task 2 (CP035-002): Document identified lessons learned so that such can be understood, assessed for applicability, and applied in future programs/projects.

Quality Eng—Basic	Quality Eng—Senior	Q Manager	Supplier QE	CSoftQE	Program QE	Contract Review	MRO QE	Spec Process QE	Met Labs QE	Production Control	Regulatory Specialist
S	S	S	S	S	P	S	S	S	S	S	P

Procedural Knowledge: Ability to follow a structured process to describe the lessons learned in a format that can be understood by users (3).

Conceptual Knowledge: Ability to describe lessons learned in terms that are relevant and meaningful to users (using various means of written and visual communication), including future practitioners of the Conceptual Phase (Output must flow into integration task)(4). Ability to understand the interrelationship of business and technical processes and to make descriptive connections (4).

Factual Knowledge: Understand the importance of descriptive narratives in relation to lessons learned (3). Know the interrelationships of business and technical processes (3). Knowledge of communication methods (3). Knowledge of how to effectively apply communication methods (4).

Task 3 (CP035-003): Integrate documented lessons learned into organizational lessons learned libraries.

Quality Eng—Basic	Quality Eng—Senior	Q Manager	Supplier QE	CSoftQE	Program QE	Contract Review	MRO QE	Spec Process QE	Met Labs QE	Production Control	Regulatory Specialist
S	S	S	S	S	P	S	S	S	S	S	P

Procedural Knowledge: Ability to follow a defined and structured process to integrate the documented lessons learned of the Conceptual Phase into the organizational lessons learned libraries (3).

Conceptual Knowledge: Understanding of how and where lessons learned are to be appropriately integrated into organizational lessons learned libraries (5). Understand the interrelationship of business and technical processes (4). Ability to conceptualize linkages in business and technical processes (4).

Factual Knowledge: Know the interrelationships of business and technical processes (3). Understand the lessons learned library structure of the organization (3).

Process: Apply Lessons Learned—Design Phase

Process Purpose: Based on the lessons learned in previous programs/projects, review, identify applicable, and apply lessons learned to ensure previous issues are not repeated.

Knowledge and Skill Requirements: Ability to access lessons learned libraries and apply the experience gained from previous programs/projects to the new program/projects.

Task 1 (DP005-001): Review lessons learned libraries and determine applicability based on the program/project requirements.

Quality Eng—Basic	Quality Eng—Senior	Q Manager	Supplier QE	CSoftQE	Program QE	Contract Review	MRO QE	Spec Process QE	Met Labs QE	Production Control	Regulatory Specialist
S	P	P	S	S	P	P	S	S	S	S	P

Procedural Knowledge: Ability to follow structured process to access existing lessons learned libraries and identify applicable lessons learned for the program/project. (Libraries may reside in and outside of the organization) (3).

Conceptual Knowledge: Ability to evaluate existing lessons learned and identify which may be applicable to the new program/project (5). Assess existing lessons learned libraries and determine if additional sources of information may be required (4).

Factual Knowledge: Knowledge of lessons learned methods (3). Knowledge of existing lessons learned libraries (3). Knowledge of existing program/project requirements (3).

Task 2 (DP005-002): Assess current program/project against existing lessons learned libraries and determine applicability of application. Determine which lessons learned should be applied to the new program/project. Ensure identified lessons learned are documented so that they can be integrated into the program/project plan.

Quality Eng—Basic	Quality Eng—Senior	Q Manager	Supplier QE	CSoftQE	Program QE	Contract Review	MRO QE	Spec Process QE	Met Labs QE	Production Control	Regulatory Specialist
S	P	P	S	S	P	P	S	S	S	S	P

Procedural Knowledge: Ability to follow a structured process in performing an applicability assessment of lessons learned to the new program/project (3). Ability to follow a structured process for measuring and documenting potential risk associated with not implementing identified lessons learned (3).

Conceptual Knowledge: Ability to evaluate existing lessons learned to determine which should be applied to the new program/project (5). Ability to understand and recognize commonality of business and technical processes (4). Ability to analyze and measure risks associated with not implementing lessons learned (4). Ability to communicate risk factors into program/project planning (3).

Factual Knowledge: Understanding of lessons learned processes (3). Knowledge of applicable technologies (4). Understanding of business and technical process relationships (4). Understanding of process interrelationships (4). Knowledge of risk measurement and management processes (3).

Task 3 (DP005-003): Implement lessons learned into program/project planning and/or document initial risk management practices for risk associated with not implementing lessons learned.

Quality Eng—Basic	Quality Eng—Senior	Q Manager	Supplier QE	CSoftQE	Program QE	Contract Review	MRO QE	Spec Process QE	Met Labs QE	Production Control	Regulatory Specialist
S	P	P	S	S	P	P	S	S	S	S	P

Procedural Knowledge: Ability to follow a structured process in the implementation of applicable lessons learned into a new program/project (3).

Conceptual Knowledge: Ability to understand where lessons learned are most appropriately applied (3). Ability to define and quantify risks associated with not applying lessons learned (3).

Factual Knowledge: Knowledge of lessons learned processes (3). Knowledge of applicable technologies (3). Knowledge of interrelatedness of business and technical processes (3). Knowledge of risk assessment methods (3).

Process: Capture, Document, and Preserve Lessons Learned—Design Phase

Process Purpose: To recognize, understand, and document lessons learned during the application of the previous processes in the Design Phase so that such may be used in future programs/projects.

Knowledge and Skill Requirements: Ability to identify and document lessons learned and integrate such into organizational lessons learned libraries so that the experience gained from previous processes in the Design Phase can be applied to future program/projects.

Task 1 (DP110-001): Review the processes applied during the Design Phase and identify lessons learned for future use.

Quality Eng—Basic	Quality Eng—Senior	Q Manager	Supplier QE	CSoftQE	Program QE	Contract Review	MRO QE	Spec Process QE	Met Labs QE	Production Control	Regulatory Specialist
S	P	P	S	S	P	P	S	S	S	S	P

Procedural Knowledge: Ability to follow a structured process to ensure a post-Design Phase review is performed and potential lessons learned are identified (4).

Conceptual Knowledge: Ability to apply evaluation to the previous processes applied during the Design Phase and to recognize events and issues that may qualify as lessons learned (5). Ability to evaluate events and issues and determine root cause (5). Ability to create the lessons learned (new knowledge) for future use (6).

Factual Knowledge: Knowledge of the lessons learned identification process (3). Knowledge of Design Phase processes (3). Understand how lessons learned for future use are documented (2). Knowledge of root cause analysis techniques (3).

Task 2 (DP110-002): Document identified lessons learned so that such can be understood, assessed for applicability, and applied in future programs/projects.

Quality Eng—Basic	Quality Eng—Senior	Q Manager	Supplier QE	CSoftQE	Program QE	Contract Review	MRO QE	Spec Process QE	Met Labs QE	Production Control	Regulatory Specialist
S	P	P	S	S	P	P	S	S	S	S	P

Procedural Knowledge: Ability to follow a structured process to describe the lessons learned in a format that can be understood by users (2).

Conceptual Knowledge: Ability to describe lessons learned in terms that are relevant and meaningful to users (using various means of written and visual communication), including future practitioners of the Design Phase (Output must flow into integration task)(4). Ability to understand the interrelationship of business and technical processes and to make descriptive connections (4).

Factual Knowledge: Understand the importance of descriptive narratives in relation to lessons learned (2). Know the interrelationships of business and technical processes (3). Knowledge of communication methods (3). Knowledge of how to effectively apply communication methods (4).

Task 3 (DP110-003): Integrate documented lessons learned into organizational lessons learned libraries.

Quality Eng—Basic	Quality Eng—Senior	Q Manager	Supplier QE	CSoftQE	Program QE	Contract Review	MRO QE	Spec Process QE	Met Labs QE	Production Control	Regulatory Specialist
S	P	P	S	S	P	P	S	S	S	S	P

Procedural Knowledge: Ability to follow a defined and structured process to integrate the documented lessons learned of the Design Phase into the organizational lessons learned libraries (2).

Conceptual Knowledge: Understanding of design phase business processes. Understanding of how and where lessons learned are to be appropriately integrated into organizational lessons learned libraries (4). Understand the interrelationship of business and technical processes (3). Ability to conceptualize linkages in business and technical processes (4).

Process: Apply Lessons Learned—Pre-Production Organizational

Process Purpose: Based on the lessons learned in previous programs/projects, apply lessons learned to ensure previous issues are not repeated.

Knowledge and Skill Requirements: Ability to access lessons learned libraries and apply the experience gained from previous programs/projects to the new program/projects.

Task 1 (PO005-001): Review lessons learned libraries and determine applicability based on the program/project requirements.

Quality Eng—Basic	Quality Eng—Senior	Q Manager	Supplier QE	CSoftQE	Program QE	Contract Review	MRO QE	Spec Process QE	Met Labs QE	Production Control	Regulatory Specialist
P	P	P	S	P	P	S	S	S	S	S	P

Procedural Knowledge: Ability to follow structured process to access existing lessons learned libraries and identify applicable lessons learned for the program/project (3). (Libraries may reside in and outside of the organization).

Conceptual Knowledge: Ability to evaluate existing lessons learned and identify which may be applicable to the new program/project (5). Assess existing lessons learned libraries and determine if additional sources of information may be required (4).

Factual Knowledge: Knowledge of lessons learned methods (3). Knowledge of existing lessons learned libraries (2). Knowledge of existing program/project requirements (3).

Task 2 (PO005-002): Assess current program/project against existing lessons learned libraries and determine applicability of application. Determine which lessons learned should be applied to the new program/project. Ensure identified lessons learned are documented so that they can be integrated into the program/project plan.

Quality Eng—Basic	Quality Eng—Senior	Q Manager	Supplier QE	CSoftQE	Program QE	Contract Review	MRO QE	Spec Process QE	Met Labs QE	Production Control	Regulatory Specialist
P	P	P	S	P	P	S	S	S	S	S	P

Procedural Knowledge: Ability to follow a structured process in performing an applicability assessment of lessons learned to the new program/project (3). Ability to follow a structured process for measuring and documenting potential risk associated with not implementing identified lessons learned (3).

Conceptual Knowledge: Ability to evaluate existing lessons learned to determine which should be applied to the new program/project (5). Ability to understand and recognize commonality of business and technical processes (4). Ability to analyze and evaluate risks associated with not implementing lessons learned (5). Ability to communicate risk factors into program/project planning (3).

Factual Knowledge: Understanding of lessons learned processes (3). Knowledge of applicable technologies (3). Understanding of business and technical process relationships (3). Understanding of process interrelationships (3). Knowledge of risk measurement and management processes (3).

Task 3 (PO005-003): Implement lessons learned into program/project planning and/or document initial risk management practices for risk associated with not implementing lessons learned.

Quality Eng—Basic	Quality Eng—Senior	Q Manager	Supplier QE	CSoftQE	Program QE	Contract Review	MRO QE	Spec Process QE	Met Labs QE	Production Control	Regulatory Specialist
P	P	P	S	P	P	S	S	S	S	S	P

Procedural Knowledge: Ability to follow a structured process in the implementation of applicable lessons learned into a new program/project planning and schedules (3).

Conceptual Knowledge: Ability to evaluate where lessons learned are most appropriately applied (5). Ability to analyze and quantify risks associated with not applying lessons learned (4).

Factual Knowledge: Knowledge of lessons learned processes (3). Knowledge of applicable technologies (3). Knowledge of interrelatedness of business and technical processes (3). Knowledge of risk assessment methods (3).

Process: Capture, Document, and Preserve Lessons Learned—Pre-Production Organizational

Process Purpose: To recognize, understand, and document lessons learned during the application of the previous processes in the Pre-Production Manufacturing/Maintenance Phase so that such may be used in future programs/projects.

Knowledge and Skill Requirements: Ability to identify and document lessons learned and integrate such into organizational lessons learned libraries so that the experience gained from previous processes in the Pre-Production Manufacturing/Maintenance Phase can be applied to future program/projects.

Task 1 (PO170-001): Review the processes applied during the Pre-Production Manufacturing/Maintenance Phase and identify lessons learned for future use. Document identified lessons learned so that such can be understood, assessed for applicability, and applied in future programs/projects.

Quality Eng—Basic	Quality Eng—Senior	Q Manager	Supplier QE	CSoftQE	Program QE	Contract Review	MRO QE	Spec Process QE	Met Labs QE	Production Control	Regulatory Specialist
P	P	P	S	P	P	S	S	S	S	S	P

Procedural Knowledge: Ability to follow a structured process to ensure a post-Pre-Production Manufacturing/Maintenance Phase review is performed and potential lessons learned are identified (3). Ability to follow a structured process to describe the lessons learned in a format that can be understood by users (3).

Conceptual Knowledge: Ability to apply evaluation to the previous processes applied during the Pre-Production Manufacturing/Maintenance Phase and to recognize events and issues that may qualify as lessons learned (5). Ability to analyze events and issues and determine root cause (5). Ability to create the lessons learned for future use (6). Ability to understand the interrelationship of business and technical processes and to make descriptive connections (4). Ability to describe lessons learned in terms that are relevant and meaningful to users (using various means of written and visual communication), including future practitioners of the Pre-Production Manufacturing/Maintenance Phase (Output must flow into integration task) (4).

Factual Knowledge: Knowledge of the lessons learned identification process (3). Knowledge of Pre-Production Manufacturing/Maintenance Phase processes (3). Understand how lessons learned for future use are documented (3). Understanding of organizational knowledge/memory libraries (3) Understand the importance of descriptive narratives in relation to lessons learned (2). Know the interrelationships of business and technical processes (3). Knowledge of communication methods (3). Knowledge of how to effectively apply communication methods (4).

Task 2 (PO170-002): Integrate documented lessons learned into organizational lessons learned libraries.

Quality Eng—Basic	Quality Eng—Senior	Q Manager	Supplier QE	CSoftQE	Program QE	Contract Review	MRO QE	Spec Process QE	Met Labs QE	Production Control	Regulatory Specialist
P	P	P	S	P	P	S	S	S	S	S	P

Procedural Knowledge: Ability to follow a defined and structured process to integrate the documented lessons learned of the Pre-Production Manufacturing/Maintenance Phase into the organizational lessons learned libraries (2).

Conceptual Knowledge: Evaluate how and where lessons learned are to be appropriately integrated into organizational lessons learned libraries (in evaluation review where the project/program specific lessons learned may be integrated into other unrelated projects/programs) (5). Understand the interrelationship of business and technical processes (4). Ability to conceptualize linkages in business and technical processes (4).

Factual Knowledge: Know the interrelationships of business and technical processes (3). Understand the lessons learned library structure of the organization (3).

Process: Apply Lessons Learned—Pre-Production Supplier

Process Purpose: Based on the lessons learned in previous programs/projects, apply lessons learned to ensure previous issues are not repeated.

Knowledge and Skill Requirements: Ability to access lessons learned libraries and apply the experience gained from previous programs/projects to the new program/projects.

Task 1 (PP005-001): Review lessons learned libraries and determine applicability based on the program/project requirements.

Quality Eng—Basic	Quality Eng—Senior	Q Manager	Supplier QE	CSoftQE	Program QE	Contract Review	MRO QE	Spec Process QE	Met Labs QE	Production Control	Regulatory Specialist
S	P	P	P	S	P	S	S	S	S	S	P

Procedural Knowledge: Ability to follow structured process to access existing lessons learned libraries and identify applicable lessons learned for the program/project (3). (Libraries may reside in and outside of the organization.)

Conceptual Knowledge: Ability to evaluate existing lessons learned and identify that may be applicable to the new program/project (5). Assess existing lessons learned libraries and determine if additional sources of information may be required (4).

Factual Knowledge: Knowledge of lessons learned methods (3). Knowledge of existing lessons learned libraries (2). Knowledge of existing program/project requirements (3).

Task 2 (PP005-002): Assess current program/project against existing lessons learned libraries and determine applicability of application. Determine which lessons learned should be applied to the new program/project. Ensure identified lessons learned are documented so that they can be integrated into the program/project plan.

Quality Eng—Basic	Quality Eng—Senior	Q Manager	Supplier QE	CSoftQE	Program QE	Contract Review	MRO QE	Spec Process QE	Met Labs QE	Production Control	Regulatory Specialist
S	P	P	P	S	P	S	S	S	S	S	P

Procedural Knowledge: Ability to follow a structured process in performing an applicability assessment of lessons learned to the new program/project (3). Ability to follow a structured process for measuring and documenting potential risk associated with not implementing identified lessons learned (3).

Conceptual Knowledge: Ability to evaluate existing lessons learned to determine which should be applied to the new program/project (5). Ability to understand and recognize commonality of business and technical processes (4). Ability to analyze and evaluate risks associated with not implementing lessons learned (5). Ability to communicate risk factors into program/project planning (3).

Factual Knowledge: Understanding of lessons learned processes (3). Knowledge of applicable technologies (3). Understanding of business and technical process relationships (3). Understanding of process interrelationships (3). Knowledge of risk measurement and management processes (3).

Task 3 (PP005-003): Implement lessons learned into program/project planning and/or document initial risk management practices for risk associated with not implementing lessons learned.

Quality Eng—Basic	Quality Eng—Senior	Q Manager	Supplier QE	CSoftQE	Program QE	Contract Review	MRO QE	Spec Process QE	Met Labs QE	Production Control	Regulatory Specialist
S	P	P	P	S	P	S	S	S	S	S	P

Procedural Knowledge: Ability to follow a structured process in the implementation of applicable lessons learned into a new program/project planning and schedules (3).

Conceptual Knowledge: Ability to evaluate where lessons learned are most appropriately applied (5). Ability to analyze and quantify risks associated with not applying lessons learned (4).

Factual Knowledge: Knowledge of lessons learned processes (3). Knowledge of applicable technologies (3). Knowledge of interrelatedness of business and technical processes (3). Knowledge of risk assessment methods (3).

Process: Capture, Document, and Preserve Lessons Learned—Pre-Production Supplier

Process Purpose: To recognize, understand, and document lessons learned during the application of the previous processes in the Pre-Production Supplier Phase so that such may be used in future programs/projects.

Knowledge and Skill Requirements: Ability to identify and document lessons learned and integrate such into organizational lessons learned libraries so that the experience gained from previous processes in the Pre-Production Supplier Phase can be applied to future program/projects.

Task 1 (PP170-001): Review the processes applied during the Pre-Production Manufacturing/Maintenance—Supplier Focus Phase and identify lessons learned for future use. Document identified lessons learned so that such can be understood, assessed for applicability, and applied in future programs/projects.

Quality Eng—Basic	Quality Eng—Senior	Q Manager	Supplier QE	CSoftQE	Program QE	Contract Review	MRO QE	Spec Process QE	Met Labs QE	Production Control	Regulatory Specialist
S	P	P	P	S	P	S	S	S	S	S	P

Procedural Knowledge: Ability to follow a structured process to ensure a post-Pre-Production Manufacturing/Maintenance Phase review is performed and potential lessons learned are identified (3). Ability to follow a structured process to describe the lessons learned in a format that can be understood by users (3).

Conceptual Knowledge: Ability to apply evaluation to the previous processes applied during the Pre-Production Manufacturing/Maintenance—Supplier focus Phase and to recognize events and issues that may qualify as lessons learned (5). Ability to analyze events and issues and determine root cause (5). Ability to create the lessons learned for future use (6). Ability to understand the interrelationship of business and technical processes and to make descriptive connections (4). Ability to describe lessons learned in terms that are relevant and meaningful to users (using various means of written and visual communication), including future practitioners of the Pre-Production Manufacturing/Maintenance—Supplier focus Phase (Output must flow into integration task) (4).

Factual Knowledge: Knowledge of the lessons learned identification process (3). Knowledge of Pre-Production Manufacturing/Maintenance—Supplier focus Phase processes (3). Understand how lessons learned for future use are documented (3). Understanding of organizational knowledge/memory libraries (3) Understand the importance of descriptive narratives in relation to lessons learned (2). Know the interrelationships of business and technical processes (3).

Knowledge of communication methods (3). Knowledge of how to effectively apply communication methods (4).

Task 2 (PP170-002): Integrate documented lessons learned into organizational lessons learned libraries.

Quality Eng—Basic	Quality Eng—Senior	Q Manager	Supplier QE	CSoftQE	Program QE	Contract Review	MRO QE	Spec Process QE	Met Labs QE	Production Control	Regulatory Specialist
S	P	P	P	S	P	S	S	S	S	S	P

Procedural Knowledge: Ability to follow a defined and structured process to integrate the documented lessons learned of the Pre-Production Manufacturing/Maintenance—Supplier focus Phase into the organizational lessons learned libraries (2).

Conceptual Knowledge: Evaluate how and where lessons learned are to be appropriately integrated into organizational lessons learned libraries (in evaluation review where the project/program specific lessons learned may be integrated into other unrelated projects/programs) (5). Understand the interrelationship of business and technical processes (4). Ability to conceptualize linkages in business and technical processes (4).

Factual Knowledge: Know the interrelationships of business and technical processes (3). Understand the lessons learned library structure of the organization (3).

Process: Apply Lessons Learned—Production/Maintenance Product Focused Phase

Process Purpose: Based on the lessons learned in previous programs/projects, apply lessons learned to ensure previous issues are not repeated.

Knowledge and Skill Requirements: Ability to access lessons learned libraries and apply the experience gained from previous programs/projects to the new program/projects.

Task 1 (PF005-001): Review lessons learned libraries and determine applicability based on the program/project requirements.

Quality Eng—Basic	Quality Eng—Senior	Q Manager	Supplier QE	CSoftQE	Program QE	Contract Review	MRO QE	Spec Process QE	Met Labs QE	Production Control	Regulatory Specialist
P	P	P	S	P	P	S	P	P	P	P	P

Procedural Knowledge: Review lessons learned libraries and determine applicability based on the program/project requirements.

Conceptual Knowledge: Ability to analyze existing lessons learned and identify which may be applicable to the new program/project (4). Ability to understand and recognize commonality of business and technical processes (5). Evaluate existing lessons learned libraries and determine if additional sources of information may be required (5).

Factual Knowledge: Knowledge of lessons learned methods (3). Knowledge of existing lessons learned libraries (3). Knowledge of existing program/project requirements (3). Knowledge of sources of additional information related to project/program lessons learned (3).

Task 2 (PF005-002): Assess current program/project against existing lessons learned libraries and determine applicability of application. Determine which lessons learned should be applied to the new program/project. Ensure identified lessons learned are documented so that they can be integrated into the program/project plan.

Quality Eng—Basic	Quality Eng—Senior	Q Manager	Supplier QE	CSoftQE	Program QE	Contract Review	MRO QE	Spec Process QE	Met Labs QE	Production Control	Regulatory Specialist
P	P	P	S	P	P	S	P	P	P	P	P

Procedural Knowledge: Ability to follow a structured process in performing an applicability assessment of lessons learned to the new program/project (3). Ability to follow a structured process for measuring and documenting potential risk associated with not implementing identified lessons learned (3).

Conceptual Knowledge: Ability to evaluate existing lessons learned to determine which should be applied to the new program/project (5). Ability to understand and recognize commonality of business and technical processes (5). Ability to analyze and measure risks associated with not implementing lessons learned (4). Ability to communicate risk factors into program/project planning (4).

Factual Knowledge: Understanding of lessons learned processes (3). Knowledge of applicable technologies (3). Understanding of business and technical process relationships (3). Understanding of process interrelationships (3). Knowledge of risk measurement and management processes (3).

Task 3 (PF005-003): Implement lessons learned into program/project planning and/or document initial risk management practices for risk associated with not implementing lessons learned.

Quality Eng—Basic	Quality Eng—Senior	Q Manager	Supplier QE	CSoftQE	Program QE	Contract Review	MRO QE	Spec Process QE	Met Labs QE	Production Control	Regulatory Specialist
P	P	P	S	P	P	S	P	P	P	P	P

Procedural Knowledge: Ability to follow a structured process in the implementation of applicable lessons learned into a new program/project (3).

Conceptual Knowledge: Ability to apply the lessons learned planning (3). Ability to analyze the risks associated with not implementing the planned lessons learned (4). Ability to document the applied lessons learned (3).

Factual Knowledge: Knowledge of lessons learned processes (3). Knowledge of applicable technologies (3). Knowledge of interrelatedness of business and technical processes (3). Knowledge of project/program documentation practices (3).

Process: Capture, Document, and Preserve Lessons Learned—Production/Maintenance Product Focused Phase

Process Purpose: To recognize, understand, and document lessons learned during the application of the previous processes in the Production/Maintenance Phase so that such may be used in future programs/projects.

Knowledge and Skill Requirements: Ability to identify and document lessons learned and integrate such into organizational lessons learned libraries so that the experience gained from previous processes in the Production/Maintenance Phase can be applied to future program/projects.

Task 1 (PF135-001): Review the processes applied during the Production/Maintenance Phase and identify lessons learned for future use.

Quality Eng—Basic	Quality Eng—Senior	Q Manager	Supplier QE	CSoftQE	Program QE	Contract Review	MRO QE	Spec Process QE	Met Labs QE	Production Control	Regulatory Specialist
P	P	P	S	P	P	S	P	P	P	P	P

Procedural Knowledge: Ability to follow a structured process to ensure a post-Production/Maintenance Phase review is performed and potential lessons learned are identified.

Conceptual Knowledge: Ability to apply evaluation to the previous processes applied during the Production/Maintenance Phase and to recognize events and issues that may qualify as lessons learned (5). Ability to analyze events and issues and determine root cause (4). Ability to create the lessons learned for future use (6).

Factual Knowledge: Knowledge of the lessons learned identification process (3). Knowledge of Production/Maintenance Phase processes (3). Understanding of how lessons learned for future use are documented (incl libraries) (3).

Task 2 (PF135-002): Document identified lessons learned so that such can be understood, assessed for applicability, and applied in future programs/projects.

Quality Eng—Basic	Quality Eng—Senior	Q Manager	Supplier QE	CSoftQE	Program QE	Contract Review	MRO QE	Spec Process QE	Met Labs QE	Production Control	Regulatory Specialist
P	P	P	S	P	P	S	P	P	P	P	P

Procedural Knowledge: Ability to follow a structured process to describe the lessons learned in a format that can be understood by users.

Conceptual Knowledge: Ability to describe lessons learned in terms that are relevant and meaningful to users (using various means of written and visual communication), including future practitioners of the Production/Maintenance Phase (Output must flow into integration task)(4). Ability to understand the interrelationship of business and technical processes and to make descriptive connections (4).

Factual Knowledge: Understand the importance of descriptive narratives in relation to lessons learned (3). Know the interrelationships of business and technical processes (3). Knowledge of communication methods (3). Knowledge of how to effectively apply communication methods (4).

Task 3 (PF135-003): Integrate documented lessons learned into organizational lessons learned libraries.

Quality Eng—Basic	Quality Eng—Senior	Q Manager	Supplier QE	CSoftQE	Program QE	Contract Review	MRO QE	Spec Process QE	Met Labs QE	Production Control	Regulatory Specialist
P	P	P	S	P	P	S	P	P	P	P	P

Procedural Knowledge: Integrate documented lessons learned into organizational lessons learned libraries.

Conceptual Knowledge: Understanding of how and where lessons learned are to be appropriately integrated into organizational lessons learned libraries (4). Understand the interrelationship of business and technical processes (4). Ability to conceptualize linkages in business and technical processes (4).

Factual Knowledge: Know the interrelationships of business and technical processes (3). Understand the lessons learned library structure of the organization (3).

Process: Apply Lessons Learned—Production System Phase

Process Purpose: Based on the lessons learned in previous programs/projects, apply lessons learned to ensure previous issues are not repeated.

Knowledge and Skill Requirements: Ability to access lessons learned libraries and apply the experience gained from previous programs/projects to the new program/projects.

Task 1 (PM005-001): Review lessons learned libraries and determine applicability based on the program/project requirements.

Quality Eng—Basic	Quality Eng—Senior	Q Manager	Supplier QE	CSoftQE	Program QE	Contract Review	MRO QE	Spec Process QE	Met Labs QE	Production Control	Regulatory Specialist
P	P	P	S	P	P	S	P	P	P	P	P

Procedural Knowledge: Ability to follow structured process to access existing lessons learned libraries and identify applicable lessons learned for the program/project. (Libraries may reside in and outside of the organization) (3).

Conceptual Knowledge: Ability to analyze existing lessons learned and identify which may be applicable to the new program/project (4). Evaluate existing lessons learned libraries and determine if additional sources of information may be required (5).

Factual Knowledge: Knowledge of lessons learned methods (3). Knowledge of existing lessons learned libraries (3). Knowledge of existing program/project requirements (3).

Task 2 (PM005-002): Ensure identified lessons learned are integrated into the program/project plan. Evaluate the risks/benefits associated with implementing lessons learned.

Quality Eng—Basic	Quality Eng—Senior	Q Manager	Supplier QE	CSoftQE	Program QE	Contract Review	MRO QE	Spec Process QE	Met Labs QE	Production Control	Regulatory Specialist
P	P	P	S	P	P	S	P	P	P	P	P

Procedural Knowledge: Ability to apply a structured approach to ensure applicable lessons learned are integrated into applicable project/program plans (3). Ability to follow a structured process for measuring and documenting potential risk/benefits associated with lessons learned (3).

Conceptual Knowledge: Ability to understand and recognize commonality of business and technical processes (4). Ability to evaluate and measure risks and benefits associated with implementing lessons learned (5). Ability to communicate risk and benefits factors into program/project planning (4).

Factual Knowledge: Understanding of lessons learned processes (3). Knowledge of applicable technologies (4). Understanding of business and technical process relationships (4). Understanding of process interrelationships (3). Knowledge of risk and benefit measurement and management processes (3).

Task 3 (PM005-003): Implement lessons learned into program/project planning and/or document initial risk management practices for risk associated with not implementing lessons learned.

Quality Eng—Basic	Quality Eng—Senior	Q Manager	Supplier QE	CSoftQE	Program QE	Contract Review	MRO QE	Spec Process QE	Met Labs QE	Production Control	Regulatory Specialist
P	P	P	S	P	P	S	P	P	P	P	P

Procedural Knowledge: Ability to follow a structured process in the implementation of applicable lessons learned into a new program/project (3).

Conceptual Knowledge: Ability to evaluate program/project requirements and planning and determine where lessons learned are most appropriately applied (5). Ability to analyze, define and quantify risks associated with not applying lessons learned (4).

Factual Knowledge: Knowledge of lessons learned processes (3). Knowledge of applicable technologies (4). Knowledge of interrelatedness of business and technical processes (4). Knowledge of risk assessment methods (3).

Process: Capture, Document, and Preserve Lessons Learned—Production System Phase

Process Purpose: To ensure knowledge of past problems, challenges, and best practices are documented in a fashion in which can be easily understood and organized so that the case is illustrated to show relationship to the various business process steps within the lifecycle. Ensure the process for maintaining and accessing lessons learned is developed to allow long-term memory to the lesson and facilitates easy retrieval.

Knowledge and Skill Requirements: Understanding of lessons learned libraries. Ability to identify and document lessons learned and integrate such into organizational lessons learned libraries so that the experience gained from previous processes in the Production/Service Phase can be applied to future program/projects.

Task 1 (PM260-001): Review the processes applied during the Production/Service Phase and identify lessons learned for future use.

Quality Eng—Basic	Quality Eng—Senior	Q Manager	Supplier QE	CSoftQE	Program QE	Contract Review	MRO QE	Spec Process QE	Met Labs QE	Production Control	Regulatory Specialist
P	P	P	S	P	P	S	P	P	P	P	P

Procedural Knowledge: Ability to follow a structured process to ensure a post-Production/Service Phase review is performed and potential lessons learned are identified.

Conceptual Knowledge: Ability to apply evaluation to the previous processes applied during the Production/Service Phase and to recognize events and issues that may qualify as lessons learned (6). Ability to analyze events and issues and determine root cause (5). Ability to create the lessons learned for future use (4).

Factual Knowledge: Knowledge of the lessons learned identification process (3). Knowledge of Production/Service Phase processes (3). Understand how lessons learned for future use are documented (2).

Task 2 (PM260-002): Document identified lessons learned so that such can be understood, assessed for applicability, and applied in future programs/projects.

Quality Eng—Basic	Quality Eng—Senior	Q Manager	Supplier QE	CSoftQE	Program QE	Contract Review	MRO QE	Spec Process QE	Met Labs QE	Production Control	Regulatory Specialist
P	P	P	S	P	P	S	P	P	P	P	P

Procedural Knowledge: Ability to follow a structured process to describe the lessons learned in a format that can be understood by users (3).

Conceptual Knowledge: Ability to describe lessons learned in terms that are relevant and meaningful to users (using various means of written and visual communication), including future practitioners of the Production/Service Phase (Output must flow into integration task)(4). Ability to understand and evaluate the interrelationship of business and technical processes and to make descriptive connections (5).

Factual Knowledge: Understand the importance of descriptive narratives in relation to lessons learned (3). Know the interrelationships of business and technical processes (3). Knowledge of communication methods (3). Knowledge of how to effectively apply communication methods (3).

Task 3 (PM260-003): Integrate documented lessons learned into organizational lessons learned libraries.

Quality Eng—Basic	Quality Eng—Senior	Q Manager	Supplier QE	CSoftQE	Program QE	Contract Review	MRO QE	Spec Process QE	Met Labs QE	Production Control	Regulatory Specialist
P	P	P	S	P	P	S	P	P	P	P	P

Procedural Knowledge: Ability to follow a defined and structured process to integrate the documented lessons learned of the Production/Service Phase into the organizational lessons learned libraries (3).

Conceptual Knowledge: Understanding of how and where lessons learned are to be appropriately integrated into organizational lessons learned libraries (4). Understand the interrelationship of business and technical processes (4). Ability to conceptualize linkages in business and technical processes (4).

Factual Knowledge: Know the interrelationships of business and technical processes (3). Understand the lessons learned library structure of the organization (3).

Process: Apply Lessons Learned—Supplier Selection Phase

Process Purpose: Based on the lessons learned in previous programs/projects, apply lessons learned to ensure previous issues are not repeated. To assess commodity-related history and evaluate the appropriateness for application in the present procurement. Apply this knowledge appropriately.

Knowledge and Skill Requirements: Ability to access lessons learned libraries and apply the experience gained from previous programs/projects to the new program/projects.

Task 1 (SS010-001): Review lessons learned libraries and determine applicability based on the program/project requirements.

Quality Eng—Basic	Quality Eng—Senior	Q Manager	Supplier QE	CSoftQE	Program QE	Contract Review	MRO QE	Spec Process QE	Met Labs QE	Production Control	Regulatory Specialist
S	P	P	P	P	P	S	S	P	P	S	P

Procedural Knowledge: Ability to follow structured process to access existing lessons learned libraries and identify applicable lessons learned for the program/project. (Libraries may reside in and outside of the organization) (3).

Conceptual Knowledge: Ability to evaluate existing lessons learned and identify which may be applicable to the new program/project (4). Assess existing lessons learned libraries and determine if additional sources of information may be required (4).

Factual Knowledge: Knowledge of lessons learned methods (3). Knowledge of existing lessons learned libraries (3). Knowledge of existing program/project requirements (3).

Task 2 (SS010-002): Assess current program/project against existing lessons learned libraries and determine applicability of application. Determine which lessons learned should be applied to the new program/project. Ensure identified lessons learned are documented so that they can be integrated into the program/project plan.

Quality Eng—Basic	Quality Eng—Senior	Q Manager	Supplier QE	CSoftQE	Program QE	Contract Review	MRO QE	Spec Process QE	Met Labs QE	Production Control	Regulatory Specialist
S	P	P	P	P	P	S	S	P	P	S	P

Procedural Knowledge: Ability to follow a structured process in performing an applicability assessment of lessons learned to the new program/project (3). Ability to follow a structured process for measuring and documenting potential risk associated with not implementing identified lessons learned (3).

Conceptual Knowledge: Ability to evaluate existing lessons learned to determine which should be applied to the new program/project (5). Ability to understand and recognize commonality of business and technical processes (5). Ability to analyze and measure risks associated with not implementing lessons learned (5). Ability to communicate risk factors into program/project planning (4). Ability to recognize new sources of potential lessons learned that may apply to the present project/program (5).

Factual Knowledge: Understanding of lessons learned processes (3). Knowledge of applicable technologies (4). Understanding of business and technical process relationships (4). Understanding of process interrelationships (4). Knowledge of risk measurement and management processes (4). Knowledge of diverse lessons learned libraries (3).

Task 3 (SS010-003): Implement lessons learned into program/project planning and/or create risk mitigation plans. Document initial risk management practices for risk associated with not implementing lessons learned.

Quality Eng—Basic	Quality Eng—Senior	Q Manager	Supplier QE	CSoftQE	Program QE	Contract Review	MRO QE	Spec Process QE	Met Labs QE	Production Control	Regulatory Specialist
S	P	P	P	P	P	S	S	P	P	S	P

Procedural Knowledge: Ability to follow a structured process in the implementation of applicable lessons learned into a new program/project (3). Ability to apply a structured process to measure potential risk associated with not implementing lessons learned (3). Ability to apply a structured process to create a risk mitigation plan (3).

Conceptual Knowledge: Ability to understand where lessons learned are most appropriately applied (4). Ability to define and quantify risks associated with not applying lessons learned (5).

Factual Knowledge: Knowledge of lessons learned processes (3). Knowledge of applicable technologies (4). Knowledge of interrelatedness of business and technical processes (4). Knowledge of risk assessment measurement methods (3). Knowledge of risk mitigation plans (3).

Process: Capture, Document, and Preserve Lessons Learned—Supplier Selection Phase

Process Purpose: To recognize, understand, and document lessons learned during the application of the previous processes in the Partner/Supplier Selection Phase so that such may be used in future programs/projects.

Knowledge and Skill Requirements: Ability to identify and document lessons learned and integrate such into organizational lessons learned libraries so that the experience gained from previous processes in the Partner/Supplier Selection Phase can be applied to future program/projects.

Task 1 (SS125-001): Review the processes applied during the Partner/Supplier Selection Phase and identify lessons learned for future use.

Quality Eng—Basic	Quality Eng—Senior	Q Manager	Supplier QE	CSoftQE	Program QE	Contract Review	MRO QE	Spec Process QE	Met Labs QE	Production Control	Regulatory Specialist
S	P	P	P	P	P	S	S	P	P	S	P

Procedural Knowledge: Review the processes applied during the Partner/Supplier Selection Phase and identify lessons learned for future use.

Conceptual Knowledge: Ability to follow a structured process to ensure a post-Partner/Supplier Selection Phase review is performed and potential lessons learned are identified (3).

Factual Knowledge: Ability to apply evaluation to the previous processes applied during the Partner/Supplier Selection Phase and to recognize events and issues that may qualify as lessons learned (4). Ability to analyze events and issues and determine root cause (4). Ability to create the lessons learned for future use (6).

Task 2 (SS125-002): Document identified lessons learned so that such can be understood, assessed for applicability, and applied in future programs/projects.

Quality Eng—Basic	Quality Eng—Senior	Q Manager	Supplier QE	CSoftQE	Program QE	Contract Review	MRO QE	Spec Process QE	Met Labs QE	Production Control	Regulatory Specialist
S	P	P	P	P	P	S	S	P	P	S	P

Procedural Knowledge: Ability to follow a structured process to describe the lessons learned in a format that can be understood by users (2).

Conceptual Knowledge: Ability to describe lessons learned in terms that are relevant and meaningful to users (using various means of written and visual communication), including future practitioners of the Partner/Supplier Selection Phase (Output must flow into integration task)(4).

Ability to evaluate the interrelationship of business and technical processes and to make descriptive connections (5).

Factual Knowledge: Understand the importance of descriptive narratives in relation to lessons learned (2). Knowledge of the interrelationships of business and technical processes (3). Knowledge of communication methods (3). Knowledge of how to effectively apply communication methods (4).

Task 3 (SS125-003): Integrate documented lessons learned into organizational lessons learned libraries.

Quality Eng—Basic	Quality Eng—Senior	Q Manager	Supplier QE	CSoftQE	Program QE	Contract Review	MRO QE	Spec Process QE	Met Labs QE	Production Control	Regulatory Specialist
S	P	P	P	P	P	S	S	P	P	S	P

Procedural Knowledge: Ability to follow a defined and structured process to integrate the documented lessons learned of the Partner/Supplier Selection Phase into the organizational lessons learned libraries (3).

Conceptual Knowledge: Understanding of how and where lessons learned are to be appropriately integrated into organizational lessons learned libraries (4). Understand the interrelationship of business and technical processes (4). Ability to conceptualize and evaluate linkages in business and technical processes (5).

Factual Knowledge: Know the interrelationships of business and technical processes (3). Understand the lessons learned library structure of the organization (3). Knowledge of how to update the organizational knowledge base (3).

VI.A. Collecting and Summarizing Data

Process: Collect Performance Metrics & Perform Data Analysis

Process Purpose: Collect data related to process performance and quality and perform analysis of data. Present data in visual formats for use in the organization.

Knowledge and Skill Requirements: Understanding of performance measurement processes. Ability to collect data related to process performance and quality. Ability of conduct data analysis. Ability to present data in visual formats for use in the organization. Ability to effectively communicate performance measurements.

Task 1 (PM020-001): Ensure the personnel within the value stream who are collecting and communicating performance measurements have the appropriate knowledge to successfully execute the process. Review value stream performance to ensure effectiveness of performance measurement selection and effectiveness of improvement/corrective actions.

Quality Eng—Basic	Quality Eng—Senior	Q Manager	Supplier QE	CSoftQE	Program QE	Contract Review	MRO QE	Spec Process QE	Met Labs QE	Production Control	Regulatory Specialist
S	P	P	P	S	P		P	S	S	P	

Procedural Knowledge: Ability to apply a structured assessment process to ensure the personnel within the value stream who are establishing performance measurements have the appropriate knowledge to successfully execute the process (3). Ability to apply a structured review process to ensure effectiveness of performance measurement selection and effectiveness of improvement/corrective actions within the value stream (3).

Conceptual Knowledge: Ability to evaluate work processes and work product to ensure that persons within the value stream who are establishing performance measurements have the appropriate knowledge to successfully execute the process (5). Ability to analyze work processes and work product to ensure effectiveness of performance measurement selection and effectiveness of improvement/corrective actions within the value stream (4).

Factual Knowledge: Understanding of performance measurements (3). Understanding of effective communication (3). Understanding of applicable measured processes (3). Understanding of program/project requirements (3). Understanding of requirements management tools (3). Understanding of audit practices (3). Understanding of value stream structure and where work is being performed (3). Understanding of personnel training records (3).

Task 2 (PM020-002): Collect performance data related to process and product performance and quality.

Quality Eng—Basic	Quality Eng—Senior	Q Manager	Supplier QE	CSoftQE	Program QE	Contract Review	MRO QE	Spec Process QE	Met Labs QE	Production Control	Regulatory Specialist
P	P		P	P	S		P	S	S	P	

Procedural Knowledge: Ability to apply a structured process to collect performance data related to process and product performance and quality (3).

Conceptual Knowledge: Ability to collect performance data related to process and product performance and quality (3).

Factual Knowledge: Understanding of data collection methods (3). Understanding of production/servicing work processes (3). Understanding of measurement processes (3).

Task 3 (PM020-003): Conduct data analysis to determine process performance.

Quality Eng—Basic	Quality Eng—Senior	Q Manager	Supplier QE	CSoftQE	Program QE	Contract Review	MRO QE	Spec Process QE	Met Labs QE	Production Control	Regulatory Specialist
P	P		P	S	S		S	S	S	S	

Procedural Knowledge: Ability to apply a structured approach to conducting data analysis to determine process performance (3).

Conceptual Knowledge: Ability to conduct data analysis to determine process performance (4).

Factual Knowledge: Understanding of data collection methods (3). Understanding of data analysis (3). Understanding of production/servicing work processes (3). Understanding of measurement processes (3).

Task 4 (PM020-004): Effectively communicate performance measurements to applicable parties (e.g., management, customer, and value stream). (Note: This task is for all types of performance measurement communication—internal and external.)

Quality Eng—Basic	Quality Eng—Senior	Q Manager	Supplier QE	CSoftQE	Program QE	Contract Review	MRO QE	Spec Process QE	Met Labs QE	Production Control	Regulatory Specialist
P	P	P	P	P	P	P	P	P	P	P	P

Procedural Knowledge: Ability to apply a structured process to effectively communicate performance measurements (3).

Conceptual Knowledge: Ability to effectively communicate performance measurements so that it may understood and action taken (4).

Factual Knowledge: Understanding of performance measurements (3). Understanding of effective communication (3). Understanding of applicable measured processes (3). Understanding of program/project requirements (3).

VI.F. Statistical Process Control (SPC)

Process: Monitor Process Certification/Statistical Process Control

Process Purpose: To ensure statistical techniques and inspection sampling plans are appropriately and effectively applied.

Knowledge and Skill Requirements: Understanding of applicable regulatory and industry standard requirements related to sampling and application of statistical process control methods. Understanding of statistics and statistical process controls. Ability to assess the correctness of statistical techniques and inspection sampling plans that are being applied in the supplier/partner base. Ability to verify that sampling and SPC records are effectively maintained throughout the value stream.

Task 1 (PF065-001): Ability to assess the correctness of statistical techniques and inspection sampling plans that are being appropriately applied.

Quality Eng—Basic	Quality Eng—Senior	Q Manager	Supplier QE	CSoftQE	Program QE	Contract Review	MRO QE	Spec Process QE	Met Labs QE	Production Control	Regulatory Specialist
P	P	S	P		S	S	P	S	S	P	S

Procedural Knowledge: Ability to apply a defined and structured review and assessment process to verify that established statistical techniques and sampling plans are controlled and in compliance with appropriate sampling standards and/or other applicable requirements (3).

Conceptual Knowledge: Ability to analyze statistical control plans to verify that they are in compliance with appropriate requirements and follow valid statistical models (4). Ability to analyze sampling plans to verify that they are in compliance with appropriate sampling standards (4). Ability to assess established control processes to verify that sampling plans are controlled (3).

Factual Knowledge: Understanding of statistical process control (4). Understanding of applied statistics (4). Understanding of sampling plans (4). Understanding of regulatory requirements related to sampling plans (3). Understanding of contractual and project/program requirements related to sampling plans (3). Understanding of industry standards related to sampling plans (3). Understanding of audit practices (3). Understanding of the value stream structure and where work is performed (3). Understanding of requirements management tools (3).

Task 2 (PF065-002): Verify that sampling and SPC records are effectively maintained.

Quality Eng—Basic	Quality Eng—Senior	Q Manager	Supplier QE	CSoftQE	Program QE	Contract Review	MRO QE	Spec Process QE	Met Labs QE	Production Control	Regulatory Specialist
P	P	S	P		S	S	P	S	S	P	S

Procedural Knowledge: Ability to apply a structured and defined assessment process to verify that sampling and SPC records are effectively maintained (3).

Conceptual Knowledge: Ability to assess the compliance of verifying that sampling and SPC records are effectively maintained (3).

Factual Knowledge: Understanding of statistical process control (3). Understanding of sampling plans (3). Understanding of regulatory requirements related to sampling plans (3). Understanding of contractual and project/program requirements related to sampling plans (3). Understanding of industry standards related to sampling plans (3). Understanding of audit practices (3). Understanding of the value stream structure and where work is performed (3). Understanding of requirements management tools (3).

Appendix A: Definition of Key Terms & Acronyms

Advisory Circular (AC): A document produced by the FAA to provide guidance for compliance to Federal Aviation Regulations.

Aircraft Certification Office (ACO): The engineering office within the FAA that oversees the certification of new and changed FAA-approved designs.

Airworthiness Directive (AD): A rule promulgated by the FAA requiring corrective or preventative action on an FAA-approved product, part, or component.

Aerospace and Defense Learning Institute (ADLI): The organization within the aviation, space, and defense industry that is responsible for developing and sustaining domain PBOE. The AS&D industry focused forum within the Center for Professional Knowledge.

Aftermarket: In the civil aviation industry, products or services that are provided for incorporation into product that already has an original airworthiness certificate issued. PMA parts are intended for aftermarket installations and are inherently aftermarket. Modification to OEM products after initial delivery and airworthiness certification is an aftermarket activity. In the defense industry, aftermarket includes break-out parts and services.

Airworthiness (Airworthy): That condition of an aircraft, engine, propeller, or component part thereof, where the article conforms to its approved design and is in a condition for safe operation.

American Society for Quality (ASQ): The global professional society for quality practitioners.

Approved Supplier List (ASL): A controlled listing of an organization's approved suppliers.

Aviation, Space, and Defense (AS&D) Industry: Includes designers, manufacturers, suppliers, repairers, operators, and policymakers of products and services that are aviation, space, and/or defense related (ITA, 2002).

Body of Knowledge (BoK): A body of knowledge is an all-inclusive term that describes the sum of knowledge within a specific profession. Specific to this study, a body of knowledge is the specialized skills and knowledge that a professional must possess in order to execute the processes for which they are responsible (Gruber, 1993).

Business Process: An activity that is performed to create value in the performance of a service or creation of a product (Womack and Jones, 1996; Olalla, 2000).

Capability and Maturity Model Integration (CMMI): A structured measurement methodology for measuring the maturity of organizational processes.

Center for Professional Knowledge (CPK): The global organization that fosters the development and sustainment of industry focused PBOE.

Civil Aviation Administration of China (CAAC): The civil aviation authority in China.

Civil Aviation Administration of Israel (CAAI): The civil aviation authority in Israel.

Computer Produced Design (CPD): A product design that is drafted completely within electronic media and may only have spatial geometric tolerances (e.g., CAD/CAM, CATIA, etc.).

Conceptual Knowledge: The knowledge of more complex, organized knowledge forms (Bloom 1956). It includes knowledge of classifications and categories; principles and generalizations; and theories, models, and structures (Anderson and Krathwohl 2001).

Continuous Improvement (CI): An ongoing effort to improve products, services, or processes.

Corrective Action (C/A or CA): The process of investigating a nonconformance or noncompliance to identify its root cause and establish actions to prevent its recurrence.

Corrective Action Report/Request (CAR): A document that can be utilized to document the corrective and preventative action activities.

Department of Defense (DoD): The military department of the United States.

Designated Airworthiness Representative (DAR): An individual delegated by the FAA to perform certain airworthiness determinations and acceptance activities.

Designated Engineering Representative (DER): An individual delegated by the FAA to perform certain FAA engineering functions.

Designated Manufacturing Inspection Representative (DMIR): A delegation issued to a member of a manufacturer's inspection organization to perform conformity inspections on behalf of the FAA.

Domain Ontology: Represents a specific field of knowledge, such as quality, production, engineering, mathematics, computer science, medicine, etc. (Noy and McGuiness, 2001; Stojanovic et al., 2002; Spyns et al., 2002).

Engineering Order (EO): An engineering writing created to document engineering changes.

Engineering Source Approval (ESA): The process of validating and verifying a service or manufacturing process by an organization's engineering function.

Escape: A nonconformance to design, process, or procedure that leaves an organization's established quality management system.

European Aviation Safety Agency (EASA): EASA is the civil aviation authority within the European Union.

Explicit Knowledge: Knowledge that has been memorialized in such a way that it can be transferred independent of a human.

Factual Knowledge: The knowledge of facts (e.g., the sky is blue, the hardness values of metals, etc.).

Failure Mode and Effects Analysis (FMEA): A procedure in product development and operations management for analysis of potential failure modes within a system, product, or service for classification by the severity and likelihood of the failures.

Federal Aviation Administration (FAA): The FAA is the civil aviation authority in the United States.

Federal Aviation Regulations (FAR): The statutory rules of performance within the United States civil aviation industry.

First Article Inspection (FAI): The documented inspection of a product, part, or component that substantiates conformity to design and production requirements. (Reference AS/EN/SJAC 9102.)

Flight Standards District Office (FSDO): The office within the FAA that is responsible for the monitoring the maintenance and operational safety of operational civil aircraft under the rules of the United States.

Functional Configuration Audit (FCA): The formal examination of functional characteristics of a configuration item, or system, to verify that the item has achieved the requirements specified in its functional and/or allocated configuration documentation. (Reference MIL-HDBK-61A.)

Industry-Controlled Other Party (ICOP): The IAQG process for certification of organizations to the 9100 series of AQMS standards.

Integrated Product Development (IPD): The conceptual framework of integrated teams focused around a specific product, process, or value stream.

Integrated Product/Process Team (IPT): The team that represents the combined functional disciplines and business interests of a specific product or process.

Integrated Product/Program Management Team (IPMT): The highest level of leadership over a product line and/or program, which utilizes the IPD methodology.

International Aerospace Quality Group (IAQG): An international forum of aerospace and defense leaders, representing member companies, who have come together to self-manage the quality interests of the AS&D industry.

Just-In-Time (JIT): A term that relates to an organization's business processes that ensure material and/or services are provided when needed and are not held in reserve, such as in inventory, and is a characteristic of a *pull system.*

Knowledge Capability: The demonstrated capability of an individual with regard to an element of a knowledge requirement (reference Bloom's six levels of cognition).

Knowledge Management (KM): Knowledge management is concerned with the representation, organization, acquisition, creation, usage, and evolution of knowledge in its many forms (Polanyi, 1966).

Knowledge Management Process (KMP): The knowledge management process elicits, transforms, and diffuses knowledge throughout an enterprise so that it can be shared and reused (Mach and Owoc, 2001).

Lifecycle: The entire cycle of a project, business process, and/or product that is defined by how such is planned, controlled, executed, and monitored from its inception to its completion (Womack and Jones, 1996).

Lifecycle Phases: The discrete phases of a product's lifecycle. Traditional lifecycle phases include conceptual, design, production, and service. (See figure 3 for the modified lifecycle phases for this BoK.)

Material Requirements Planning (MRP): A production or servicing planning and inventory control system used to manage manufacturing/servicing processes. Also referred to as material resource planning.

Macro Process Mapping: The mapping of organizational business processes at the highest level of work design. It may not include the actual execution instruction of work but will capture the major work processes in an organization's business process lifecycle.

Maintenance Repair and Overhaul (MRO): The maintenance, repair, and overhaul (MRO) aspects of the AS&D that is also referred to as "aftermarket." MRO work is that which is performed on product that has already been provided for operational use.

Material Control Laboratory (MCL): The quality activity within an organization that engages advanced technologies and methods in the destructive and nondestructive evaluation of product and services.

Measurement System Analysis (MSA): A specially designed experiment that seeks to identify the components of variation in the measurement.

Metacognitive Knowledge: Knowledge about cognition in general as well as awareness of knowledge about one's own cognition (Bloom 1956). It encompasses strategic knowledge; knowledge about cognitive tasks, including contextual and conditional knowledge; and self-knowledge (Anderson and Krathwohl 2001).

Micro Process Mapping: The detailed mapping of organizational processes to the actual level of work execution.

Manufacturing Inspection District Office (MIDO): The office within the FAA that oversees FAA production approval holders within a prescribed geographic region.

Material Review Board (MRB): An authorized review team within the organization that establishes the proper disposition for product or service discrepancies.

National Aeronautics and Space Administration (NASA): The governmental organization in the United States responsible for space exploration. NASA also performs an appellate role in FAA administrative matters.

Nondestructive Testing/Evaluation (NDT/NDE): A special process used to evaluate product/service conformity without physically damaging the specimen.

National Institute of Standards and Technology (NIST): The governmental agency in the United States responsible for the development and maintenance of national standards of measure.

Online Aerospace Supplier Information System (OASIS): The IAQG database that maintains information on AS/EN/JQIS 9100 and 9110 organizational QMS registration, auditor certification, and accredited certification bodies.

Ontological Commitment: An agreement of consistency with respect to queries and assertions using the vocabulary defined in the ontology (Mach and Owac, 2001).

Ontology: Ontology provides a shared vocabulary, which can be used to model a domain of knowledge, that is, the type of objects and/or concepts that exist, and their properties and relations. It allows for common understanding and communication between people and computers (Mennis et al., 2000; Lubit, 2001; Mach and Owac, 2001).

Ontology Engineering (OE): A methodology for the design of a knowledge base that includes a strict definition of meaning, identification of relationships, and common vocabulary (Sayed and Zeid, 2003).

Other Party Management Team (OPMT): The IAQG global, industry-controlled, oversight committee chartered to manage the Industry-Controlled Other Party (ICOP) process for certification of organizations to the 9100 series of AQMS standards.

Organizational Memory (OM): The collective knowledge of an organization, both tacit and explicit, structured and unstructured (Abecker et al., 1998; Anand and Manz, 1998; Attipoe, 1999; Vasconcelos et al, 2002; Weinberger et al., 2003).

Original Equipment Manufacturer (OEM): In the aviation industry, the organization that holds a civil aviation authority design and production approval for an engine, propeller, or airframe. In the defense industry, the organization that was originally approved by the government customer for both the design and production of a useable end-item.

Parts Manufacturer Approval (PMA): A production approval issued by the FAA to qualified manufacturers of parts and components. As a PMA is not required for parts that are incorporated into an airframe, engine, or propeller for new OEM production, PMA parts and components are inherently for aftermarket use, such as replacement parts and modifications.

Physical Configuration Audit (PCA): The formal examination of the "as-built" configuration of a configuration item against its technical documentation to establish or verify the configuration item's product baseline.

Preliminary Review: The initial assessment conducted on a product or service discrepancy to determine severity, as part of the MRB process.

Preventative Action (P/A or PA): The process of evaluating processes for the identification of actions that may possibly prevent a nonconformance/noncompliance from occurring.

Principal Inspector (PI): The primary FAA representative assigned to a PAH for certificate management.

Principal Maintenance Inspector (PMI): The primary FAA representative assigned to an FAA repair station or operator for certificate management.

Production Certificate (PC): The FAA production approval issued to manufacturers of aircraft, engines, and propellers.

Procedural Knowledge: The knowledge of how to do something (Bloom 1956). It includes knowledge of skills and algorithms, and techniques and methods, as well as, knowledge of the criteria used to determine and/or justify when to do what within specific domains and disciplines (Anderson and Krathwohl 2001).

Process Based Ontological Engineering (PBOE): The application of process mapping as a structuring methodology to engineer professional domain ontology.

Process Mapping: The process of analyzing work or business processes for description of objective and/or chronological order (Womack and Jones, 1996).

Production Approval Holder (PAH): Within the FAA's system of production approval, a PAH may be a Production Certificate (PC), Parts Manufacturing Approval (PMA), or a Technical Standard Order Authorization (TSOA).

Pull System: When there is little or no inventory in the work environment, a structured methodology is required to ensure materials and services are provided when required.

Purchase Order (PO): A controlled document released to a supplier to procure a product or service.

Quality Assurance (QA): The planned and systematic activities implemented in a quality system so that quality requirements for a product or service will be fulfilled (ASQ 2012).

Quality Assurance Manual (QAM): The core quality document within an organization.

Quality Assurance Procedure (QAP): A supplemental document relating to the quality assurance activities for a specific process.

Quality Control (QC): The observation techniques and activities used to fulfill requirements for quality (ASQ 2012).

Quality Management Process (QMP): A business process that defines and controls an element of the Quality Management System of an organization.

Quality Management System (QMS): the organizational structure, procedures, processes, and resources needed to implement quality management.

Request for Quotation (RFQ): A document released to potential suppliers identifying the requirements for a desired product or service and requesting a bid to be submitted.

Requirements Management Tool (RMT): A software tool that facilitates the documentation, linkage, tracking, and configuration management of requirements and compliance to requirements (e.g., Cradle, DOORS, Polarion, Requisite Pro, etc.).

Safety Management System (SMS): A SMS provides a systematic way to identify hazards and control risks while maintaining assurance that these risk controls are effective.

Service Bulletin (SB): A document provided by a PAH or a supplier of a PAH to provide information (e.g., alerts, suggested repairs, suggested modifications, etc.) to owners and operators of civil aviation products.

Six Sigma: A structured measurement methodology for understanding and controlling variation in processes.

Statistical Process Control (SPC): The application of statistical methods to the monitoring and control of a process to ensure that it operates at within desired parameters to produce conforming product and/or services.

Subject Matter Expert (SME): A person that is recognized by his/her peers as being expert in a particular subject matter.

Supplemental Type Certificate (STC): The design approval issued by civil aviation authorities (e.g., FAA, EASA, CAAC, CAAI, etc.) for changes in approved designs.

Supply Chain Management Handbook (SCMH): The SCMH is provided by the IAQG to assist with the advancement of quality practices within the AS&D value stream.

Tacit Knowledge: The knowledge that resides in the minds of individuals and not memorialized for transference.

Technical Standard Order (TSO): An FAA design specification for specific equipment or components.

Technical Standard Order Authorization (TSOA): The FAA design and production approval for products compliant with a TSO.

Type Certificate (TC): A design approval issued by civil aviation authorities (e.g., FAA, EASA, CAAC, CAAI, etc.) for designs of product (e.g., aircraft, engines, and propellers.)

Value stream: A value stream is made up of the collective value-adding activities that are required to bring a product or service to a customer. A complete value stream includes all tiers of suppliers, internal business processes, and servicing activities.

Appendix B: Delta of ASQ QE BoK vs AS&D Quality BoK

The following is a gap analysis between the the ASQ Quality Engineer BoK* and the AS&D Quality BoK.

	ASQ Major Knowledge Categories (QE BoK)		**AS&D Quality BoK Major Knowledge Categories**
I	**Management and Leadership**	I	**Management and Leadership**
I.A.	Quality Philosophies and Foundations	I.A.	Quality Philosophies and Foundations
I.B.	The Quality Management System (QMS)	I.B.	The Quality Management System (QMS)
I.B.1.	Strategic planning	I.B.1.	Strategic planning
I.B.2.	Deployment techniques	I.B.2.	Deployment techniques
I.B.3.	Quality information system (QIS)	I.B.3.	Quality information system (QIS)
I.C.	ASQ Ethics for Professional Conduct	I.C.	Ethics for Professional Conduct
I.D.	Leadership Principles and Techniques	I.D.	Leadership Principles and Techniques
I.E.	Facilitation Principles and Techniques	I.E.	Facilitation Principles and Techniques
I.F.	Communication Skills	I.F.	Communication Skills
I.G.	Customer Relations	I.G.	Customer Relations
		I.G.1.	*Contracts*
I.H.	Supplier Management	I.H.	Supplier Management
I.I.	Barriers to Quality Improvement	I.I.	Barriers to Quality Improvement
		I.J.	*Knowledge Management*
		I.K.	*AS&D General Requirements*
		I.L.	*Systems Integration*
		I.M.	*Material Resource Planning (MRP)*
II	**Quality Management System**	II	**Quality Management System**
II.A.	Elements of the Quality System	II.A.	Elements of the Quality System
II.B.	Documentation of the Quality System	II.B.	Documentation of the Quality System
II.C.	Quality Standards and Other Guidelines	II.C.	Quality Standards and Other Guidelines
		II.C.1.	*AS9100C*
		II.C.2.	*Government Contract Requirements*
		II.C.2.a.	*Single Process Initiative*
		II.C.3.	*Civil Aviation Regulatory Requirements*
		II.C.3.a.	*Manufacturing (Part 21)*
		II.C.3.b.	*Maintenance (Part 145)*
		II.C.3.c	*Operations (Part 121)*
II.D.	Quality Audits	*II.D.*	Quality Audits *(AS9101)*
II.D.1.	Types of audits	II.D.1.	Types of audits
II.D.2.	Roles and responsibilities in audits	II.D.2.	Roles and responsibilities in audits
II.D.3.	Audit planning and implementation	II.D.3.	Audit planning and implementation
II.D.4.	Audit reporting and follow up	II.D.4.	Audit reporting and follow up
		II.D.5.	*OASIS Database Audit Reporting & Use*
II.E.	Cost of Quality (COQ)	II.E.	Cost of Quality (COQ)
II.F.	Quality Training	II.F.	Quality Training
		II.G	*Earned Value Management*
		II.H.	*Control of customer/government material*
		II.I.	*equipment/property*
		II.J.	*Capability/CapacityMeasurement*
III	**Product and Process Design**	III	**Product and Process Design**
III.A.	Classification of Quality Characteristics	III.A.	Classification of Quality Characteristics
III.B.	Design Inputs and Review	III.B.	Design Inputs and Review
		III.B.1.	*Workmanship Standards*
		III.B.2.	*Produceability Review*

		III.B.3.	*Inspectability Review*
		III.B.4.	*Design Validation*
III.C.	Technical Drawings and Specifications	III.C.	Technical Drawings and Specifications
		III.C.1.	*Approved data*
		III.C.2.	*Computer Produced Design (CPD) Control*
III.D.	Design Verification	III.D.	Design Verification
III.E.	Reliability and Maintainability	III.E.	Reliability and Maintainability
III.E.1.	Predictive and preventive maintenance tools	III.E.1.	Predictive and preventive maintenance tools
III.E.2.	Reliability and maintainability indices	III.E.2.	Reliability and maintainability indices
III.E.3.	Bathtub curve	III.E.3.	Bathtub curve
III.E.4.	Reliability / Safety / Hazard Assessment Tools	*III.E.4.*	Reliability / Safety / Hazard Assessment Tools *(Incl all FMEA)*
		III.F.	*Critical Systems/Components*
		III.G	*Configuration Management*
		III.G.1.	*Configuration Baseline*
		III.G.2.	*Tech Data Management (Incl. Export Control)*
IV	**Product and Process Control**	**IV**	**Product and Process Control**
IV.A.	Tools	*IV.A.*	*Quality* Tools
		IV.A.1	*Key Characteristics (AS9103)*
		IV.A.2.	*First Article Inspections (AS9102)*
		IV.A.3.	*Process/Mfg/Svc/Quality Planning*
		IV.A.3.a	*Mandatory Inspections*
		IV.A.3.b	*Delegated Functions (incl OpCert)*
IV.B.	Material Control	IV.B.	Material Control
IV.B.1.	Material identification, status, and traceability	IV.B.1.	Material identification, status, and traceability
		IV.B.1.a	*Part marking requirements*
IV.B.2.	Material segregation	IV.B.2.	Material segregation
IV.B.3.	Classification of defects	IV.B.3.	Classification of defects
IV.B.4.	Material review board (MRB)	IV.B.4.	Material review board (MRB)
		IV.B.4.a	*Escape Containment*
		IV.B.4.b	*Industry/Government notification of escapes and suspected escapes*
		IV.B.4.c	*Customer in MRB*
		IV.B.4.d	*Government in MRB*
		IV.B.4.e	*Positive control of scrap*
IV.C.	Acceptance Sampling	IV.C.	Acceptance Sampling *(Non-SPC)*
IV.C.1.	Sampling concepts	IV.C.1.	Sampling concepts *(Non-SPC)*
IV.C.2.	Sampling standards and plans	IV.C.2.	Sampling standards and plans *(Non-SPC)*
IV.C.3.	Sample integrity	IV.C.3.	Sample integrity *(Non-SPC)*
IV.D.	Measurement and Test	IV.D.	Measurement and Test
IV.D.1.	Measurement tools	IV.D.1.	Measurement tools
IV.D.2.	Destructive and nondestructive tests	IV.D.2.	Destructive and nondestructive tests
		IV.D.3.	*Measuring Techniques*
IV.E.	Metrology	IV.E.	Metrology
IV.F.	Measurement System Analysis (MSA)	IV.F.	Measurement System Analysis (MSA)
		IV.F.1	Significant-out-of-tolerance (SOT) processes
		IV.G.	*Manufacturing Processes*

V Continuous Improvement

V.A. Quality Control Tools

V.B. Quality Management and Planning Tools

V.C. Continuous Improvement Techniques

V.D. Corrective Action

V.E. Preventive Action

IV.G.1. Shop Assist

IV.H. Maintenance Processes

IV.I. Control of Tooling

IV.J. Control of life limited materials

IV.K. Supplier Management

IV.K.1 Risk Based Acquisition Management

IV.K.2. Statement of Work

IV.K.3. Flow Down of Requirements (Incl PO)

IV.K.4. Deliverable Data

IV.K.5. Pre-award/Post-award Evaluations

IV.K.6. Supplier Approvals (ASL)

IV.K.7. Sub-tier Supplier Control

IV.K.8. Supplier Delegations

IV.K.9. Direct/Drop Shipments

IV.K.10. Supplier support resources

IV.K.11. Monitoring of Suppliers

IV.K.12. Termination of Suppliers

IV.K.13. Supplier Quality Plans

IV.L. Transfer of Work

IV.M. Special process control

IV.M.1. Monitoring of Second & Third Party Certifications (e.g., Nadcap)

IV.M.2. Special process operator certification

IV.M.3 Place holder for various special processes

IV.N. Software Controls

IV.N.1. Deliverable software

IV.N.2. Nondeliverable software

IV.O. Packaging, Handling & Shipping

IV.O.1. Cleanliness & Contamination

IV.O.2. Foreign Object Debris

IV.O.3 Electro Static Discharge

IV.P. Product Pedigree Assurance

IV.Q. Suspected unapproved parts

IV.R. Certification of Flight Readiness/Airworthiness Certification

IV.R.1. Civil aviation

IV.R.2. Military aviation

IV.R.3. Space (Launch Processing & Servicing)

V Continuous Improvement

V.A. Quality Control Tools

V.B. Quality Management and Planning Tools

V.C. Continuous Improvement Techniques

V.D. Corrective Action

V.E. Preventive Action

V.F. Corrective Action Requests

V.F.1. Letters of Investigation

V.F.2 Government CARs

V.G. Lessons Learned

		V.H	*Cost of Quality*
		V.I.	*Process Effectiveness Measures*
VI	**Quantitative Methods and Tools**	**VI**	**Quantitative Methods and Tools**
VI.A.	Collecting and Summarizing Data	VI.A.	Collecting and Summarizing Data
VI.A.1.	Types of data	VI.A.1.	Types of data
VI.A.2.	Measurement scales	VI.A.2.	Measurement scales
VI.A.3.	Data collection methods	VI.A.3.	Data collection methods
VI.A.4	Data accuracy	VI.A.4	Data accuracy
VI.A.5	Descriptive statistics	VI.A.5	Descriptive statistics
VI.A.6.	Graphical methods for depicting relationships	VI.A.6.	Graphical methods for depicting relationships
VI.A.7.	Graphical methods for depicting distributions	VI.A.7.	Graphical methods for depicting distributions
VI.B.	Quantitative Concepts	VI.B.	Quantitative Concepts
VI.B.1.	Terminology	VI.B.1.	Terminology
VI.B.2.	Drawing statistical conclusions	VI.B.2.	Drawing statistical conclusions
VI.B.3.	Probability terms and concepts	VI.B.3.	Probability terms and concepts
VI.C.	Probability Distributions	VI.C.	Probability Distributions
VI.C.1.	Continuous distributions	VI.C.1.	Continuous distributions
VI.C.2.	Discrete distributions	VI.C.2.	Discrete distributions
VI.D.	Statistical Decision-Making	VI.D.	Statistical Decision-Making
VI.D.1.	Point estimates and confidence intervals	VI.D.1.	Point estimates and confidence intervals
VI.D.2.	Hypothesis testing	VI.D.2.	Hypothesis testing
VI.D.3.	Paired-comparison tests	VI.D.3.	Paired-comparison tests
VI.D.4.	Goodness-of-fit tests	VI.D.4.	Goodness-of-fit tests
VI.D.5.	Analysis of variance (ANOVA)	VI.D.5.	Analysis of variance (ANOVA)
VI.D.6.	Contingency tables	VI.D.6.	Contingency tables
VI.E.	Relationships Between Variables	VI.E.	Relationships Between Variables
VI.E.1.	Linear regression	VI.E.1.	Linear regression
VI.E.2.	Simple linear correlation	VI.E.2.	Simple linear correlation
VI.E.3.	Time-series analysis	VI.E.3.	Time-series analysis
VI.F.	Statistical Process Control (SPC)	VI.F.	Statistical Process Control (SPC)
VI.F.1.	Objectives and benefits	VI.F.1.	Objectives and benefits
VI.F.2.	Common and special causes	VI.F.2.	Common and special causes
VI.F.3.	Selection of variable	VI.F.3.	Selection of variable
VI.F.4.	Rational subgrouping	VI.F.4.	Rational subgrouping
VI.F.5.	Control charts	VI.F.5.	Control charts
VI.F.6.	Control chart analysis	VI.F.6.	Control chart analysis
VI.F.7.	PRE-control charts	VI.F.7.	PRE-control charts
VI.F.8.	Short-run SPC	VI.F.8.	Short-run SPC
VI.G.	Process and Performance Capability	VI.G.	Process and Performance Capability
VI.G.1.	Process capability studies	VI.G.1.	Process capability studies
VI.G.2.	Process performance vs. specifications	VI.G.2.	Process performance vs. specifications
VI.G.3.	Process capability indices	VI.G.3.	Process capability indices
VI.G.4.	Process performance indices	VI.G.4.	Process performance indices
VI.H.	Design and Analysis of Experiments	VI.H.	Design and Analysis of Experiments
VI.H.1.	Terminology	VI.H.1.	Terminology
VI.H.2.	Planning and organizing experiments	VI.H.2.	Planning and organizing experiments
VI.H.3.	Design principles	VI.H.3.	Design principles
VI.H.4.	One-factor experiments	VI.H.4.	One-factor experiments
VI.H.5.	Full-factorial experiments	VI.H.5.	Full-factorial experiments
VI.H.6.	Two-level fractional factorial experiments	VI.H.6.	Two-level fractional factorial experiments

Appendix C: Recognition of Contributors

Throughout the development of the AS&D Quality BoK, members of industry and academia made professional contributions of time and intellect.

The Subject Matter Expert (SME) Panel

SME were employed throughout this study to validate the direction of the research, the results of the organizational process mapping, identification of tasks associated with the processes, and the analysis of knowledge and skill scoring. The following AS&D professionals participated in various phases of this research.

Name	*Organization*
Robert Ellison	NASA
Robert Faticanti	Textron
Frank Hutto	Northrop Grumman
Maria Havenhill	SAIC/NASA
Ray Kacmar	NASA
Harry Keane	Northrop Grumman
Ryan Nowosielski	The IPL Group, LLC
Dr. Jaye Omberg	Lockheed Martin
Francois Pageau	General Dynamics-Canada
Kevin Treanor	Northrop Grumman

The Examination Committee

As the AS&D Quality BoK was a product of a broader research project, and strict academic protocols were maintained. The following thought leaders examined the research methodology, research application, analysis of data, results, and conclusions and accepted such as meeting the highest levels of rigor.

Name	*Organization*
Dr. Neal E. Chalofsky	The George Washington University
Dr. Michael J. Marquardt	The George Washington University
Dr. Chetan (Chet) Date	Honeywell Aerospace
Dr. Lynne Thompson	National Defense University
Dr. Daniel Krieger	NASA Goddard Space Flight Center

Final Manuscript Reviewers

Prior to the publishing of this book, the manuscript was circulated to top leaders within the AS&D industry for review and comment. Their time and effort in this endeavor is greatly appreciated, and they are to be commended for their dedication to the AS&D industry.

Name	*Organization*
Lloyd Barker	Alcoa
John Clay	Harris Corp
Robert Cook	Federal Aviation Administration
Dr. Chetan (Chet) Date	Honeywell Aerospace
Jack Fletcher	NASA (Retired)
Peter Friedkin	Parker Aerospace
Rayner Hutchinson	AAR Corp
LTC Robert W. Jackson II	United States Air Force
Steve Kerr	Meggitt Aircraft Braking Systems
Dr. Jesse Mangual	Textron Systems Corporate
Dr. Alan Stolzer	Embry-Riddle Aeronautical University

References

Abecker, A., A. Bernardi, K. Hinkelmann, O. Kuhn, and M. Sintek (1998). Toward a technology for organizational memories. IEEE Intelligent Systems & Applications (3); 40-48.

Anand, V. and C. C. Manz (1998). An organizational memory approach to information management. Academy of Management Review (4); 796-809.

Anderson, L. and D. Krathwohl (2001). A taxonomy for learning, teaching, and assessing. New York: Longman.

Attipoe, A. (1999). Knowledge structuring for corporate memory. Markup Languages: Theory & Practice (4); 27-36.

Bertalanffy, L. (1972). General system theory: Foundations, development, applications. London: Allen Lane.

Bloom, B. and R. Paige (1995). Transformational design and implementation of a new efficient solution to the ready simulation problem. Science of Computer Programming (3); 189.

Bloom, B. S., Ed. (1956). Taxonomy of educational objectives: The classification of educational goals, Handbook 1: Cognitive domain. New York: Longman-Green.

Chan, C. and B. Lewis (2002). A basic primer on data mining. Information Systems Management (4); 56-60.

Dreikorn, M. J. (2009). Exploring the application of process mapping in the ontological engineering of a body of knowledge for the supplier quality engineer profession. The School of Education. The George Washington University. Washington, DC.

Fishburn, P. C. (1964). Decision and value theory. New York: Wiley.

Gruber, T. R. (1993). Toward principles for the design of ontologies used for knowledge sharing. Palo Alto, California: Stanford University: Knowledge Sharing Laboratory.

Hassell, L. (2003). A philosophical perspective on knowledge management. College of Information Science and Technology. Philadelphia, PA: Drexel University.

Hofstadter, D. (1989). Godel, Escher, Bach: An eternal golden braid. New York; Vintage Books.

ITA (2002). Final report of the commission on the future of the United States aviation Industry. Arlington, VA: International Trade Administration.

Kimble, C., F. Li, and A. Barlow (2000). Effective virtual teams through communities of practice. Glasgow: University of Strathclyde, Strathclyde Business School.

Laszlo, E. (1972). The systems view of the world: The natural philosophy of the New developments in the sciences. New York: G. Braziller.

Lin, F. and S. Lin (2001). A conceptual model for virtual organizational learning. Journal of Organizational Computing and Electronic Commerce (3); 155-178.

Lindgren, B. W. (1971). Elements of decision theory. New York: Macmillan.

Lubit, R. (2001). Tacit knowledge and knowledge management: The keys to sustainable competitive advantage. Organizational Dynamics (4); 164-178.

Mach, M. and M. Owoc (2001). Validation as the integral part of knowledge management process. Prague, Poland: University of Economics.

Maxwell, J. A. (1996). Qualitative research design: An interactive approach. Thousand Oaks, CA: Sage.

Mennis, J., D. Puequet, and L. Qian (2000). A conceptual framework for incorporating principles into geographical database representation. International Journal of Geographical Information Science (6); 501-520.

Nonaka, I. (1994). A dynamic theory of organizational knowledge creation. Organizational Science (1); 14-37.

Nonaka, I. and T. Takeuchi (1995). The Knowledge Creating Company. New York: Oxford University Press.

Noy, N. F. and D. L. McGuiness (2001). Ontology development 101: A guide to creating your first ontology. Stanford University.

Olalla, M. F. (2000). Information technology in business process reengineering. International Advances in Economic Research (3); 581-588.

Polanyi, M. (1966). The tacit dimension. New York: Doubleday.

Sayed, E. and A. Zeid (2003). What can ontologies learn from knowledge management? Journal of Computer Information Systems (3); 109-117.

Spyns, P., R. Meersman, and M. Jarrar (2002). Data modeling versus ontological engineering. ACM SIGMOD Record (2); 12-17.

Stojanovic, L., N. Stojanovic, S. Handschuh (2002). Evolution of the Metadata in the ontology-based knowledge management systems. Karlsruhe, Germany: University of Karlsruhe, FZI Research for Information Technologies, Institute AIFB.

Sure, Y., S. Staab, R. Studer (2002). Methodology for development and employment of ontology based knowledge management applications. Karlsruhe, Germany: University of Karlsruhe.

Vasconcelos, J., Gouveia, F., and Kimble, C. (2002). An organizational memory information system using ontologies. 3rd Conference of the Association of Portuguese Systems Information. Portugal: University of Coimbra.

Vasconcelos, J., Kimble, C., Gouveia, F., and Kudenko, D. (2000). A group memory system for corporate knowledge management: An ontological approach. 1st European Conference on Knowledge Management. Slovenia: Bled School of Management.

Weinberger, H., D. Te'eni, J. Frank (2003). Ontologies of organizational memory as a basis for evaluation. 11th European Conference on Information Systems: Naples, Italy.

Womack, J. P. and D. T. Jones (1996). Lean thinking: Banish waste and create wealth in your corporation. New York: Simon and Schuster.

www.ingramcontent.com/pod-product-compliance
Lightning Source LLC
LaVergne TN
LVHW080011210926
848481LV00011B/2995